Progress in Scientific Computing
Vol. 6

Edited by
S. Abarbanel
R. Glowinski
G. Golub
P. Henrici
H.-O. Kreiss

Birkhäuser
Boston · Basel · Stuttgart

Progress and Supercomputing in Computational Fluid Dynamics

Proceedings of U.S.-Israel Workshop, 1984

Earll M. Murman
Saul S. Abarbanel
editors

1985

Birkhäuser
Boston · Basel · Stuttgart

Editors

Earll M. Murman
Department of Aeronautics and Astronautics
Massachusetts Institute of Technology, Rm. 33–217
Massachusetts Avenue
Cambridge, MA 02139 (USA)

Saul S. Abarbanel
Tel-Aviv University
Department of Applied Mathematics
Ramat-Aviv, Tel-Aviv
Israel

Library of Congress Cataloging in Publication Data

U.S.-Israel Workshop (1984 : Jerusalem)
Progress and supercomputing in computational fluid dynamics.

(Progress in scientific computing ; vol. 6)
1. Fluid dynamics -- Data processing -- Congresses.
2. Supercomputers -- Congresses. I. Murman, Earll M., 1942– . II. Abarbanel, Saul S., 1931– III. Title. IV. Series: Progress in scientific computing ; v. 6.
QA911.U2 1984 532'.05'028551 85–13420
ISBN 0-8176-3321-9

CIP-Kurztitelaufnahme der Deutschen Bibliothek

Progress and supercomputing in computational fluid dynamics : proceedings of US Israel workshop, 1984 / Earll M. Murman ; Saul S. Abarbanel, ed. – Boston ; Basel ; Stuttgart : Birkhäuser, 1985.
(Progress in scientific computing ; Vol. 6)
ISBN 3-7643-3321-9 (Stuttgart . . .)
ISBN 0-8176-3321-9 (Boston . . .)

NE: Murman, Earll M. [Hrsg.]; GT

Printed in Germany
ISBN 0-8176-3321-9
ISBN 3-7643-3321-9

TABLE OF CONTENTS

PREFACE

The present volume, with the exception of the introductory chapter, consists of papers delivered at the workshop entitled "The Impact of Supercomputers on the Next Decade of Computational Fluid Dynamics." The workshop, which took place in Jerusalem, Israel during the week of December 16, 1984, was initiated by the National Science Foundation of the USA (NSF), by the Ministry of Science and Development, Israel (IMSD), and co-sponsored by the National Aeronautics and Space Administration (NASA), the Office of Scientific Research of the U.S. Air Force (AFOSR), Tel Aviv University and Massachusetts Institute of Technology.
The introductory chapter attempts to summarize what transpired at the workshop.

The genesis of the workshop was an agreement between NSF and IMS, signed in the spring of 1983, to conduct a series of bi-national workshops and symposia. This workshop represented the first activity sponsored under the agreement. The undersigned were selected by their respective national bodies to act as co-coordinators and organizers of the workshop.

The first question that we faced was to decide upon a topic. In the past few years the field of CFD has mushroomed and consequently there have been many meetings, symposia, workshops, congresses, etc. dealing with all aspects of CFD. The smaller workshops and symposia have by and large been specialists' meetings. Among them they have covered the whole spectrum--from the mathematical theory of hyperbolic conservation laws through numerical simulation of turbulence to numerical boundary conditions in CFD. We thought that the time was ripe to pause and present to the community a sort of "State of the CFD-Nation" report consisting of two elements: technical papers by leading researchers and an attempt to assess where the field is going. It was decided that the technical papers should represent diverse areas and in fact we relished the idea of having theoreticians and practitioners (such as code-developers) meet together. We had two reasons for this--the first was that this is hardly ever done in small meetings and secondly the diversity was necessary for our second aim of assessing where are we going. The theme of this assessment, "The Impact of Supercomputers on the Next Decade of CFD," was selected since, from among all the factors driving the field and progress in our understanding of it, the coming Supercomputers are particularly important at this juncture.

The undersigned have enjoyed their involvement in this undertaking. All participants were very cooperative in meeting deadlines, being present at all sessions, and in presenting work that is at the cutting edge of their research. We would like to thank them all for their splendid cooperation. As with all such meetings, the "nitty-gritty" of organizing the event determines the welfare of the undertaking. Many people contributed to the success of the workshop--we would like to thank them especially; Ms. Ellen Mandigo at MIT, who coordinated the travel of the U.S. participants and the manuscripts; Ms. Sara Marcus at Tel Aviv University, who coordinated the arrangements for the workshop on the Israeli side; the Kennes company, which was in charge of the hotel and meeting facilities arrangements; and the Birkhauser Book Publishing Co. of Boston.

Saul S. Abarbanel
Earll M. Murman

LIST OF PARTICIPANTS

Saul ABARBANEL
Tel Aviv University

Achi BRANDT
Weizmann Institute of Science

Sidney FERNBACH
Consultant

Joel FERZIGER
Stanford University

Roland GLOWINSKI
University of Paris

David GOTTLIEB
Tel Aviv University

Moshe ISRAELI
Technion

Antony JAMESON
Princeton University

Heinz-Otto KREISS
California Institute of Technology

Robert MacCORMACK
University of Washington

Daniel MICHELSON
Hebrew University

Earll M. MURMAN
M. I. T.

Steven A. ORSZAG
Princeton University

Stanley OSHER
U. of California, Los Angeles

Yirmiahu RIMON
Ministry of Defence

Michael SEVER (MOCK)
Hebrew University

Joseph STEGER
NASA Ames Research Center

Pierre L. SULEM
Tel Aviv University

Eitan TADMOR
Tel Aviv University

W. Tilton THOMPKINS
M. I. T.

Eli TURKEL
Tel Aviv University

Micha WOLFSHTEIN
Technion

Progress in Scientific Computing, Vol. 6
Proceedings of U.S.-Israel Workshop, 1984

IMPACT OF SUPERCOMPUTERS ON THE NEXT DECADE OF COMPUTATIONAL FLUID DYNAMICS

Earll M. Murman
Saul S. Abarbanel

Introduction

A small group of CFD researchers from the United States and Israel gathered in Jerusalem during December 1984 at a workshop entitled "The Impact of Supercomputers on the Next Decade of Computational Fluid Dynamics." The background of the workshop attendees ranged from CFD code developers to applied mathematicians to computer experts. During the workshop the participants presented and discussed results of their current research. They then engaged in discussion of the workshop theme. This article attempts to summarize their observations and speculations on what the impact of supercomputers will be on CFD during the next decade. First, however, we briefly summarize the papers in these proceedings and the current status of CFD.

The Present

Supercomputers and CFD have affected every aspect of fluid dynamics to some degree during the past decade. Perhaps the area which has experienced the most dramatic impact is the field of attached flow aerodynamics, typical of design point conditions for transport aircraft. In this situation the fluid flow is well behaved by design. Separated and unsteady flow are avoided. The turbulent flow models applicable to attached boundary layers are acceptable (though not perfect). There are no chemical reactions or phase changes taking place. The major challenges lie in solving the nonlinear inviscid flow equations (primarily transonic) and dealing with the complex geometry. During the past decade, the capability of the computers and algorithms has developed enormously in this one particular subdiscipline. In some instances they are as much utilized as wind tunnel testing, although by no means supplementing them.

In most other fields of science and engineering, many of the more difficult fluid dynamic phenomena which are absent in attached flow aerodynamics are of paramount importance. For example, turbomachines are dominated by three-dimensional viscous and unsteady phenomena which affect heat transfer and performance. In many devices for propulsion and chemical processes, multicomponent chemical reactions and turbulent mixing must be modeled. High performance aircraft and helicopters are strongly influenced by vortical and unsteady effects. Low drag bodies are dominated by the prediction of transition. Separated, unsteady wakes of automobiles influence their fuel consumption and handling capabilities. In large scale geophysical fluid dynamics, coriolis forces and stratification effects are dominant. These lead to multiple time scale wave phenomena. Unstable stratification produces turbulent, buoyant mixing. Turbulent flow is present in virtually every situation, yet can only be adequately modeled for the simplest of flows like attached boundary layers and jets. This does not exhaust the list, but the point made here is quite clear; only the tip of the iceberg has been seen by the progress made in attached flow aerodynamics. The biggest challenges are yet to come. The papers in the proceedings give one assessment of where the field stands in this respect.

The Papers

Fernbach's paper gives a comprehensive overview of the current performance of supercomputers and what is on the horizon. This field is now very active following a dormant period in the 70's. The basic message is that computer speed and main memory will both increase by about two orders of magnitude in the next decade. Also, all future supercomputers will be a combination of vector (pipeline) and parallel (multiprocessor) architectures. Algorithms will have to adapt to and exploit these architectural features to achieve the stated machine performance. Thompkins' paper demonstrates that supercomputing does not necessarily have to be done on supercomputers. It makes an interesting case for the need of personal-sized supercomputers with speeds about one order of magnitude slower than the mainframes, but with comparable memories. The idea of using higher level languages for multiprocessor applications is brought out by Thompkins. Navier-Stokes results are also presented for turbomachine cascades, illustrating the state of the art for CFD applied to these flows. The paper by Jameson, Leicher, and Dawson demonstrates one way that the current generation of algorithms for inviscid external flows can be modified for multi-processor architectures.

The paper by Steger and Buning provides an overview of current issues regarding the computation of inviscid and viscous aerodynamic flows. In addition to a number of interesting results which are presented, the experience of these authors using the current generation of supercomputers is recorded. Vectorization of implicit algorithms is explained. The need for good graphical postprocessing of large data bases turns out to be mandatory. A similar message is given in the paper by Murman, Rizzi and Powell, which compares two independently obtained solutions for leading edge vortex flows for delta wings. This paper also illustrates that this class of compressible flows is relatively unexplored compared to shock wave dominated flows.

Several papers present new algorithms for the Euler and the incompressible or compressible Navier-Stokes equations. Since the solution of these equations will become more frequent with the higher power of computers, this is an important topic for the future. The paper by Walters and Dwoyer introduces an upwind differencing line relaxation algorithm for the Euler equations. McCormack presents a new algorithm for the compressible Navier-Stokes equations which has some similarities to the algorithm of Walters and Dwoyer. Turkel presents methods for accelerating the convergence to a steady state solution of the Euler or Navier-Stokes equations by using preconditioning to alter the time consistency of the equation set. The paper by Glowinski considers finite element methods for the incompressible Navier-Stokes equations and presents a number of results concerning entry to ducts and the subsequent internal flow. Remarks are also included on finite element methods for compressible Navier-Stokes equations and applications on supercomputers. Israeli gives an algorithm for the parabolized Navier-Stokes (PNS) equations. Brandt and Ta'asan present the latest multi-grid algorithms for quasi-elliptic systems which arise from discrete approximations to the Navier-Stokes and related equations. The importance of algorithms, such as multigrid methods, which have convergence rates (spectral radius) independent of the number of mesh points will be mentioned later.

Perhaps no problem is more central to fluid mechanics than the prediction of transition and turbulent flows. Three papers deal with this topic. Brachet, Metcalfe, Orszag and Riley present new results for instability of free shear flows based upon direct computations of the Navier-Stokes equations. Numerical experiments such as these can lead to new theoretical understanding of instability of rotational flows. Ferziger's paper gives a comprehensive overview of current and future approaches to turbulent flow computations using direct simulations of the Navier-Stokes equations, large eddy simulation, and turbulence models. Wolfshtein considers the latter topic in much greater detail. These two papers point out the capabilities and shortcomings of current turbulence models. The importance of having accurate algorithms is stressed by both authors.

Papers by Sulem and by Michelson illustrate how numerical results can be used to understand the nature of the solutions to partial differential equations. The use of spectral methods for problems which require high accuracy is receiving increased interest. The presentations by Abarbanel and Gottlieb, and Gottlieb and Tadmor consider some basic issues regarding the resolution of extreme gradients by spectral methods. The paper by Browning and Kreiss illustrates that many fluid problems with multiple time and length scales are exceedingly difficult to compute, even with "unlimited" computer power. It is important to understand that the powerful new supercomputers will only yield useful results if the mathematical and numerical analysis formulation is carefully done. The paper by Sever is another illustration of this.

The Next Decade

During the next decade supercomputer power will increase dramatically. The directly addressable high speed memory capacity will increase by about two orders of magnitude, from 2-16 Mwords to 256 Mwords or 2 million words to 256 million more. Processor speed will increase an order of magnitude from about 100 MFLOPS to 1000 MFLOPS, or perhaps more. It is likely that the corresponding parameters of smaller computers will increase by similar factors. These estimates are important because history has shown that whenever an important parameter is varied by an order of magnitude, new discoveries are made. The difficulty is to have some feeling as to what those discoveries might be. The participants realized, of course, that forecasting the future is more of a "guestimate" than an exact science. It is interesting to note, however, that during the panel discussion almost everyone subscribed to the idea that the new supercomputers will not only allow tackling bigger problems, but will also lead to a better understanding of the physics of some complex problems such as turbulence, vortical flows, and chemically reacting flows.

In the remainder of this article we summarize the feelings expressed by the attendees concerning four questions which were posed by the panel.

Impact of Supercomputers on CFD

In the field of aerodynamics, the preliminary design of transport aircraft will primarily be done on supercomputers. The modeling and computing capability will be basically in place. Unlike earlier predictions that computers will make wind tunnels obsolete, few people subscribe to that viewpoint now. What is more likely to happen is that the use of wind tunnels by researchers and design engineers will change. Less and less of the exploratory design will be done by tests as predictive methods become more reliable. This has already happened in several instances with the current generation of computers. The next generation will provide enough resolution and speed that a realistic model of an actual cruising transport aircraft can be computed.

The capability to model "off-design" or "unclean" aerodynamic flows will increase. These are flows which are separated, unsteady, vortex dominated, and the like. Such flows are of great importance for maneuvers of high performance aircraft or for emergency situations for transport aircraft. The loads developed in these regimes often determine the required strength of the aircraft components. The same phenomena often dominate rotary wing and rotating machinery aerodynamics. Capacity of computers and algorithms up to now has not been adequate to support a frontal assault on this class of problems. The payoff for analysis of unsteady, separated, vortical flows will be much greater than for the clean flows representing cruise aerodynamics. This is because little theory has ever been developed for them.

The complexity of problems which the researcher and the engineer will be dealing with will grow in some proportion to the new computer power. This will have a number of impacts on the daily life of the fluids mechanician. Problems under investigation will have many length and time scales. Analysis of results will be more challenging. It will be harder to understand the solution due to the number of interacting physical phenomena present. This is illustrated by the paper of Brachet, Metcalfe, Orszag, and Riley. New methods of analyzing and presenting results will have to emerge in order to deal with this. Graphical output is crucial, and maybe artificial intelligence types of technology will help out.

One difficulty which can be foreseen is the problem of verifying the accuracy or fidelity of a computed result. Up to now, it has generally been possible to compare computed results with theory for limiting conditions. For example, a transonic wing calculation can be compared with linear wing theory for low Mach numbers, or a Navier-Stokes solution can be compared with a laminar boundary layer. But as the computations move into more nonlinear flows, the past theoretical framework will become less and less applicable. Comparison with experiment is essential, and independent computations of the same problem by different researchers will be necessary. Perhaps a renewed interest in theory will result from this need.

Impact of Supercomputers on Basic Sciences

As one participant stated, the great masters of fluid mechanics in the past solved all the linear problems and left us with only the nonlinear ones. Since most fluid mechanic problems are nonlinear, we can speculate that the ability to model highly nonlinear problems with powerful computers will lead to many new discoveries. Another participant thought that the impact of supercomputers will influence the basic way we think about physical problems. New information will be discovered from numerical experiments and provide insight for modeling. In this sense, computational experiments are akin to laboratory experiments which have provided insight and ideas throughout the history of fluid mechanics and other scientific disciplines.

In the past, computational methods have made a major impact on our ability to compute and understand potential flows and inviscid flows dominated by shockwaves. One can conclude that these classes of flows are well understood both from the physical and algorithmic points of view. Although the ability to analyze shock dominated flows has been a major step forward in fluid mechanics, much is left to be done. For example, only limited CFD studies have been done for vortical flows, and little is understood about the algorithm requirements for inviscid rotational flows. Many studies have been done for two-dimensional separated flows, but only limited studies for three-dimensional separated flows. Although efficient algorithms for steady flows are under development, indications are that the flows these algorithms are to be applied to may be unsteady in nature. See for example Thompkins or Murman, Rizzi and Powell.

Perhaps no area is more tempting to speculate on than the field of turbulence. This is an area in which progress has been relatively stagnant since Reynolds introduced all the unknowns without introducing any new equations. In the past decade, computational models and laboratory experiments have opened a new look at turbulence. The idea of organized or coherent structure has emerged. On the other hand, mathematicians have shown that solutions to fairly simple dynamical systems have chaotic behavior. An interesting question which was posed is "What will be the resolution of the speculation that there is both determinism and chaos in nonlinear equations?" Computational experiments could provide a framework for helping to answer this question.

Another area which will probably be strongly influenced by more powerful computational approaches is the coupling of chemical reactions and heat release to fluid flow problems. Even fairly "simple" reactions involve many species with many time scales of reactions. In the past, computers simply were not large enough to tackle many of these problems. Rate constants are always an uncertain factor in such calculations. Perhaps being able to model the experimental conditions under which the rate constants are measured will lead to more accurate measurements of their values.

One issue on which there was quite a difference of opinion is the degree to which modeling will be required prior to computing. On the one hand, many participants felt that the time was upon us to tackle the full three-dimensional Navier-Stokes equations, possibly adding models only for subgrid scale turbulence. Others felt that the past practice of selecting simplified sets of equations such as inviscid or parabolized viscous will still be prudent. It is likely that some level of modeling will always be required as computer capability will never be big enough to solve a complex problem from first principles. In fact, for most problems this is unnecessary. The question is, will the type of modeling appropriate for the future be different from that used in the past when computer memory, speed, and accessibility were much more limited?

Impact of Supercomputers on Algorithms and Languages

An important issue regarding algorithms arises from the multiprocessor and vector architectures of supercomputers. Algorithms which cannot be efficiently used on these architectures will be of limited utility. Many fluid mechanic problems are solved using time dependent integration procedures for initial boundary value problems. Both explicit and implicit methods are utilized. In general, explicit algorithms are easier to vectorize than implicit ones. The latter usually involve recursive steps in the matrix inversion. However, the paper by Steger and Buning demonstrates an effective vectorization strategy for simultaneous inversion of a large number of tridiagonal matrices applicable to approximate factorization methods. Explicit algorithms are also easier to adapt to multiprocessor architectures as the solution domain can be subdivided without introducing difficulties in the algorithm. The paper by Jameson, Leicher and Dawson reports a strategy and results which illustrate this. There is the need to find effective strategies for adapting implicit algorithms for multiprocessor architectures. Participants generally agreed that the real payoff is in designing algorithms which can work effectively on tens or hundreds of parallel processors, not just on two, four, or eight.

It is clear that algorithm developers must be cognizant of the advantages and constraints which non-Von Neumann architectures will place on supercomputing. The paper by Thompkins is an indicator of what will come. In addition to the conceptual changes in algorithms due to multiprocessors and vector processors, efficiency limitations arise in the speed at which main memory can be accessed from various processors or the speed at which data can be exchanged between processors. The personal-size supercomputer reported by Thompkins involves two processors (a scalar host and an attached array). Movement of the data must be carefully managed to avoid bottlenecks. In the future algorithm development must take into consideration not only traditional numerical analysis, but also computer science aspects.

Another algorithm issue arises from the shear size of the main memory of supercomputers. Problems with very fine meshes will be possible. With the exception of multigrid algorithms for elliptic equations, the asymptotic convergence rate (spectral radius) of iterative methods is dependent on the number of mesh points. Recent estimations done by researchers at NASA Langley indicate that the time required to reach convergence for three-dimensional Navier-Stokes equations on a 256 megaword machine is excessive to the point of being unrealistic. This indicates a real need for finding algorithms which have asymptotic convergence rates which are either mesh independent or vary (at worst) slowly with the number of grid points. Such algorithms must work for problems with widely varying length and time scales, not just model problems or model equations (see Jameson et al).

There was considerable discussion on the need for higher level languages that will make it easier to construct a solution approach for a new problem, as well as make it easier to utilize the new architecture. The general strategy for solving a problem by CFD is more or less common. A grid is generated, discretization of spatial derivatives and boundary conditions is done, an iteration or time integration method is selected, and various outputs are required. It was suggested that assembling these tools and manipulating them for embedded subdomains and the like would lend itself to a higher, and therefore simpler, language. FORTRAN is the language of the CFD community to date. The paper by Thompkins introduces some higher level constructs for managing the solution process in a multiprocessor environment.

Another area in which algorithm innovation may be required is in preprocessing (grid generation) and post-processing (data base analysis). The papers by Steger and Buning and Murman, Rizzi and Powell indicate that graphical analysis is imperative, but other ways of manipulating the data bases would be desirable. Maybe knowledge base programs ("expert systems") will be helpful in finding the important features in a given solution. Or perhaps pattern recognition approaches will be needed.

Impact on Subsystems

The workshop attendees for the most part represented users of supercomputers, and not hardware specialists. However, with the large data bases which will be generated, the participants felt that several of the supporting subsystems might well be inadequate to match the power of the high speed processors. Participants who have had experience with supercomputers felt that strong graphics capability is a number one requirement for analyzing results. As discussed above, some new graphics algorithms may well need to be devised to analyze complicated flow fields. But powerful processing and graphical display capabilities are also required. As the paper by Thompkins points out, a researcher will typically spend much of his or her time performing graphical analysis which requires a processor about an order of magnitude smaller than the supercomputer. A large number of operations are required to construct the output quantities which typically are different combinations of the dependent variables than those which are stored in the data base. Many researchers currently think that for these reasons the plot files will be created on the supercomputer itself. However, super graphical processors, which could be much cheaper and therefore more available to the users, should be developed. Typically, algorithms for creating the graphical data base are easily vectorized and could be done on array processors attached to most processors. These could lead to very powerful and inexpensive graphics workstations. Fortunately the technology is developing rapidly for medium to high resolution color display devices with interactive capability.

Most of the attendees at the workshop fall in the category of "remote" users. They are not located at the same site as the supercomputer. There was significant concern that remote communications will be a real bottleneck. Regular dial-up capability is barely adequate at present for editing files or transmitting small output files to remote users. It certainly would be impossible to do interactive graphics processing from a remote site or to transmit the entire data base for onsite analysis using even dedicated data lines currently available. The best way to communicate with remote facilities at present is to transmit magnetic tapes via express mail services. This inevitably leads to delays and slow turnaround. The impact of supercomputers on researchers who are not co-located with the machines will be minor if high bandwidth communications are not available.

Conclusions

The impact of supercomputers on the next decade of computational fluid dynamics will be substantial. With processor speeds and high speed memory increasing by two orders of magnitude, many changes will take place. The difficulty lies in accurately forecasting what those changes will be. The above discussion presents the thinking of one group of active CFD researchers. Perhaps their viewpoints will serve to help others to become aware of, and think about, these changes as they take place. The next decade has already started, twenty years ago!

Progress in Scientific Computing, Vol. 6
Proceedings of U.S.-Israel Workshop, 1984

CURRENT STATUS OF SUPERCOMPUTERS AND WHAT IS YET TO COME

Sidney Fernbach
Alamo, CA

We have lived through six generations of very high performance computers - the so-called supercomputers. The next generation is due in a year or so. The attached chart, Table 1.1, shows my definition of the generations.

ETA, Cray Research, and Denelcor have their experts busily engaged in planning their respective entries, namely the CYBER 250 (or GF10), the Cray-3 (and X-MP (n)), and the HEP-2. These will all be multiprocessors. The first two manufacturers seem to be aiming at having eight (maybe 16) processors in their systems, and the last for an even larger number.

What do we know about multiprocessor systems? Unfortunately, not enough. At the present time there are numerous dual processors (e.g., IBM 3081, CDC CYBER 875, Cray X-MP, Fujitsu 382, etc.). There are also computers with attached processors (e.g., CDC + MAP IV (a STAR Technology product), NEC ACOS 1000, Hitachi 200 + IAP, Univac + a Datawest AP etc.), and there are even four processor systems (e.g., DEC VAX-4, Denelcor HEP-1, ELXSI). Some of these can grow to eight or more systems. In addition, there are already in existence a number of "home-built" multiprocessors (MIDAS at LBL and a host of others in various stages of planning or construction.

Up until this generation, we made do with uniprocessors (except for a few experimental machines, notably Illiac 4), with vector instructions added if we wanted much greater effectiveness. Having used uniprocessors for well over 30 years now, we believe that we can use them quite efficiently (except that we resort to higher level languages and lose some performance thereby). Vector processors were introduced with the CDC STAR 100 and Texas Instruments ASC in the 70's. These were followed by the Cray and CYBER 200s, and now the Fujitsu and Hitachi entries (Table 1.2). While we have had some experience over the past ten years, we still have a long way to go to gain real performance from these systems.

Computer Scientists throughout the world tell us that the best way to gain performance is to multiplex our CPUs (scalar or vector). At the same time it is becoming more and more difficult to get higher speed electronic components. At least the vendors are more eager to sell the millions and millions of slower devices for the ever expanding personal and small computer market, than invest in higher speed

SUPERCOMPUTER GENERATIONS

GENERATION	TYPICAL SYSTEM	APPROXIMATE DATE
THE BEGINNING	UNIVAC I	1950
I	IBM 700s	1955
II	IBM 7000s	1960
	EXPERIMENTAL MACHINES:	
	IBM STRETCH	
	UNIVAC LARC	
III	CDC 6600	1965
IV	CDC 7600	1970
V	THE DOLDRUMS	
	EXPERIMENTAL MACHINES:	
	BURROUGHS ILLIAC IV	1975
	CDC STAR 100	
	TEXAS INSTRUMENTS ASC	
VI	CRI CRAY-1/CDC CYBER 205	1980
	(FUJITSU VP200)	
	(HITACHI S810-20)	
VII	CRAY 2,3/ETA GF10 or CDC CYBER 250/DENELCOR HEP2	1985
	(NEC SX2)	

TABLE 1.1

CURRENT SUPERCOMPUTERS

Organization	Fujitsu	Hitachi	CRAY	CDC
Model	VP-200	S-810/20	X-MP	205
announcement (or project start)	July 1982	Aug. 1982	Aug. 1982	June 1981
availability	Dec. 1983	Oct. 1983	June 1983	Feb. 1982
architecture	vector IBM compatible	vector IBM compatible	vector multi-processor	vector
maximum performance (MFLOPS)	500	630	630	400
maximum main memory size MB	256	256	32	128
Technology				
o Logic (gate delay)	ECL (350 ps)	ECL (350 ps)	ECL (.5-1 ns)	ECL (700 ps)
o main memory	64K MOS static (55 ns)	16K MOS static (40 ns)	4K ECL (25 ns)	MOS (45 ns)
Input/Output				
o No. paths (total)	32	32	19	16
o max data rate path (MB/sec)	3	3	1,000	25
o max aggregate data rate (MB/sec)	900	1,100	1,550	400

TABLE 1.2

device research that would create components needed in limited quantities by a few. Because of these two reasons, we are seeing these major efforts by the "few" on multiprocessor vector machines. Table 1.3 lists these efforts as well as performance goals, and expected delivery dates. Denelcor, of course, is not going into this vector regime, but rather into a greater number of CPUs.

The academic community and some select government laboratory research projects are aiming mostly at the multiprocessor schemes to achieve the next level of performance desired. Table 1.4 shows some 40 such efforts mostly in the U.S. These are basically research projects supported primarily by government agencies. We know even less about optimizing the use of numerous processors operating on a single problem, than we know about vector processing. There are studies, however, which could aid us in our designs. Presumably memory size and I/O capabilities have to be significantly enlarged to deal with the vast quantity of data to be processed in so short a time.

The current supercomputer generation, Class VI, the one which we are now learning to use, has been with us for a while, especially the Cray-1 which was first delivered in 1976. All the others are essentially of the '80s vintage. See Figures 1.1-1.3. Current and projected supercomputers, their performance and completion dates are shown in Figure 1.4. Most manufacturers are looking ahead to Class VII. The first of these, the Cray-2 is to be delivered in 1985. This machine has a four nanosecond clock, consists of four processors, each similar in structure to the Cray-1. Fortunately, in this system, addressable memory can be much larger. The first systems already spoken for will have at least 16 million words of main memory. In this instance, a new operating system based on UNIX is being worked on by Cray Research, Inc. The first customer, the Magnetic Fusion Energy Computer Center (MFECC), is writing its own operating system based on the Cray Time Sharing System (CTSS) being used at this center on the Cray-1.

Seymour Cray, himself, is designing the Cray-3 while the finishing touches are being put on the Cray-2. The Cray-3 will have a 1ns clock, using GaAs instead of silicon for chips. A special production facility has been built to manufacture these GaAs chips. The number of processors will also be increased, possibly to 8 or 16. Here, then, we see efforts to move ahead even to Class VIII Computers.

ETA Systems, the supercomputer off-shoot of Control Data Corporation is working on its version of Class VII. This has variously been called the GF10 (Gigaflops 10) or the CYBER 250. It will be essentially identical to the CYBER 205 in logical structure, except that eight modules will operate cooperatively on a single problem. This will be a silicon based machine. The strategy is to keep costs and price down by going to high density chips. This machine also will be looking at a UNIX derivative for its operating system.

ETA Systems also claims that it could easily build a 30 Gigaflops machine. This would get us into the Class VIII regime, but seems to be somewhat further off than the GF10.

SUPERCOMPUTER NOW IN DESIGN

Organization	CRAY	CRAY	ETA	Denelcor	NEC
Model	2	3	GF10	HEP-2	SX-2
announcement (or project start)	none yet	none yet	Sep. 1983	May 1983	Apr. 1983
availability	Apr. 1985	1986-7	1986-7	1986-7	1985
architecture	vector multi-processor	vector multi-processor	vector multi-processor	scalar	vector
maximum performance (MFLOPS)	1,000	NA	10,000	4,000	1,300
maximum main memory size MB	512	NA	2,048	4,000	256
Technology					
o Logic (gate delay)	ECL	ECL. GaAs	CMOS (0.5-1 ns)	ECL	NA
o main memory	MOS	MOS	MOS	ECL/MOS	64K MOS static
Input/Output			16		32
o No. paths (total)	NA	NA		NA	
o max date rate path (MB/sec)			50-60		1.5
o max aggregate data rate (MB/sec)	NA	NA	1,000	NA	1,350

TABLE 1.3

FORTY UNIVERSITY RESEARCH COMPUTER EFFORTS

Project	University
Ultra	MYU
Cedar	Illinois
Homogeneous Machine	CalTech
Pumps	Rice/Purdue
Trac	Texas
Dataflow (Static)	MIT
Dataflow (Tagged Token)	MIT
DDMI	Utah
Dataflow	Chicago
Manchester Dataflow	Manchester
Nonvon	Columbia
DADO	Columbia
Hypertree	Berkeley
Tree Machine	CMU
HM^2P	RPI
Mago	NCU
MUKET	MIT
Wavefront Array Proc	USC
PACS	Tsukuba
NNCP	CalTech
Database Machine	Wisc
EMPRESS	ETH Zurich
SERFRE	CNAM Paris
ZMOB	Maryland
MAP	Lille
AMP-1	Illinois
Polyp	Heidelberg
Parallel Hetrarchical Machine (LISP)	Mexico
PASM	Purdue
MP/C	Princeton
Parallel Speech Processor	Purdue
Midas	UCB
Manip	Purdue
Multiprocessor Reduction Machine	Newcastle
Connection Machine	MIT
Boolean Vector Machine	Duke
Tree Machine	CalTech
Blue Chip	Purdue
Systolic Processor	Kung
ELI	Yale

TABLE 1.4

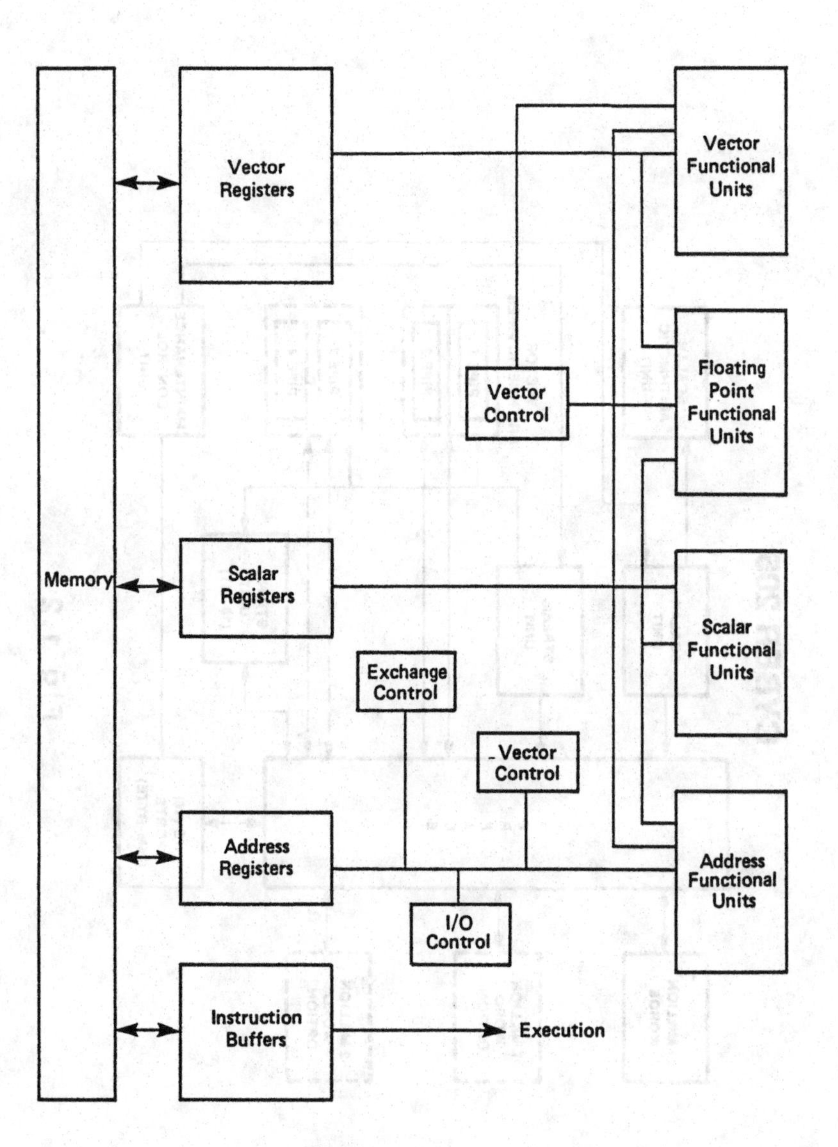
Memory
Vector
Registers
Scalar
Registers
Address
Registers
Instruction
Buffers
Execution
Vector
Control
Exchange
Control
Vector
Control
I/O
Control
Vector
Functional
Units
Floating
Point
Functional
Units
Scalar
Functional
Units
Address
Functional
Units

CRAY - 1

Fig 1.1

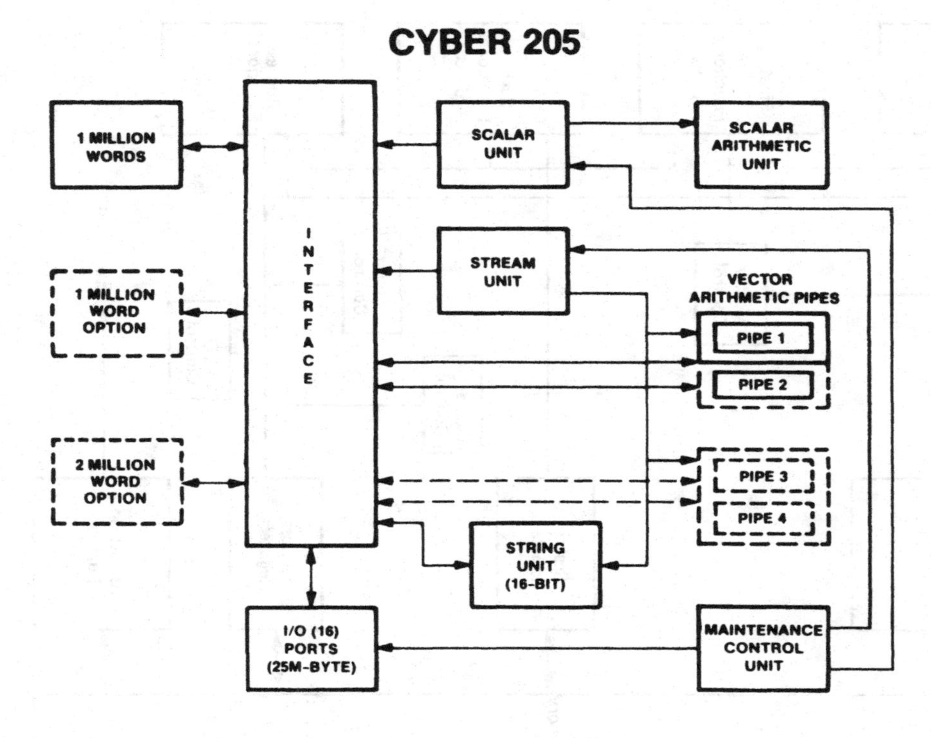

CYBER 205
1 MILLION WORDS
1 MILLION WORD OPTION
2 MILLION WORD OPTION
INTERFACE
SCALAR UNIT
SCALAR ARITHMETIC UNIT
STREAM UNIT
VECTOR ARITHMETIC PIPES
PIPE 1
PIPE 2
PIPE 3
PIPE 4
STRING UNIT (16-BIT)
I/O (16) PORTS (25M-BYTE)
MAINTENANCE CONTROL UNIT

Fig 1.2

DENELCOR
Heterogeneous Element Processor
(HEP - 1)

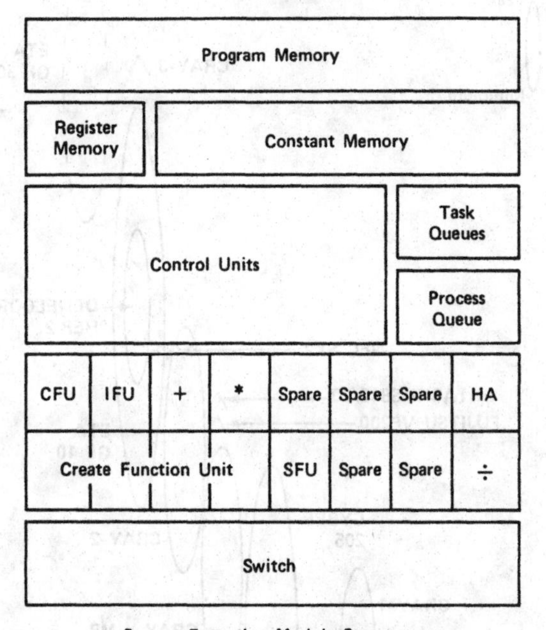

Process Execution Module Structure

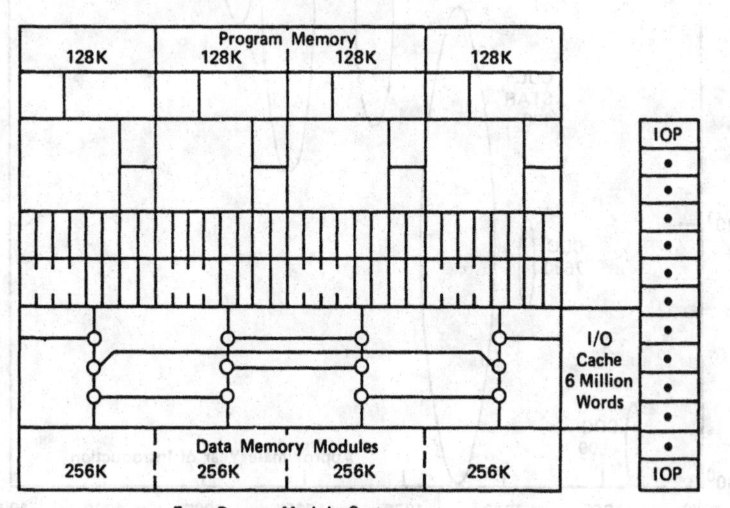

Four Process Module System

Fig 1.3

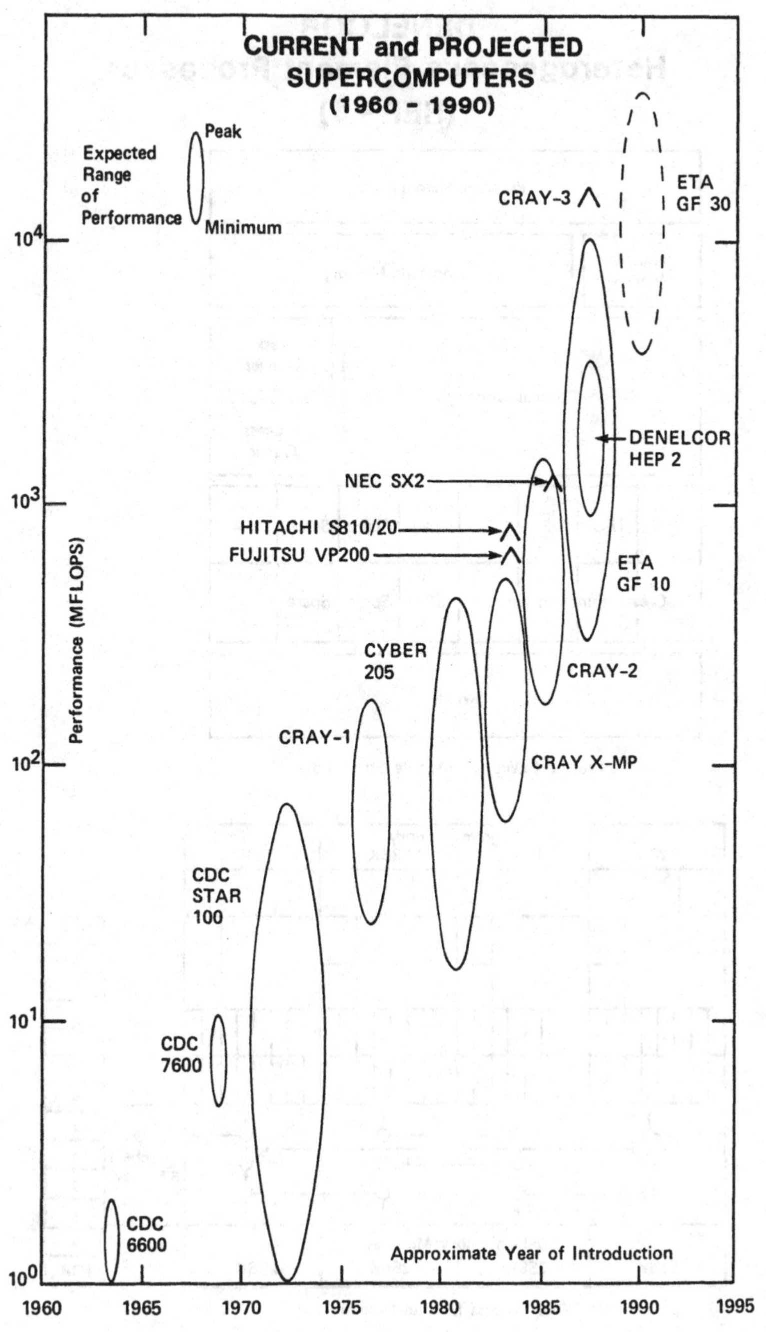
CURRENT and PROJECTED
SUPERCOMPUTERS
(1960 - 1990)
Expected
Range
of
Performance
Peak
Minimum
10^4
10^3
10^2
10^1
10^0
Performance (MFLOPS)
CRAY-3
ETA
GF 30
DENELCOR
HEP 2
NEC SX2
HITACHI S810/20
FUJITSU VP200
ETA
GF 10
CYBER
205
CRAY-2
CRAY-1
CRAY X-MP
CDC
STAR
100
CDC
7600
CDC
6600
Approximate Year of Introduction
1960
1965
1970
1975
1980
1985
1990
1995

Fig 1.4

Other possible U.S. contenders for the supercomputer market will be Denelcor, the AP manufacturers, and possibly IBM. The fact that IBM is more quiet than ever concerning its plans at the high end of their computer line, indicates that there may be an imminent announcement (within the next year), about its plans. At the same time, Enrico Clementi in IBM's research lab is using a number of Floating Point Systems Array Processors to solve his large chemistry problems. Also there seems to be a desire on the part of IBM to construct the NYU Ultra computer to test this concept.

Denelcor having had a fair amount of success with its HEP-1 is now designing the HEP-2. This will be more of the same; that is, it will consist of a larger number of processors and make use of higher performance components. The HEP-1 is fairly slow, but does gain considerable potential performance as processors are added. The largest system yet delivered is the 4-CPU system at Aberdeen Proving Grounds. The Denelcor machines also will use UNIX as the operating system. The performance being sought for the HEP-2 is of the order of five Gigaflops.

Some of the U.S. Scientists are pushing the concept of using a large number of array processors in a single system. This would make for a fairly inexpensive "supercomputer" if it can be assembled and made to work well. As mentioned above, Clementi of IBM is already using such a system. It is also being pushed by Professor Ken Wilson of Cornell University who recently was awarded the Nobel prize for his work in Lattice Gauge Theory. Floating Point Systems is, of course, interested in pursuing this model. Others, e.g. Univac, have tried such systems, but the scientific world at large is not too impressed with this design for a supercomputer.

One much overlooked multiprocessor system is the Control Data CYBER-PLUS. It originated as a fixed point special purpose system, built for a customer to do image processing. At that time it was called the AFP (Advanced Flexible Processor, see Figure 1.5). An eight processor system, front ended by a DEC VAX was actually delivered (1983). Since that time a floating point unit has been added, as well as some additional much-needed software, such as a Fortran compiler. This machine uses the same components as the CYBER 205 and has a 20ns clock. If it also were to be upgraded with the components going into the CYBER 250, it could be an excellent Class VII system.

The other supercomputer manufacturers in the world are all in Japan. Hitachi, Fujitsu and NEC either have delivered one or more systems or (NEC) are completing their first model. Hitachi built an IBM plug-compatible 200 model to which an array processor, the IAP had been added with satisfying results. This was followed by a more highly integrated system, the HAP-1, now called the S 810/10 or S 810/20 (Figure 1.6). The higher performing system, the 20, is supposedly capable of 630 MFLOPS at peak performance. One has been installed at the University of Tokyo. This machine has a 15ns clock. I would suspect that it can easily be followed up with a higher performance machine. This will depend on how big the market is for these systems in Japan. No plans are known for its being marketed in the U.S.

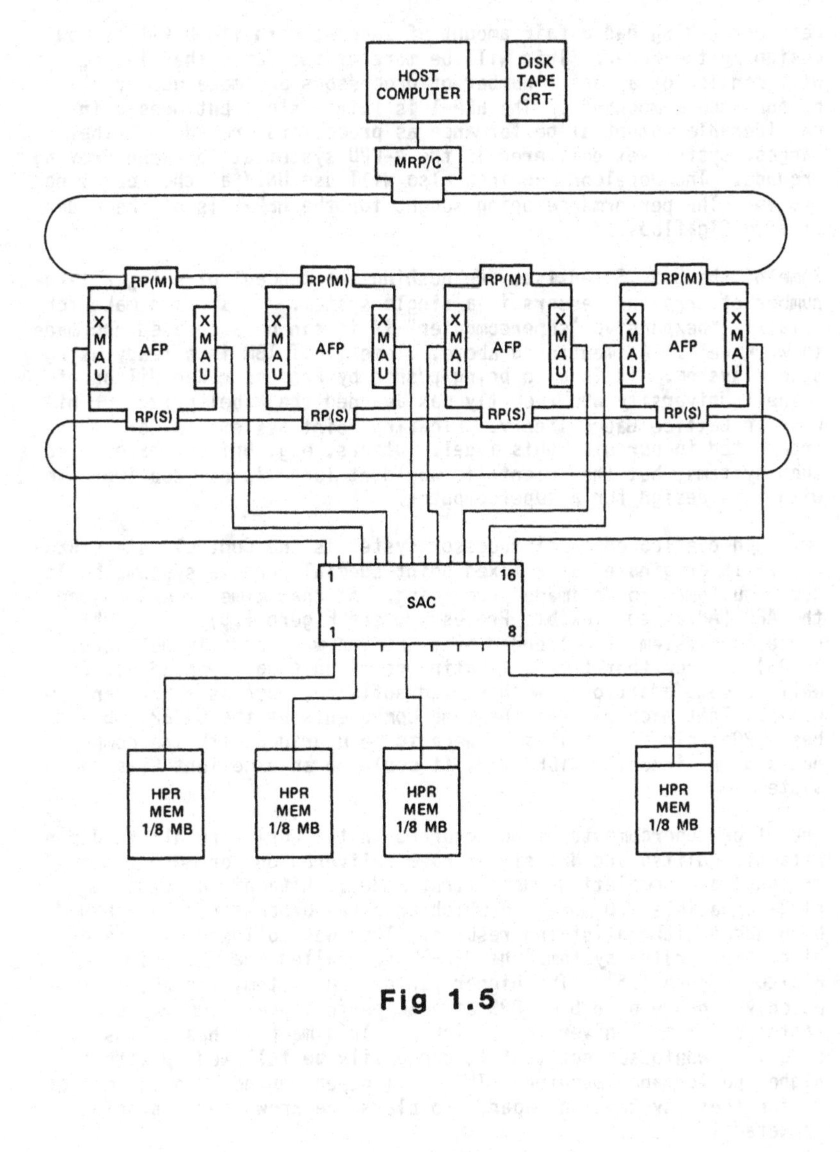

Advanced Flexible Processor
HOST COMPUTER
DISK TAPE CRT
MRP/C
RP(M)
XMAU
AFP
RP(S)
SAC
1
16
8
HPR MEM 1/8 MB

Fig 1.5

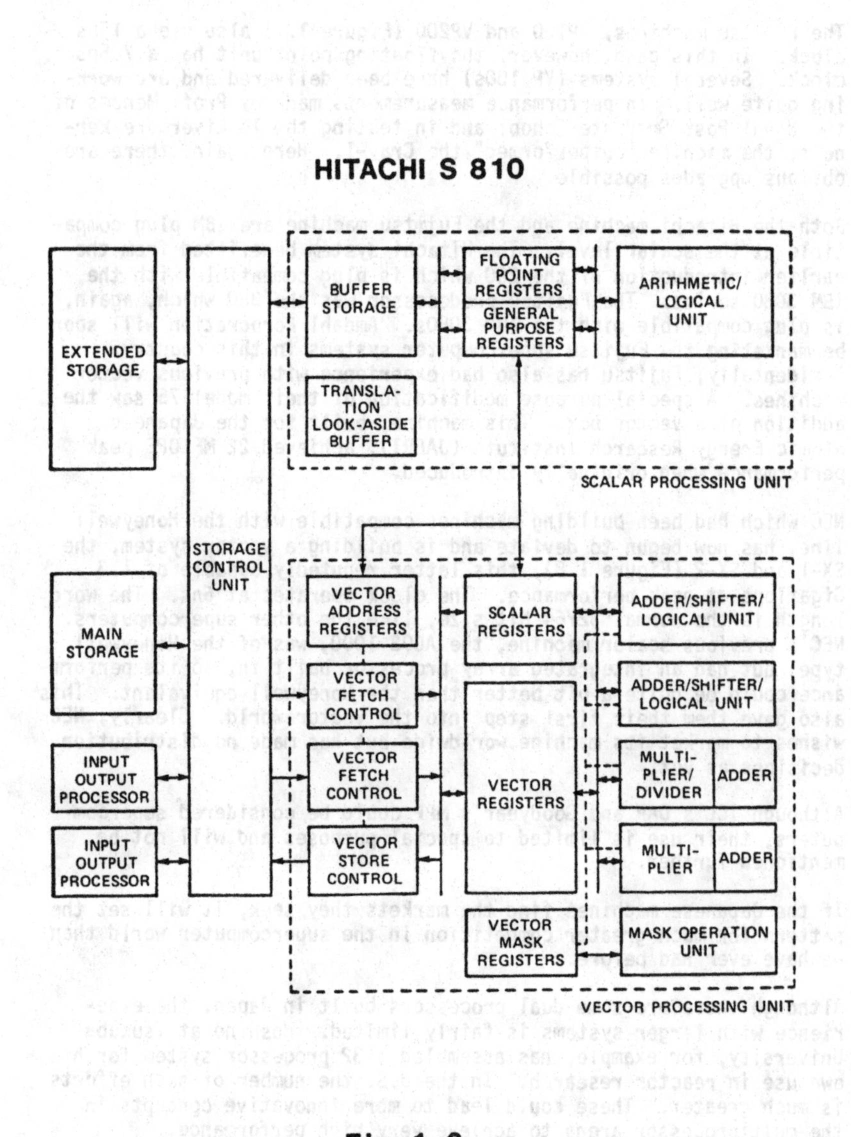
HITACHI S 810
EXTENDED STORAGE
BUFFER STORAGE
FLOATING POINT REGISTERS
GENERAL PURPOSE REGISTERS
ARITHMETIC/ LOGICAL UNIT
TRANSLA- TION LOOK-ASIDE BUFFER
SCALAR PROCESSING UNIT
STORAGE CONTROL UNIT
MAIN STORAGE
VECTOR ADDRESS REGISTERS
SCALAR REGISTERS
ADDER/SHIFTER/ LOGICAL UNIT
VECTOR FETCH CONTROL
ADDER/SHIFTER/ LOGICAL UNIT
INPUT OUTPUT PROCESSOR
VECTOR FETCH CONTROL
VECTOR REGISTERS
MULTI- PLIER/ DIVIDER
ADDER
INPUT OUTPUT PROCESSOR
VECTOR STORE CONTROL
MULTI- PLIER
ADDER
VECTOR MASK REGISTERS
MASK OPERATION UNIT
VECTOR PROCESSING UNIT

Fig 1.6

The Fujitsu machines, VP100 and VP200 (Figure 1.7) also use a 15ns clock. In this case, however, the floating point unit has a 7.5ns clock. Several systems (VP 100s) have been delivered and are working quite well. In performance measurements made by Prof. Mendes of the Naval Post Graduate School and in testing the 14 Livermore kernels, the machine "outperformed" the Cray-1. Here again, there are obvious upgrades possible.

Both the Hitachi machine and the Fujitsu machine are IBM plug compatible at the scalar level. The Hitachi system benefitted from the earlier introduction of the 280 which is plug compatible with the IBM 3080 series. The Fujitsu predecessor was the 380 which, again, is plug compatible with the IBM 3080s. Amdahl Corporation will soon be marketing the Fujitsu supercomputer systems in this country. Incidentally, Fujitsu has also had experience with previous vector machines. A special purpose modification of their model 75 saw the addition of a vector box. This machine, built for the Japanese Atomic Energy Research Institute (JAERI), achieved 22 MFLOPS peak performance when originally introduced.

NEC which had been building machines compatible with the Honeywell line, has now begun to deviate and is building a vector system, the SX-1 and SX-2 (Figure 1.8), this latter reputedly capable of 1.3 Gigaflops at peak performance. The clock operates at 6ns. The word length is the normal 32/64 bit size, like the other supercomputers. NEC's previous scalar machine, the ACOS 1000, was of the Honeywell type, but had an integrated array processor built in, so its performance could be quite a bit better than the Honeywell equivalent. This also gave them their first step into the vector world. Clearly, NEC wishes to market its machine worldwide but has made no distribution decisions as yet.

Although ICL's DAP and Goodyear's MPP could be considered supercomputers, their use is limited to special purposes and will not be mentioned further.

If the Japanese machines find the markets they seek, it will set the pattern for much greater competition in the supercomputer world than we have ever had before.

Although there are some dual processors built in Japan, the experience with larger systems is fairly limited. Hoshino at Tsukuba University, for example, has assembled a 32 processor system for his own use in reactor research. In the U.S. the number of such efforts is much greater. These could lead to more innovative concepts in the multiprocessor arena to achieve very high performance.

Jack Schwartz, one of the designers of the Ultra Systems at NYU has put together a taxonomy on the efforts worldwide (Table 1.5). He has also listed some 40 of the systems in accordance with this Taxonomy. If any of these systems are agressively pursued and completed (including software), the world of supercomputing could rapidly change. Today we seem to live in a vector world. Tomorrow it might be all multiprocessing.

FUJITSU VP 200

FACOM VECTOR PROCESSOR BLOCK DIAGRAM

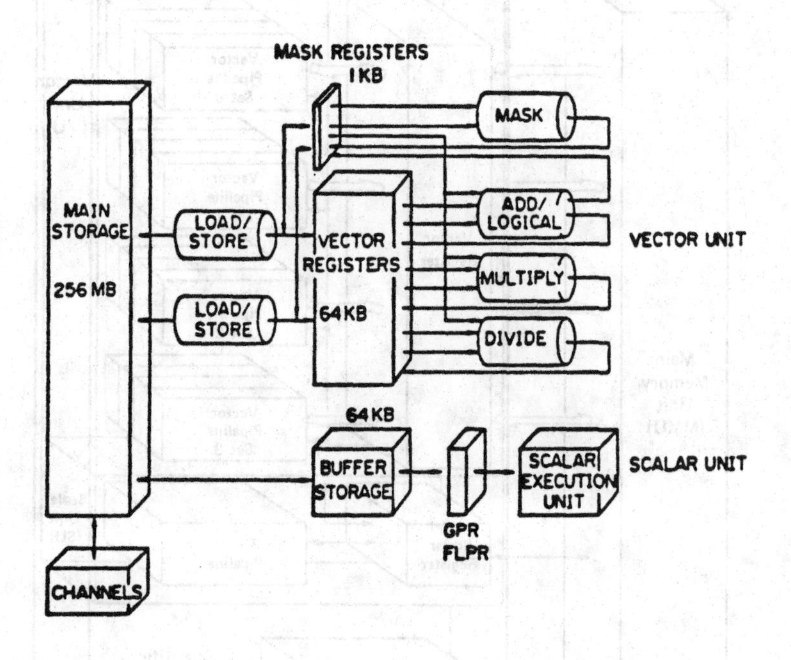

Fig 1.7

NEC SX 2

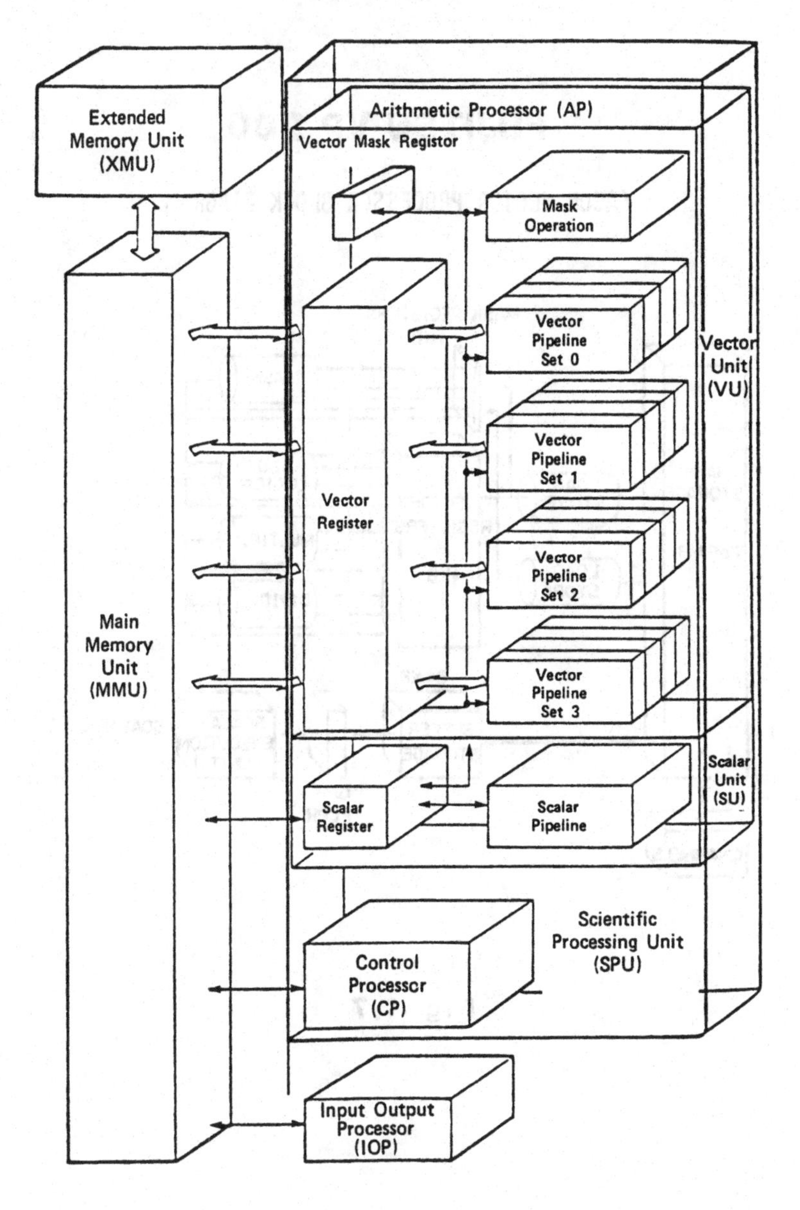

Fig 1.8

TAXONOMIC TABLE OF PARALLEL COMPUTERS

By JACK SCHWARTZ

I. "COARSE GRAINED" DESIGNS

- A. PROCEDURALLY ORIENTED, OMEGA NETWORK OR CUBE DESIGN
  - 1. PACKET SWITCHING
  - 2. CIRCUIT SWITCHING
- B. DATAFLOW
- C. TREE STRUCTURED MACHINES
- D. NEAREST NEIGHBOR MACHINES
- E. CROSSBAR DESIGNS
- F. RING STRUCTURED MACHINES
- G. BUS STRUCTURED MACHINES
- H. MISCELLANEOUS

II. "FINE GRAINED" DESIGNS

- A. BITWISE CUBE AND SHUFFLE
- B. NEAREST NEIGHBOR BITWISE PROCESSING
- C. TREE MACHINES
- D. CIRCUIT SWITCHING RECONFIGURABLE
- E. SYSTOLIC SPECIAL PURPOSE CHIPS

III. MICRO-OVERLAPPED SERIAL PROCESS OR DESIGN

TABLE 1.5

The National Security Agency (NSA) in the U.S. has decided to explore some of these systems, particularly as applied to Cryptology, nuclear and space research. A new supercomputer Research Laboratory will be opened sometime in 1985 in the Baltimore-Washington area to carry out this work. Eighteen million dollars have been budgeted for 1985. This number is to double in the following year. Presumably a few of the better designs will be pursued to completion.

It is not clear that we are ready for dataflow or systolic computers as yet, but presumably research in this area will continue. There have been but a few data flow machines actually built. Perhaps we will learn from experience with them when this technology becomes most useful.

Areas that have been neglected over the years with respect to the design and implementation of supercomputer systems is in peripherals and software. The vast peripheral marketplace is designed to take care of the small systems and now, personnel computers - primarily because of the huge volume (as compared with perhaps 100 supercomputers).

Software also has been neglected. It is left to the user to develop his own systems, utilities, etc. There would be many more users for supercomputer systems if they were as easy to use as some of the lesser systems.

In the U.S. the supercomputer manufacturers are small and specialize in mainframe hardware only. Japan, on the other hand, has highly integrated vertical organizations capable of solving the peripheral and software problems and providing complete system integration. In addition, however, the Japanese government has established a National Supercomputer project to assist its manufacturers in their studies of new components and multiprocessing. It may be that Japan will dominate the entire supercomputer industry for that reason alone. Sufficient to say that the U.S. needs extensive support for their supercomputer manufacturers if they are to survive.

Incidentally, France also has a national project on supercomputers (see Table 1.6).

NATIONAL SUPERCOMPUTER PROJECTS

Organization	Japan (MITI Project)	France (Min. de la Defense DRET Projects)		
Model		ISIS	MARIANNE	MARISIS
announcement (or project start)	(Jan. 1981)	(1980) (BULL)	(1981) (SINTRA)	(1982)
availability	1989	1985/86	1986	1988
architecture	vector multi-processor	vector	multi-processor	vector multi-processor
maximum performance (MFLOPS)	10,000	50-100	low	200
maximum main memory size MB	1,000	256	NA	NA 800
Technology o Logic (gate delay)	JJ/GaAs (10-30 ps)	NA	NA	NA
o main memory	16K Ga As (10 ns)			
Input/Output o No. paths (total) o max date rate path (MB/sec)	NA	NA	NA	NA
o max aggregate data rate (MB/sec)	1,500	1,200	NA	NA

TABLE 1.6

Progress in Scientific Computing, Vol. 6
Proceedings of U.S.-Israel Workshop, 1984

EXPERIENCE WITH A PERSONAL SIZED SUPERCOMPUTER IMPLICATIONS FOR ALGORITHM DEVELOPMENT

W. T. Thompkins, Jr.

ABSTRACT

While the use of supercomputers for computational fluid dynamics, CFD, has received considerable attention, alternatives to batch use of supercomputers for CFD have received little discussion. In the current paper, a class of systems which is inexpensive enough that they may be dedicated full time to single projects is discussed to demonstrate that such systems are a useful alternative to batch supercomputers for CFD research projects. Low-cost, dedicated systems that can perform computations of complexity equivalent to supercomputers in reasonable time will be termed personal sized supercomputers. A research project that has successfully constructed a personal-sized supercomputer from minicomputer system components will be discussed along with the computational model and message-passing algorithm representation required for implementation on the minicomputer system and future multiprocessing systems. Several other projects for developing personal-sized systems exist, and the successful use of such systems for significant computational projects will depend in large measure on the ability to efficiently implement the same numerical algorithm on several different computers or architectures. The message-passing style developed for the minicomputer system is suggested as a sound programming tool to achieve the required degree of architecture independence.

INTRODUCTION

The problems posed by computational aerodynamics for external flow configurations have received considerable discussion, and the computational demands of this discipline are widely perceived as a principal driver for supercomputer technology. Aerodynamic and heat transfer predictions for internal flow configurations, especially turbomachinery, have received considerably less publicity, but are also of extreme commercial importance. The technical problems of internal flow computations include all those associated with external flows as well as the presence of flow unsteadiness in most important applications. Because of the intrinsic design features of turbomachinery, few important examples of traditional analytical problems like potential flows or two-dimensional, steady, inviscid flows exist. Such flow models are of course used, but only as models for a more complicated flow situation. For turbomachinery flows a capability for solving the Navier-Stokes equations is required to have the same impact that potential flow solvers have had on external aerodynamics design. This fact requires that very large computational capacity be allocated for these computations.

In 1980 a research computing project was begun in the MIT Gas Turbine Laboratory to study the internal fluid dynamics of high speed, axial flow turbine stages. This project requires the development of new algorithms for compressible Navier-Stokes flows as well as detailed analysis of some specific three-dimensional geometries. Accurate heat transfer rate predictions are a major goal of the project which translated into finite difference grid sizes of 250,000 to 500,000 nodes for the geometries of interest. Since problems with this node size were, and remain, too large in memory requirements to be processed efficiently on commercially available supercomputers, this project was one of the first to rely completely on low cost, dedicated computational facilities rather than remote supercomputer access.

Dedicated computer facilities are considered to be those which deliberately trade off processing speed for other considerations and then are dedicated full time to a particular task rather than shared in order to make up the processing speed differential. In the turbine research project, processing speed was traded for an increased number of finite difference nodes by using a minicomputer as a host for an array processor and a large attached, solid state memory device, see reference [1]. Since this project was successful in providing computational results for three-dimensional, viscous flows before suitable commercial supercomputers were available, a natural question is what, if any, role special purpose machines of this type will play in future scientific computing research. A discussion of this question requires that a few common terms of reference be defined:

commercial supercomputer - term describing commercially available machines designed for highest possible processing speed on scientific computations. Outstanding examples in 1984 are the CRAY-1 series and the CDC 205.

batch supercomputer - pejorative term emphasizing that while commercial supercomputers are fast, they are expensive to maintain and run, they must be shared among many projects, and they offer less than an ideal computing environment.

personal-sized supercomputer - term for machines that trade processing speed for low purchase price, are inexpensive to maintain and run, can be dedicated to single projects, and offer a more balanced computing environment than batch machines. To qualify as a personal-sized supercomputer, a machine must be capable of computations of a complexity equivalent to commercial supercomputers in a reasonable time frame.

research computing - class of projects characterized by a few big CPU time cases, much algorithm development and job setup time, and detailed analysis of computational results.

engineering computing - class of projects characterized by many jobs whose run time is chosen for quick turnaround, fixed algorithms, minimal setup time, and minimal analysis of computational results.

The distinction made between engineering computing and research computing is critical to a discussion of personal-sized supercomputers and is perhaps best shown by this notion of detailed analysis of computational results. By its very nature, detailed analysis implies that one does not really know exactly what one is looking for or where to look for it. Processing to be conducted is ill-defined, one of a kind in type, and has no obvious parallelism. These qualities are, of course, the antithesis of problem types which "fit" on any commercial supercomputer, and such tasks require considerable personal attention of the research personnel. In the case of the turbine flow project, one flow solution may be analyzed for two to four weeks before an understanding of the flow phenomena emerges. In such environments, the urgency for processing speed that engineering computing environments generate is greatly diminished. If results of equivalent fidelity and resolution are available in a time scale short compared to a time scale for absorbing the results, how are very high processing speed rates valued?

The notion of shared resources is also critical to discussions of research computing and personal-sized machines. Research computing jobs are often very long by engineering computing standards and suffer significant slowdowns in batch computing environments. If only 1/20 of the capacity of a batch supercomputer is available to a research group, usually a very generous estimate, then that group could get just as many FLOPS from a personal-sized supercomputer. For the truly remote access to batch supercomputers that is typical of academic computing, an average usage of 1 CPU hour per day for any real length of time is essentially impossible to maintain due to logistics of large data base transfer and resource quotas. If a group actually uses a batch supercomputer 7 CPU hours per week at rate of 100 MFLOPS (millions of floating point operations per second), the average rate is a very modest 4 MFLOPS. True computational capacity is a much more complex and subjective issue than just peak or average FLOP rate.

These rather simplistic examples illustrate that different computing groups may rationally value FLOP rates quite differently given their tasks and responsibilities and that shared access to a batch supercomputer is not the same as unrestricted access to a commercial supercomputer. Since many different requirements exist for scientific computing, it should not be expected that any single performance metric like FLOP rate is adequate to judge all applications against. Thus it should be expected that machines with different performance levels and special capabilities will emerge to "fill out" the performance space.

It is the position of this paper that personal-sized supercomputers with a computation rate of order 1/20 that of commercial supercomputers are viable choices for many research computing projects if they can in fact perform calculations of equivalent fidelity and resolution. This rather arbitrary computation rate was chosen as a reference point because a batch computer user will typically get "overnight" service on long run time jobs. To illustrate that this performance level can be obtained with commercially available minicomputer system components, some computational aspects of the turbine research project will be discussed along with the algorithm expression constructs that were required to achieve this performance level. Other potential users of dedicated machines should then be in a better position to evaluate whether personal-sized supercomputer configurations will be suitable for their own needs or whether other architectures must be chosen.

RESEARCH PROBLEM STATEMENT AND ANALYSIS

The major scientific goals for the turbine computation project are aerodynamic and heat transfer predictions for high speed, axial flow turbine stages. These three-dimensional, unsteady, compressible, viscous flow fields are of considerable practical importance and at the same time tractable from an analytical viewpoint to only a modest degree. A full computational understanding of these flows would require a coupled solution of the unsteady flow over several hundred interacting airfoils which will clearly be beyond supercomputer computational capacity for some time to come.

Computations for the turbine project concentrate on the flow over a single airfoil with the effect of other airfoils accounted for through periodic boundary conditions. This approximation produces a model flow which is roughly analogous to that of a typical external flow analysis but has different far-field boundary conditions. A typical computational domain spans the flow area bounded by one blade passage, and the flow over every other blade in that same row is assumed to be identical. This model problem corresponds to a quasi-steady type approximation which ignores much of the interaction of one airfoil on another. For brevity, descriptions of algorithms and computed examples will be reported for two-dimensional examples, but similar flow computations in three dimensions are well underway.

The equations to be modeled are the time-dependent Navier-Stokes equations:

$$\frac{\partial \vec{q}}{\partial t} = \frac{\partial \vec{f}}{\partial x} + \frac{\partial \vec{g}}{\partial y} \tag{1}$$

where $\vec{q}$, $\vec{f}$ and $\vec{g}$ are the state and flux vectors;

$$\vec{q} = \begin{pmatrix} \rho \\ \rho u \\ \rho v \\ \rho e_t \end{pmatrix}, \quad \vec{f} = \vec{r} - \vec{f}_{iv}, \quad \vec{g} = \vec{s} - \vec{g}_{iv}$$

$$\vec{f}_{iv} = \begin{pmatrix} \rho u \\ \rho u^2 + p \\ \rho uv \\ u(\rho e_t + p) \end{pmatrix} \qquad \vec{g}_{iv} = \begin{pmatrix} \rho v \\ \rho uv \\ \rho v^2 + p \\ v(\rho e_t + p) \end{pmatrix}$$

$$\vec{r} = \begin{pmatrix} 0 \\ \tau_{xx} \\ \tau_{xy} \\ r_4 \end{pmatrix} \qquad \vec{s} = \begin{pmatrix} 0 \\ \tau_{xy} \\ \tau_{yy} \\ s_4 \end{pmatrix} \qquad \begin{aligned} r_4 &= u\tau_{xx} + v\tau_{xy} + k\partial T/\partial x \\ s_4 &= u\tau_{xy} + v\tau_{yy} + k\partial T/\partial y \end{aligned}$$

$$\tau_{xx} = \left(\frac{\partial u}{\partial x} + \frac{\partial v}{\partial y}\right) + 2\mu\,\frac{\partial u}{\partial x} \qquad \tau_{xy} = \mu\left(\frac{\partial u}{\partial y} + \frac{\partial v}{\partial x}\right)$$

$$\tau_{yy} = \lambda\left(\frac{\partial u}{\partial x} + \frac{\partial v}{\partial y}\right) + 2\mu\,\frac{\partial v}{\partial y}$$

Turbulent flow effects are accounted for by using algebraic models of the Cebeci-Smith type, see [2].

In general, some type of body-fitted grid is used to represent the solution to the Navier-Stokes equations, and the resulting grid is not suitable for direct evaluation of the flowfield derivatives in the x- and y-directions. In order to facilitate the evaluation of these derivatives, the physical domain, (x,y) space, is transformed to an evenly spaced computational domain, (ξ,η) space, and physical space derivatives are evaluated in the transformed domain. This process and the solution procedure using approximate factorization techniques is well documented in the literature, see references [3,4,5], and will be reviewed here only to the degree necessary for discussions of programming considerations for the minicomputer system to be described. The algorithm description will also ignore considerations of boundary conditions and smoothing.

The transformed equations are:

$$\frac{\partial(\vec{q}/J)}{\partial t} = \frac{\partial \vec{F}}{\partial \xi} + \frac{\partial \vec{G}}{\partial \eta} \tag{2}$$

where $\vec{F} = (\vec{f}\xi,_x + \vec{g}\xi,_y)/J$, $\vec{G} = (\vec{f}\eta,_x + \vec{g}\eta,_y)/J$ and J is the Jacobian of the mapping from (x,y) coordinates to (ξ,η) coordinates.

The numerical solution algorithm is written so that either time-accurate integration techniques or steady state acceleration techniques can be used. We specify the basic finite difference iteration scheme for each cell as:

$$(\vec{q}/J)^{n+1}_{j,k} = (\vec{q}/J)^{n}_{j,k} + \Delta t/2\left(\alpha_1(\mathrm{LNS}\,\vec{q}^{\,n})_{j,k} + \alpha_3(\mathrm{LNS}\,\vec{q}^{\,n+1})_{j,k}\right) \tag{3}$$

where $(\mathrm{LNS}\,\vec{q})^{n}_{j,k}$ is the finite volume approximation to the steady state Navier-Stokes equations, i.e. the finite difference approximation to the right hand side of equation (2). Common time marching schemes are recovered by choices of α_1 and α_3, for example the Backward Euler scheme by $\alpha_1 = 0$ and $\alpha_3 = 1$.

An implicit iteration scheme like this one cannot be solved without some linearization step which we choose to be a local Newton-Raphson step:

$$(LNS\ \vec{q}^{\,n+1}) = (LNS\ \vec{q}^{\,n}) + \frac{\partial(LNS)}{\partial\vec{q}}(\Delta\vec{q}^{\,n}) \quad (4)$$

where $\Delta\vec{q}^{\,n} = \vec{q}^{\,n+1} - \vec{q}^{\,n}$. This linearization coupled with the usual approximations and factorizations, see reference [4], yields the standard form:

$$(I - \overline{\overline{LX}})(I - \overline{\overline{LY}})(\Delta\vec{q}^{\,n}/J)_{j,k} = (\alpha_1 + \alpha_3)(\Delta t/2)((LNS\ \vec{q}^{\,n})_{j,k}) \quad (5)$$

The calculation order in this form is:

step 1

$$(I - \overline{\overline{LY}})(\Delta\vec{q}^{\,*}/J)_{j,k} = (\alpha_1 + \alpha_3)\Delta t/2((LNS\ \vec{q}^{\,n})_{j,k}) = (RHS\ \vec{q}^{\,n})_{j,k} \quad (6)$$

step 2

$$(I - \overline{\overline{LX}})(\Delta\vec{q}^{\,n}/J)_{j,k} = (\Delta\vec{q}^{\,*}/J)_{j,k} \quad (7)$$

step 3

$$\vec{q}^{\,n+1}_{j,k} = \vec{q}^{\,n}_{j,k} + \Delta\vec{q}^{\,*}_{j,k} \quad (8)$$

Myriad algorithm details essential for accurate results using a scheme of this type have been omitted in order to focus on computational representation of generic approximate factorization schemes. Readers interested in practical algorithm details should review references [3,4,5].

MINICOMPUTER SYSTEM DESCRIPTIONS

The configuration chosen for the turbine project computer experiment was a general purpose minicomputer coupled with a peripheral array processor [PAP] and an external memory device. The conceptual block diagram for this configuration is illustrated in Figure 1. The host CPU performs control structure decisions, routine housekeeping tasks, some general purpose calculations and acts as a large data buffer between the PAP and the external memory. The PAP performs the repetitive or vector floating point calculations which make up much of the operation count in typical fluid dynamics codes. The external memory is optimized for low cost and large size, and it is best considered to operate as a solid state disk.

The host minicomputer used is a 32 bit minicomputer produced by Perkin Elmer Corporation, the PE 3242. Important architecture features of the 3240 machine series are: 1) a high rate internal memory bus, up to 10 million 32 bit words/second transfer rate; 2) four DMA (direct memory access) ports, up to 2.5 million words/sec transfer rate on each port; and 3) a cache memory system well suited to internal data shuffle operations. Main memory access time is 500 ns, the cache cycle time is 50 ns, and an average floating point multiply time is approximately 2 μs for 32 bit single-precision results.

The essential characteristics distinguishing a PAP from an internal ALU are that it is programmable independently of its host, that it contains and executes its own instruction set under its own clock rate and that its instruction set is a relatively limited one optimized for specific arithmetic operations. Conceptually it matters little if a PAP is actually an external add-on peripheral or is physically located inside the host CPU. Several commercial vendors manufacture suitable processors which are optimized to perform fast Fourier transforms. The peripheral array processor used is the model AP-120B processor produced by Floating Point Systems, Inc. of Oregon. The AP-120B processor consists of a number of synchronous, parallel functional units, each operating under stored program control. A block diagram of the AP-120B is shown in Figure 2.

The parallel functional units of the AP-120B allow the overhead of array indexing, loop counting and data fetching from memory to be performed simultaneously with arithmetic operations on the data. Stored programs and data each reside in separate, independently addressable memories to reduce memory accessing conflicts. Independent floating point multiply and adder pipelines both allow operations to be initiated every AP-120B clock cycle or 167 ns. Address indexing and counting functions are performed by an independent integer arithmetic unit. For certain computations, such as a Fast Fourier Transform, the computation rate is near that of a floating point multiply and add result every clock cycle or 12 MFLOPS. For more general operations, the maximum computation rate is usually 2 to 3 MFLOPS. The AP-120B is connected to a PE3242 DMA port with a maximum data transfer rate of 0.6 MWORDS/sec. I/O operations from the host computer interface can also be concurrent with arithmetic operations as long as no data memory conflicts occur. The processor chosen for the turbine project contains 2048 (2K) words of program storage memory and 32K words of 333 ns cycle time main data memory.

The external memory device or bulk memory device (BMD) is manufactured by DataRam Corporation of New Jersey and is organized for our purposes as a register file of 8 million 39 bit words. Seven bits are used for error detection and correction. In the present configuration the BMD communicates by an interface which allows a maximum transfer rate of approximately 1.2 MWords/sec. Data transfer operations are initiated by setting up a starting address register, a word count register, and a read/write flag.

The present purchase cost of the minicomputer system including a disk drive, a tape drive, 2 MBYTES memory, terminals and software is approximately $400,000. This system provides all the computation and graphics capabilities for the turbine research project and four graduate student projects. Systems of less cost and capability could be constructed using a less expensive host computer and using disk storage to replace the BMD. Systems with more capability can be constructed using more powerful array processors.

MEMORY HIERARCHY AND I/O SYSTEM BOTTLENECKS

Since no data is permanently stored in the PAP and no direct path between the BMD and the PAP exists, the present structure demands considerable I/O system bandwidth to function efficiently. The host computer selected, the Perkin Elmer 3200 series, is probably unique in a commercially available minicomputer for its I/O system capacity. The 3242 configuration can support an aggregate system data rate 5.0 MWords per second. Device interfaces for the PAP and the BMD are limited to a 1.25 MWords per second data rate.

The AP-120B can perform block data transfers at the rate of 1 word every clock cycle or 6 MWords per second when the fast or 167 ns memory is used. With the 333 ns memory presently being used, the maximum AP-120B data rate is then 3 MWords per second. The AP-120B to host interface converts from 32 bit host floating point formats to 38 bit internal format on the fly. This potential rate is much larger than the rate which the interface that the AP-120B sits under can accommodate, and data transfer rates are limited by this interface to about 0.6 MWords per second. A typical block size transferred to or from the AP-120B is about 1000 words, and we currently incur a host system overhead time of about 1 millisecond to initiate a transfer operation. The system overhead time is in processing for the software interface between the host computer and the AP-120B.

For implicit Navier-Stokes solvers, the calculation time per data block is long compared to data transfer time, and most of the transfer time can be hidden or absorbed by concurrent calculation and I/O operations. A simple double buffering scheme is sufficient to hide the transfer time. For explicit solvers, calculation time per data block is approximately equal to data transfer and initiation time, and a complex scheme for buffering and overlap of the host and AP-120B calculations is required to hide these inefficiencies.

The situation is similar for the external bulk memory device, but a smaller system software overhead is incurred because the interface was custom designed for the application and because fewer but larger, contiguous data blocks are moved. Block lengths range from 20,000 to 45,000 words. Smaller data blocks moved to the AP-120B are assembled from larger data blocks stored in host memory. Simple double buffering schemes are sufficient to hide transfer time for the BMD.

PROGRAMMING CONSIDERATIONS

Programming is the art of converting an abstract algorithm, such as equations (6) through (8), which is machine independent to a form suitable for execution on a particular machine. Efficient computations using the minicomputer system require careful attention to computational load balancing between the host and the PAP, to data movement operations and to temporary result storage. While equations (6) to (8) completely define the algorithm from an algebraic representation, information on these other requirements is supplied only by inference. To more directly supply this information, an alternate object-oriented representation is used. This representation also serves as a vehicle to implement a particular abstract computational model which is well suited for scientific computations on multiple processor systems.

The computational model implemented is that of a collection of communicating processes, see reference [6]. A process is a set of instructions which may include instances of synchronization calls to other processes, message-passing to other processes, and computational processing. A process is not confined to execute inside a single processing node. Processes may execute concurrently, whether by virtue of being in different processing nodes or by being scheduled within a node.

In order to represent algebraic algorithms in this computational model, a control structure language has been developed. This language was originally conceived as an aid for representing algorithms, but interpreters and compliers for this language are being actively developed, reference [7]. Evaluation rules for this language are generally the same as those of a LISP interpreter, and are designed to highlight parallel operations, to eliminate explicit assignment of temporary variables and to minimize effort in writing complex control structures. For representation of an approximate factorization algorithm only two types of procedures are used:

1) Compute procedures which apply a fixed computational procedure to vector sets of input data and produce vector sets of output.
2) Gather/scatter data movement procedures which are used to move data from storage location to storage location, to transform data structures from computational procedures.

Computational processes represent the floating point operation dominated processing characteristic of CFD. These procedures transform data structures previously collected into intermediate data structures, and are assumed to have been coded in a sequential programming language which does not naturally support multiprocessing concepts although compliers for the language may support pipeline type operations where appropriate. These intermediate data structures are the messages passed between procedures. Their format is:

(computational-operator input-data-structure) => output-object

The essence of a computational operator is the computational processing performed rather than any possible data structure manipulations. A typical operator is illustrated by equation (6) which represents solving for the output vector, $\Delta\vec{q}^*$, using the technique of upper-lower factorization.

On the other hand, the essence of a gather/scatter type operation is the data structure transformation or information movement that it produces. A gather operation produces an intermediate level structure from a permanent structure, while a scatter operation produces a permanent structure from an intermediate level structure. A gather operation has the form:

(gather-op data-structure-access keys
    input-data-structure-name) => intermediate object

and a scatter operation has the form:

(scatter-op output-data-structure-name data-access keys
intermediate-object) => permanent object

For convenience, the special case of a scatter operation which produces no data structure transformation is called a move operation.

In this scheme the permanent data structures store state information about the simulations. Procedures communicate with the data structures using messages which are the data-access-keys, and information is returned via the intermediate objects. Message-passing between objects or procedures is also the mechanism that the control structure language uses to implement its function and its communications with the processor operating system. This use of message-passing as a mechanism to implement control structures is of course not new, and the best source on this idea is reference [8].

To implement the approximate factorization algorithm, we first define a computational operator, rhs-j-k, which transforms the state vector star $\vec{U}^n_{j,k}$ $\vec{U}^n_{j+1,k}$ $\vec{U}^n_{j-1,k}$ $\vec{U}^n_{j,k+1}$ $\vec{U}^n_{j,k}$ into the right hand side value of equation (6). Here $\vec{U}^n_{j,k}$ is a generalized state vector that includes not only the flow quantities, $\vec{q}$, but also any geometric quantities associated with cell j,k. Implementation of this procedure depends only on the manner in which the state vector star is passed and is independent of the details of how the state vector data structure is organized or even what device the data structure is located on. The procedure which contains knowledge of the data structures is called gather-star which returns the required values of $\vec{U}^n$ for a single cell. A right hand side calculation for a single cell is expressed as:

(rhs-j-k (gather-star cell-j-k $\vec{U}^n$)) => RHS-j-k

A set of right hand side values needed for a step 1 calculation, see equation (6), is obtained from the procedure rhs-j-k operating on set of input stars along a sweep-line:

(rhs-j-k (gather-rhs-star-vector sweep-line-j $\vec{U}^n$)) => RHS-j

A composed gather may also be used to represent this operation as:

(gather-rhs-j sweep-line-j $\vec{U}^n$) => RHS-j

$\Delta\vec{U}^*$ is the vector result of a computational procedure, see equation (6), operating on a vector of RHS values or the result of the composed procedure above:

(LX-OP (gather-rhs-j sweep-line-j $\vec{U}^n$)) => $\Delta\vec{U}^*$

The nature of approximation factorization algorithms is such that the temporary $\Delta\vec{U}^*$ values must be held until all step 1 sweeps are complete. This storage process is represented as:

(scatter-op $\Delta\vec{U}^*$ sweep-line-j (LX-OP (gather-rhs-sweep-line-j $\vec{U}^n$)))

All operations on all sweep-lines are parallel, and all gather/scatter memory operations are independent. Using these facts the complete algorithm is expressed as:

```
(parallel over- j-sweep-lines
     (scatter-op ΔU* sweep-line-j (LX-OP
          (gather-rhs sweep-line-j U^n))))
(parallel over-k-sweep-lines
     (scatter-op U^n+1 sweep-line-k
          (add (LY-OP (gather sweep-line-k ΔU*))
               (gather sweep-line-k U^n))))
```

The parallel procedure actually plays the role of defining a process for the computational model abstraction. In this representation, a process is identified with the sets of operations that can be independently carried out along grid sweep-lines. Such operations can of course be allocated to different processors in multi-processing environments, and the actual effect of the parallel procedure is to request scheduling of these processes. The control structure language interpreter is considered to be responsible for allocating storage for intermediate level objects, passing these objects from procedure to procedure, and generating constructs to loop over all j and k sweep-lines.

For single PAP systems, the task for the control structure language interpreter is not particularly difficult, but this representation is really designed for future systems which would have a number of processing nodes. Processing nodes of quite different architectures and capabilities can also be expected to be a part of future systems, i.e. inhomogeneous architectures rather than the present day style of homogeneous architectures. Algorithm representations such as those just discussed are relatively easily translated for these future systems while pure algebraic representations are not easily translated from architecture style to architecture style.

For execution on a uniprocessor machine, these simple control structures are sufficient. For execution on the minicomputer system, it is convenient to define the set of operations for a sweep-line as a process, and the individual computational or gather/scatter procedures as sub-processes that can execute on different processors. For example, the gather-rhs procedure becomes:

(AP-gather-rhs-j sweep-line-j $\vec{U}^n$)

or

(AP-compute-rhs-j (move-vector AP-MEM Buffer-i
    (gather-rhs-star-vector sweep-line-j
        (gather-star-from-BM cell-j-k $\vec{U}^n$))))

The complete algorithm would be represented as:

(parallel over-j-sweep-lines
    (scatter-to-BM $\Delta\vec{U}^*$ sweep-line-j
        (AP-LX-OP (AP-gather-rhs-j sweep-line-j $\vec{U}^n$))))
(parallel over-k-sweep-lines
    (scatter-to-BM $\vec{U}^{n+1}$ sweep-line-k
        (add (AP-LX-OP(gather-from-BM sweep-line-k $\Delta\vec{U}^*$))
            (gather-from-BM sweep-line-k $\vec{U}^n$))))

The requirements on the control structure language now are somewhat more complex since the host computer, the array processor, and the bulk memory device operate at different speeds and have different setup requirements. For the implicit Navier-Stokes solvers, the calculation time per data vector is long compared to data transfer time, and most of the transfer time can be hidden or absorbed by concurrent calculation and I/O operations. A simple double buffering scheme is sufficient to hide the transfer time. Responsibility for allocating these buffers and synchronization of computations and data transfers has not been passed to the control structure language.

SAMPLE COMPUTATIONAL RESULTS AND DISCUSSION

Since the primary purpose of this paper is to discuss dedicated computer systems for CFD applications rather than numerical algorithm performance, only a few computation examples and comparisons will be shown for a high pressure turbine cascade tested at Oxford University, see reference [8]. Figures (3) and (4) compare predicted and experimental blade surface pressure and heat transfer rate for design conditions. Agreement in surface pressure is excellent over the entire chord, and agreement in heat transfer rate is excellent over the pressure surface. Suction surface heat transfer rates are generally good except that the increase in heat transfer rate at the suction surface transition point is not well predicted. Oscillations in heat transfer rate along the suction surface seem to be generated by an unsteady separation near the leading due to overexpansion around the leading edge circle.

A typical convergence rate plot, Figure (5), illustrates one of the most important features of this example which is that a steady state solution is never reached. The periodic oscillation in residual values results from vortex shedding at the blade trailing edge. Computed velocity vectors at two different times are shown in Figures (6) and (7) which illustrate two of the vortex shedding states found during the computations. The predicted Strouhal number for these oscillations is approximately 0.20 which agrees well with general experimental results, but unfortunately the vortex shedding frequency was not directly measured in this experimental program.

Computation time for this calculation which has approximately 9000 finite difference grid nodes is 2.1 hours for 500 iterations, which is an average computation rate of about 2.5 MFLOPS. This computation rate for a machine which is not shared with other users provides adequate turnaround for algorithm development as well as flowfield studies. Significantly faster run times for a single case are of course possible on batch supercomputers, but significantly faster turnaround times over many cases are not easily available on any machine.

While two-dimensional Navier-Stokes computations are emphasized in this paper, the minicomputer system was designed for three-dimensional flow computations. At the present time three-dimensional flows using grids of size 100 x 50 x 50 are being processed. This size grid allows flowfield resolution equivalent to the two-dimensional cases shown for a three-dimensional turbine vane or rotor. A few preliminary results are available for these grids which demonstrate that computation rates of about 2 MFLOPS can be sustained. Major difficulties for these calculations are with the "scalar" portions of algorithms such as boundary conditions and turbulence modeling for complex flows rather than pure floating point computation rates, as was the case for two-dimensional examples. These "scalar" sections execute on the host CPU rather than the array processor, which degrades average processing speed much as scalar processing has a negative impact on pipeline computers. Present efforts on the three-dimensional algorithms are largely involved with reformulating these operations to forms suitable for array processor type computations.

It would appear that the best processing rate possible on the present minicomputer system for three-dimensional Navier-Stokes will be about 2 MFLOPS. For a 100 x 50 x 50 size grid, that rate translates into 24 to 48 hours runtime per 500 iterations. It must be emphasized that the state vector data base for this grid is approximately 9 MWORDS, which means a lengthy computation on any currently available computer.

Calculations of this size and runtime clearly fall into the research computing category and runtimes should be judged by those standards. For the turbine research computing project, researcher time to evaluate solutions, to modify and develop grids, or to modify algorithms still dominates these runtimes. For an engineering computing environment, overnight runtimes on batch supercomputers for problems of this size are generally all that can currently, late 1984, be achieved.

CONCLUSIONS, CONJECTURES, AND RECOMMENDATIONS

The sample results shown in this paper and in reference [3] illustrate that high quality, high resolution, computations can be carried out on dedicated machines, and that minicomputers plus attached array processors can reach performance levels that qualify as personal-sized supercomputers. Similar processing speeds can be achieved with many other configurations, and it seems a safe conjecture to suppose that other personal-sized machines will be developed. Current projects of special interest are the Cosmic cube and Hyper cube projects at California Institute of Technology.

It also seems a safe conjecture to suppose that future personal-sized machines will exploit multiple, independent processing elements to a much greater degree than will the commercial supercomputers which are emphasizing SIMD processing. This trend is evident in the personal-sized machine described in this paper which relies on having the host processor and the attached processor independently working on different tasks as well as in the Cosmic cube type configurations. Key items in efficiently utilizing even the simplest multiprocessing hardware like the minicomputer system are to develop a clear computational model that can be mapped onto the hardware and to develop algorithm representations with that computational model in mind. Such structures allow program development to proceed at appropriate abstraction levels with a minimum of attention devoted to specific hardware features.

Algorithm representation in the object-oriented or message-passing style is a natural one for multiprocessing systems because a clear separation between descriptions of process communication and descriptions of the detailed computations that a process must perform can be maintained. The representation of the approximate factorization algorithm presented uses this separation to provide a structure that can be implemented on diverse architecture styles without changing the coding of the algebraic algorithm. The control structure interpreter must of course produce different control sequences for different machines, but the burden of "understanding" the architecture has been moved to a "higher" abstraction level where generic tools for dealing with these problems can be created.

Researchers in computational methods could elect to perform a significant fraction of their computing on dedicated or personal-sized supercomputers if programs could be moved easily between different dedicated machines and to commercial supercomputers. In fact, it seems quite likely that much algorithm development, pioneering computing, and solution analysis would be performed on dedicated systems with a few very long running calculations or parametric investigations on batch supercomputers. The difficulties of generating good FORTRAN compliers for multiprocessing systems strongly indicates that algorithms for the personal-sized machines will be represented in a form such as the message-passing style. While message-passing is not traditionally considered as a useful programming style for vector processing machines, the hardware naturally requires these constructs. Efficient vector processing requires a sufficient quantity of well-defined input data for the pipelines and register or memory locations to store the intermediate results produced. These are exactly the problems addressed by message-passing styles, and it seems no more difficult to generate a suitable control structure interpreter for these machines than to generate a FORTRAN complier that "discovers" these same structures from a nested DO-LOOP programming style.

In conclusion, it does appear that a variety of personal-sized supercomputers will be developed and that these machines will heavily emphasize multiprocessing. CFD-type algorithms can easily be represented for such machines, and a significant fraction of research type computing conducted on them. If efficient schemes for translating programs designed for the personal-sized machines to commercial supercomputers can be developed, researchers can use each class of machines for their special properties rather than forcing applications to fit either style. Since future commercial supercomputers will also begin to emphasize multiprocessing architectures, considerable optimism can exist that compatible languages and algorithm representations will be developed.

ACKNOWLEDGMENTS

Major sponsors for the computer project are the Office of Naval Research under the direction of Dr. M. K. Ellingsworth, contract number N000014-81-K-0024; and Rolls-Royce Inc. I would also like to express our appreciation to Bob Haimes, without whom the minicomputer system would never have worked, and to Prof. A. H. Epstein for his help and advice throughout the project

REFERENCES

[1] Thompkins, W.T. and Haimes, R., "A Minicomputer/Array Processor/Memory System for Large-Scale Fluid Dynamic Calculations," Symposium on Impact of New Computing Systems on Computational Mechanics, ASME Winter Annual Meeting, Nov. 1983, Boston MA.

[2] Cebeci, T. and Smith, A.M.O., ANALYSIS OF TURBULENT BOUNDARY LAYERS, Academic Press, New York, 1974.

[3] Norton, R.J.G., Thompkins, W.T., and Haimes, R., "Implicit, Finite Volume Schemes with Non-simply Connected Grids: A Novel Approach," AIAA paper 84-003, presented at AIAA 22nd Aerospace Sciences Meeting, Jan. 1984, Reno, Nevada.

[4] Steger, J.L., "Implicit Finite-Difference Simulation of Flow about Arbitrary Two-Dimensional Geometries," AIAA Journal, Vol. 16, No. 7, July 1978, pp 679-686.

[5] Beam, R.M. and Warming, R.F., "An Implicit Factored Scheme for the Compressible Navier-Stokes Equations," AIAA Journal, Vol. 16, No. 4, April 1978, pp 393-402.

[6] Hoare, C.A.R., "Communicating Sequential Processes," Comm. ACM, August 1978, pp 666-677.

[7] Thompkins, W.T. and Dirks, P.W., "The Nemesis System. Simultaneous Simulation of Computer Architecture, Numerical Algorithms, and Operating Systems, MIT report CFDL-TR-84-5, September 1984.

[8] Hewitt, C., "Viewing Control Structures as Patterns of Message Passing," Artificial Intelligence, 8, 1977, 323-364.

[9] Nicholson, J.H., Forest, A.E., Oldfield, M.L.G., and Schultz, D.L., "Heat Transfer Optimised Turbine Rotor Blades--An Experimental Study Using Transient Techniques," ASME 82-GT-304, 1982.

Associate Professor
Department of Aeronautics and Astronautics
Massachusetts Institute of Technology

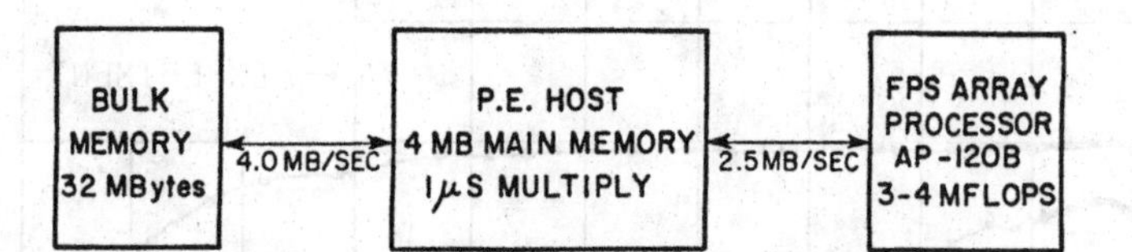

CONCURRENT OPERATIONS IN ALGORITHM PROCESSING

Vector Processing in FPS AP-120B

Scalar Processing in Host-Turbulence Model

Data Movement Between Bulk Memory ↔ Host ↔ AP-120B

Figure 1 Minicomputer System Block Diagram

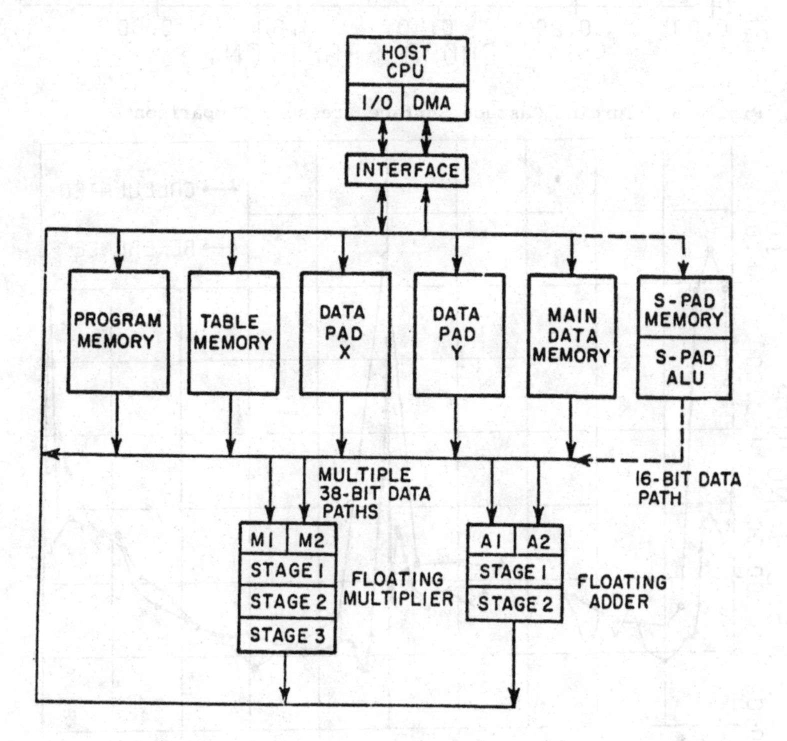

Figure 2 Array Processor Block Diagram
FPS - Model AP-120B

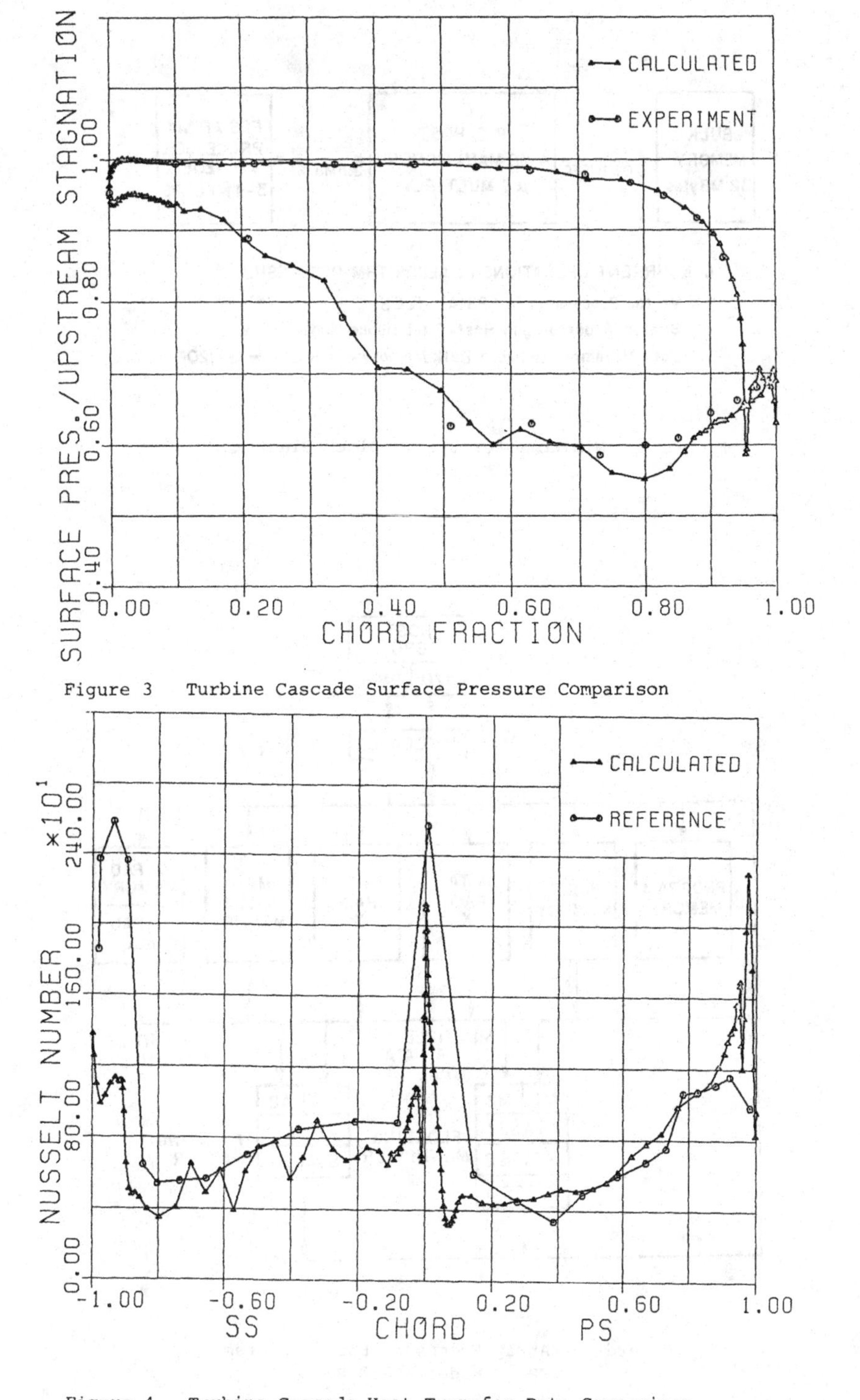

Figure 3 Turbine Cascade Surface Pressure Comparison

Figure 4 Turbine Cascade Heat Transfer Rate Comparison

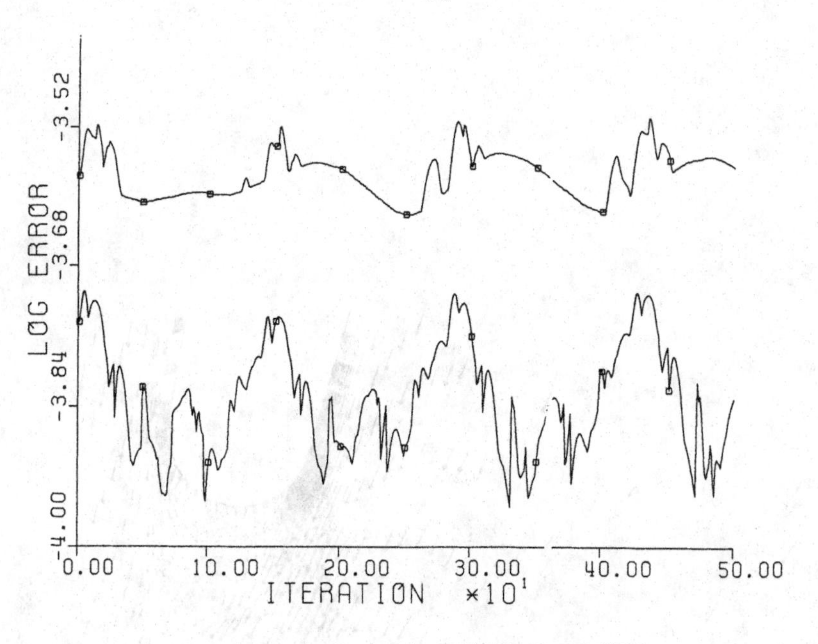

Figure 5 Typical Residual Convergence Rate

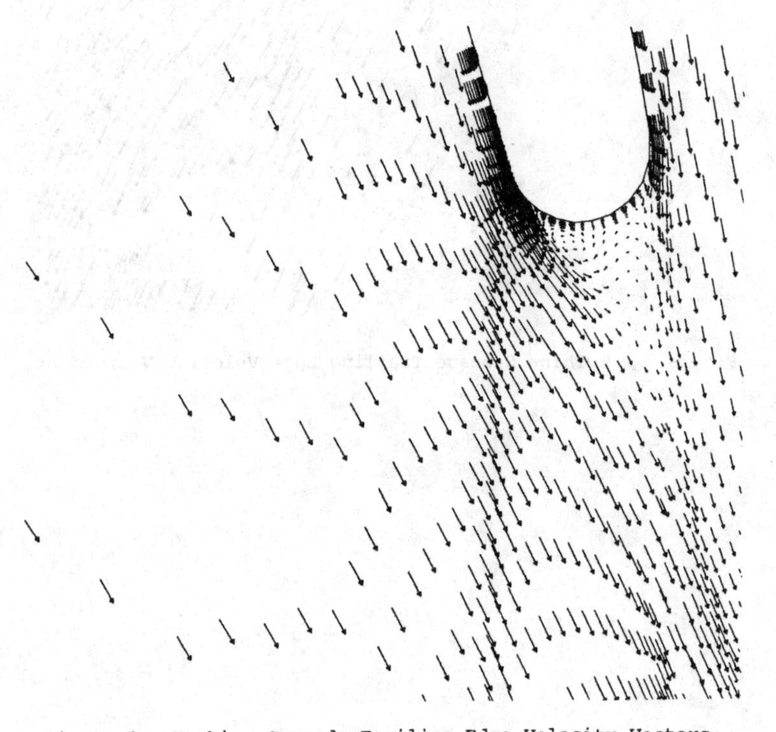

Figure 6 Turbine Cascade Trailing Edge Velocity Vectors

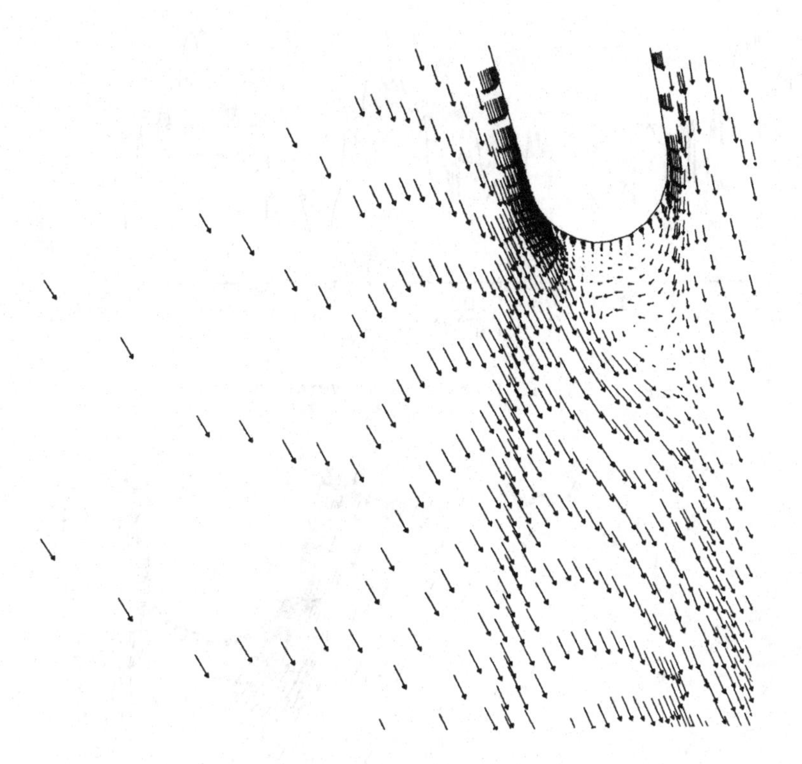

Figure 7   Turbine Cascade Trailing Edge Velocity Vectors

Progress in Scientific Computing, Vol. 6
Proceedings of U.S.-Israel Workshop, 1984

REMARKS ON THE DEVELOPMENT OF A MULTIBLOCK THREE-DIMENSIONAL EULER CODE FOR OUT OF CORE AND MULTIPROCESSOR CALCULATIONS

Antony Jameson, Stefan Leicher, Jef Dawson

1. Introduction

Our purpose in this paper is to describe some of the problems which were presented by the development of a three-dimensional Euler code with a multiblock grid structure in which only a single block at a time is held in core, and the main data base is out of core. The initial development of the code was motivated by the need to use a much denser mesh than could be accommodated in the then available supercomputers, which had one or two million words of memory, in order to provide adequate resolution of complex three-dimensional flows.

The code was based on the explicit multistage time stepping algorithm which had already been demonstrated to yield good accuracy with substantially lower computational costs than previously available algorithms [1,2]. While it was not emphasized in some of the earlier discussions of this scheme, one of the principal objectives which motivated its design had in fact been the need to find an algorithm which would perform effectively with vector, pipeline or parallel computer architectures. It seemed well worthwhile to accept some limitations on the rate of convergence to a steady state in return for factors of 20 or more in processing speed which might be realized through the use of long vectors in a pipeline machine, or by the introduction of multiple parallel processors. Accordingly, we had concentrated on extracting the maximum possible efficiency from explicit time stepping schemes, which in principle would allow simultaneous processing of every point in the entire flow field. In the event, through the introduction of measures such as variable time steps based on the local stability limit, enthalpy damping [1], residual averaging [2], and multigrid time stepping [3,4,5], it has proved possible to attain convergence sufficient for engineering predictions in about 50 cycles. Of these only the residual averaging would impose some limit on vectorization, but this restriction could be eliminated by substituting a Jacobi iterative scheme for the direct solution algorithm presently used. It has already been verified that this is effective in experiments with a scheme using a triangular mesh [6].

The present code is an adaptation of FLO57, a widely distributed code for three dimensional wing calculations. The initial multiblock scheme was programmed by Jameson during a visit to Dornier in the summer of 1982. It has subsequently been developed and refined principally by Leicher at Dornier, who also modified the scheme to allow the use of a multigrid solution algorithm in each block, and by Dawson at Cray Research. The original scheme was designed for disk storage. The consequent requirement of sequential access leads to some complications in the logical structure of the scheme which can be eliminated in situations where the data is held in a device allowing immediate access to arbitrarily located subsets. The solid state storage device of the CRAY X/MP, for example, allows blocks of data to be selected by a pointer. The same logical scheme for exchange of information across block boundaries of the grid could also be used for calculations on a multiprocessor machine with the different grid blocks assigned to different processors.

The multistage time stepping algorithm is briefly reviewed in the next section to provide a framework for a description of the logical structure of the multiblock scheme in Section 3. We discuss programming problems in Section 4, and finally we present an example of a calculation using 2.5 million mesh points in Section 5.

2. Description of the Algorithm

The algorithm is intended for use in steady state calculations of three dimensional compressible inviscid flows. It simulates the unsteady Euler equations, with modifications to accelerate convergence to a steady state. A semi-discrete approximation is first introduced by subdividing the domain into hexahedral cells. This reduces the equations to a set of ordinary differential equations, which are integrated by a multistage time stepping procedure. This separation of the space and time discretization procedure assures that the steady state is independent of the time stepping scheme.

We denote the pressure, density and three Cartesian velocity components as p, ρ, u, v and w. For a perfect gas we can express the total energy as

$$E = \frac{p}{(\gamma-1)\rho} + \frac{1}{2}(u^2 + v^2 + w^2).$$

The total enthalpy is then defined by $H = E + p/\rho$. In integral form, the Euler equations can be written as

$$\frac{\partial}{\partial t}\iiint_{\Omega} w\,d\Omega + \iint_{\partial\Omega} \underline{F} \cdot \underline{dS} = 0 \qquad (1)$$

for a fixed region Ω with boundary $\partial\Omega$. w represents the conserved quantity and $\underline{F}$ is the corresponding flux across the boundary. The x momentum equation, for example, is obtained by setting $w = \rho u$ and $\underline{F} = (\rho u^2 + p, \rho uv, \rho uw)$. The dependent variables w are assumed to be known at the center (i,j,k) of each mesh cell. A semi-discretization leads to the equation

$$\frac{d}{dt}(V_{ijk} W_{ijk}) + Q(W)_{ijk} = 0 \qquad (2)$$

Here Q_{ijk} represents the net flux out of the cell which is balanced by the rate of change of w in the cell whose volume is V. The flux is given by

$$Q_{ijk} = \sum_{\text{cell sides}} \underline{F} \cdot \underline{S}$$

where $\underline{F}$ is the flux at the center of a cell face and $\underline{S}$ denotes the cell face area. The value of $\underline{F}$ at the cell face is taken as the average of $\underline{F}$ at the cell centers on either side of the cell face.

In order to suppress the tendency for odd and even point decoupling and to capture shockwaves without any overshoots, it is necessary to add a dissipative term to equation (2). This leads to the semi-discrete equation

$$\frac{d}{dt}(V_{ijk} w_{ijk}) + Q_{ijk} - D_{ijk} = 0 \qquad (3)$$

The term D_{ijk} is constructed so that it is of third order in smooth regions of flow. For the density equation $D_{ijk}(\rho)$ has the form

$$D_{ijk} = d_{i+1/2,j,k} - d_{i-1/2,j,k} + d_{i,j+1/2,k} - d_{i,j-1/2,k} + d_{i,j,k+1/2} - d_{i,j,k-1/2}$$

where

$$d_{i+1/2,j,k} = \frac{V_{i+1/2,j,k}}{\Delta t^*} \varepsilon^{(2)}_{i+1/2,j,k} \Delta_x \rho_{i,j,k} - \varepsilon^{(4)}_{i+1/2,j,k} \Delta_x^3 \rho_{i-1,j,k}$$

and Δ_x is the forward difference operator defined by

$$\Delta_x \rho_{ijk} = \rho_{i+1,j,k} - \rho_{i,j,k}.$$

Also Δt^* is the time step corresponding to a nominal Courant number of unity, resulting in a normalization proportional to the maximum signal speed.

The coefficient $\varepsilon^{(2)}_{i+1/2,j,k}$ is made proportional to the normalized second difference of the pressure

$$\nu_{ijk} = \left| \frac{p_{i+1,j,k} - 2p_{ijk} + p_{i-1,j,k}}{p_{i+1,j,k} + 2p_{ijk} + p_{i-1,j,k}} \right|$$

in adjacent cells. This quantity is of second order except in regions containing a steep pressure gradient. The fourth differences provide background dissipation throughout the domain. In the neighborhood of a shockwave, ν_{ijk} is order one and the second differences become the dominant dissipative terms. The dissipative terms for the other equations are constructed from similar formulas with the exception of the energy equation, where the differences are of ρH rather than ρE. The purpose of this is to allow a steady state solution for which H remains constant.

The cell volume V_{ijk} is independent of time so we can write (3) as

$$\frac{dw}{dt} + R(w) = 0$$

where

$$R(w) = \frac{1}{V_{ijk}} (Q_{ijk} - D_{ijk})$$

The time integration is carried out by using a multistage scheme in conjunction with a multigrid strategy. If we let w^n and w^{n+1} represent the values of w at the beginning and end of the nth time step, then a k stage scheme for updating the fine grid values can be written as

$$w^{(0)} = w^n$$

$$w^{(q)} = w^{(0)} - \alpha_q \Delta t \, R^{(q-1,r)}, q = 1,2,\ldots k$$

$$w^{n+1} = w^{(k)}$$

In the (q+1)st stage $R^{(q,r)}$ is evaluated as

$$R^{(q,r)} = \frac{1}{V_{ijk}} (Q_{ijk}(w^{(q)}) - D_{ijk}(w^{(s)}))$$

where $s = \min(q,r)$ with $0 < r < k$.

The coefficients $\{\alpha_q \mid q = 1,\ldots, k\}$ can be chosen to generate schemes with desirable stability properties. The parameter r determines the number of stages in which the dissipative terms are re-evaluated. Thus with r equal to zero, the dissipation is evaluated once and then frozen after the first stage; with r set equal to 1 the dissipation is evaluated twice. By freezing the dissipation prior to the final stage (typically $r = 0$ or 1) it is possible to tailor the multistage scheme to provide good damping of the amplification factor at high frequencies, and thus generate a scheme well suited to a multigrid strategy.

3. Logical Structure of the Multiblock Scheme

The program is designed to use a multiblock grid structure of the type illustrated in Figure 1. Here the grid blocks are arranged in a triply indexed array with block indices IB, JB, KB, while the grid points inside each block are represented by internal indices I, J, K. The data exchange problem could be simplified by a subdivision into planes, but the present subdivision is designed to allow maximim flexibility in the grid generation procedure. It appears likely that grid generation for a complex configuration can be simplified by a subdivision into subdomains which could be different blocks of the multiblock grid. Separate grid generation procedures can then be used within each block. The topological constraints imposed by the multiplicative structure is a very slight limitation. An L-shaped region, for example, can be treated by filling out the array with empty subblocks. This procedure has been used by Fritz at Dornier in the generation of grids for the treatment of flows past cars.

The data exchange problem can be understood by referring to Figure 2, which illustrates a typical block. The grid indices inside the block range from I = 2 to IL, J = 2 to JL, K = 2 to KL. Cells external to the block are defined by index values I = 1, I2, J = 1, J2 and K = 1, K2. If the block boundary is a boundary of the whole flow field (either the surface of the configuration, or the outer boundary in the far field) then the values of the flow variables in the boundary cells must be determined by the boundary conditions. If the block boundary is an interface to a neighboring block, on the other hand, then the values in the boundary cells must be obtained from cells just inside the boundary of the next block.

In order to avoid the need to keep the neighboring blocks in core, which would require the storage of 7 blocks in the case of a fully interior block, we introduce buffers at all block boundaries. The following data structure serves this purpose. In the main database, which may be stored on a disk or a solid state storage device, we define a file MO to contain the values of the flow variables in all the blocks, and buffer files MI1, MI2, MJ1, MJ2, and MK1, MK2 to store the values at the boundaries of the blocks in the I, J and K directions. The flow variables inside a block are represented by the array W(I,J,K,N), where the index N ranges from 1 to 5 to represent the dependent variables (the density, three momentum components and the energy). To initialize W in a block for a stage of the multistage time stepping scheme we read W(I,J,K,N) for I = 2, 1L, J = 2,JL, K = 2,KL and N = 1,5 from MO. We read W(1,J,K,N) for J = 2,JL, K = 2,KL and N = 1,5 from MI1, and so on.

After updating the flow field inside the block the data is returned to the main database. In the case of disk storage, however, one would wish to read the data for the next block from the same files, and one cannot interleave read and write statements without returning to the beginning of the file. Moreover, the time stepping algorithm calls for the calculation of the correction in each cell from values in neighboring cells which have not yet been updated. Correspondingly each block should be treated using boundary values from neighboring blocks which have not yet been updated. Therefore the updated values are returned to a duplicate set of files NO, NI1, NI2, NJ1, NJ2 and NK1, NK2. The values just inside the left boundary provide the data for the right boundary of the block to the left. Thus we write W(2,J,K,N) for J = 2,JL, K = 2,KL, N = 1,5 into the file NI2. Correspondingly we write W(IL,J,K,N) for J = 2,JL, K = 2,KL, N = 1,5 into the file NI1, and the buffers on the other boundaries are treated in the same way. Finally after all the blocks have been updated we interchange the names of the files, so that on the next stage the files which have just been written as N files are read as M files.

With random access to the storage device duplicate files would not be needed for the interior data in each block, but duplicate buffer files would still be needed at the block boundaries to preserve the data flow of the algorithm. With disk storage, however, the need to read the files in the same order in which they have been written leads to an additional complication. Boundary values at the outer boundary of the entire grid have to be determined from data coming from the same block. This means, for example, that a block along the bottom boundary must provide data for its own bottom boundary which would normally be placed in NJ1, and also the data for the block immediately above it. Referring to Figure 3, we can see that if the blocks were updated in the order indicated in Figure 3(a), which is also the order in which the MJ1 file would be read, then the NJ1 file would be written in the order indicated in Figure 3(b). In the existing code this incompatibility is eliminated by using separate buffer files MI0, MI3, MJ0, MJ3 and MK0, MK3 instead of the files MI1, MI2, MJ1, MJ2 and MK1, MK2 to contain values at all outer boundaries. The result of updating the interior buffer files as they are processed is then to place the data in the proper order for these files to be read.

A final complication is caused by the topology of the C mesh which is used in the code. Referring to Figure 4, it can be seen that this leads to the need to exchange data across the cut between blocks which are not contiguously numbered. This requires the introduction of an additional set of buffer files for blocks separated by the cut.

The implementation described here does not allow for the introduction of dissipative terms using fourth differences across boundaries, and in the existing code these higher order dissipative terms are simply switched off at the block boundaries. No difficulties have been encountered, but one could include fourth differences across boundaries without too much difficulty by using buffers containing 2 planes of data points instead of only a single plane.

The same logical structure for data exchange between blocks carries over to the situation in which the grid is divided into blocks which are simultaneously updated by separate processors. If the number of blocks equals the number of processors then the interior values might be retained in the separate memories of the different processors. If there are more blocks than processors one might still wish to transfer the interior values back and forth between local memories and a second level central memory. In any case the data exchange between blocks updated by different processors can be neatly accomplished by duplicate M and N boundary buffers in the same way as in the present code. It would then only be necessary to flag the completion of the updating process in all blocks before exchanging the names of the M and N files, and proceeding to the next sweep through the blocks.

4. Remarks on Programming

In carrying out the various data transfers it is important to avoid nested read or write statements. For example the transfer of data from the file MI1 at left hand block boundary might be accomplished by

```
READ(MI1)(((W(1,J,K,N),J=2,JL), K = 2,KL),N = 1,5)
```

This results in separate input/output operations for each array element. To avoid this we can define a buffer array BUFI(J,K,M) dimensioned to J2,K2,5.

Now one can write

```
READ (MI1) BUF1
```

followed by

```
    DO 10 N = 1, 5
    DO 10 K = 2, KL
    DO 10 J = 2, JL
    W(1,J,K,N) = BUFI(J,K,N)
 10 CONTINUE
```

The block dimensions may vary from one case to the next. To avoid wasteful transfers of meaningless data it is best to use variable dimensions for the array BUFI. This can be accomplished by placing the transfer statements inside separate subroutines. The efficiency of the data transfers is also improved by using the BUFFER IN and BUFFER OUT statements to allow asynchronous transfers which can be overlapped with the calculations.

With these measures the data transfer operations incur only a small penalty in processor time, even with disk storage. The use of a disk leads, however, to a long residence time because of the time spent in data transfer during which the processor may be switched to the execution of other programs. Also, depending on the accounting system, there may be a substantial charge for the disk input/output operations themselves.

The volume of file transfer operations can be drastically reduced by modifying the algorithm so that all the stages of one time step are performed in each block before passing to the next block. Whereas the loop over the block indices IB, JB, KB would previously be inside the loop for the time stepping stages, it is now brought outside. The resulting restriction of the data flow between the blocks could have an adverse effect on the stability of the scheme. It has proved to work quite well in practice, however, incurring only a slight reduction in the rate of convergence to a steady state. One can go further and perform one complete multigrid cycle in each block before passing to the next block. This has been found to be quite effective in calculations performed by Leicher at Dornier. Experiments by Jong Yu at Boeing also indicate that in this situation, where any pretence of time accurate simulation has been abandoned, it pays to use the latest available values for the boundary data of all the blocks, so that an interior block would be treated with new values on some boundaries and old values on others, as in a Gauss-Seidel scheme. If the frequency of data exchanges between blocks is limited, it may also pay to use more complicated interface conditions taking account of the directions of wave propagation.

5. Example of a Multiblock Calculation with 2.5 Million Grid Points

The multiblock Euler code FL057 is now in routine use at Dornier for prediction of flow about wings and wing body combinations. It was adapted this summer by Leicher and Dawson for use on the CRAY X/MP with a solid state storage device (SSD) with a capacity of 128 million words. The results of a test calculation of the vortical flow past a delta wing are presented in Figure 5. A C-mesh was used, with 352 cells in the chordwise I direction, 64 cells in the normal J direction and 112 cells in the spanwise K direction, giving a total of 2,523,136 volumes. The grid was subdivided in the J and K directions to produce 16 blocks each with $352 \times 16 \times 28$ volumes. The configuration is the well known Dillner wing, at a Mach number of 1.5 and an angle of attack of 15 degrees. The figure shows pressure contours over the planform, and plots of the pressure coefficient across the span at two longitudinal stations, at 40% and 80% of the root chord. Preliminary calculations were performed using coarser meshes with $88 \times 16 \times 28$ and $176 \times 32 \times 56$ volumes, and the results on each mesh were used to provide initial values for the next mesh. A five stage time stepping scheme was used in which the dissipative terms were evaluated twice during each time step. This is the scheme which has proved most successful in multigrid calculations [4,5]. In this case residual averaging was used, allowing time steps corresponding to a Courant number of 6.5, but multigrid was not used. The complete calculation of 150 steps on the $88 \times 16 \times 28$ mesh, followed by 150 steps on the $176 \times 32 \times 56$ mesh, and 500 steps on the $352 \times 64 \times 112$ mesh, took 7.9 hours of CPU time using a single processor of the CRAY X/MP, and 8.3 hours of wall clock time. Thus the overhead incurred by the use of the SSD was about five percent.

The five stage algorithm requires 1422 floating point operations per mesh cell, including 368 for two evaluations of the dissipation, and 375 for the residual averaging. The estimation of the permissible time step before each cycle requires another 74 operations, including 3 square roots, giving a total of 1496 operations per cell. With 2,523,136 cells the number of operations to advance one time step is thus about 3.77×10^9, not counting the additional operations needed to enforce the surface and farfield boundary conditions. Allowing for the preliminary calculations on the coarse and medium meshes in addition to 500 steps on the fine mesh, the total number of operations required for the treatment of the interior cells amounts to an aggregate for the three meshes of about 1.96×10^{12}. Thus the CPU time of 7.9 hours represents a sustained computing rate in the neighborhood of 70 megaflops. This performance is a consequence of the vectorization of every inner loop in the implementation of the entire algorithm.

6. Conclusion

These experiments clearly indicate the feasibility of using supercomputers of the current generation for the prediction of transonic flow past a complete aircraft by solution of the Euler equations for inviscid flow. Using a multi-block patched grid, it should be possible to resolve the main geometric features with a grid containing about a million cells, and with the present algorithm such a calculation would require a CPU time of the order of several hours. The main remaining difficulty lies in grid generation, and attrition of the accuracy and rate of convergence induced by grid irregularities and singularities. Research is needed on the development of discretization schemes which do not require grid regularity, and methods of obtaining rapid convergence to a steady state with a multi-block structure. Looking further ahead, the inclusion of viscous effects will be the next step. This could also be accomplished with computers of the current power by the introduction of turbulence models. The development of sufficiently reliable and accurate models is likely to be the pacing item of such a development. For a more detailed prediction based on large eddy simulation we must look forward to the appearance of machines such as the CRAY 3 and ETA Systems GF10 in the vanguard of a new generation of computers.

References

1. Jameson, A., Schmidt, W., and Turkel, E., "Numerical Solution of the Euler Equations by Finite Volume Methods Using Runge-Kutta Time Stepping Schemes," AIAA Paper 81-1259, 1981.

2. Jameson, A., Baker, T.J., "Solution of the Euler Equations for Complex Configurations," Proc. AIAA 6th Computational Fluid Dynamics Conference, Danvers, 1983, pp 293-302.

3. Jameson, A., "Solution of the Euler Equations by a Multigrid Method," Applied Mathematics and Computations, 13, 1983, pp 327-356.

4. Jameson, A., and Baker, T.J., "Multigrid Solution of the Euler Equations for Aircraft Configurations," AIAA Paper 84-0093, 1984.

5. Baker, T.J., Jameson, A., Schmidt, W., "A Family of Fast and Robust Euler Codes," Proc. of CFD User's Workshop, University of Tennessee Space Institute, Tullahoma, 1984.

6. Jameson, A. and Mavripilis, D., "Finite Volume Solution of the Two-Dimensional Euler Equations on a Regular Triangular Mesh," AIAA Paper 85-0435, 1985.

Princeton University
Dornier GmbH
Cray Research, Inc.

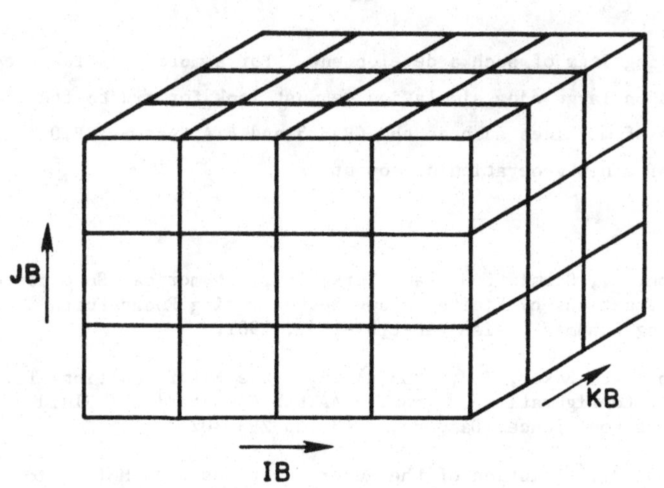

Figure 1. Multiblock grid with block indices IB, JB, KB

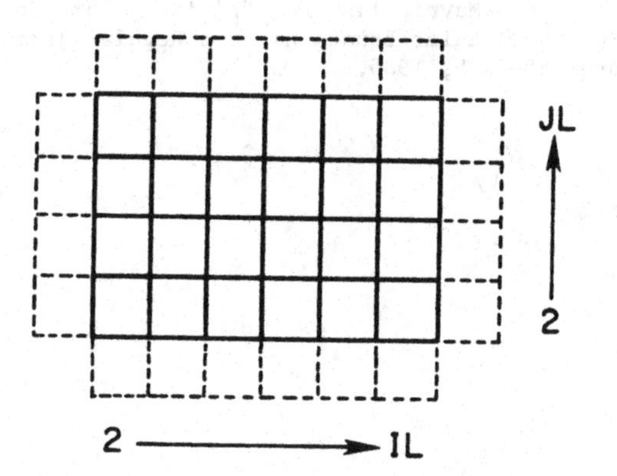

Figure 2. Data structure for a grid block in two dimensions. Interior points range over I = 2,IL, J = 2,JL

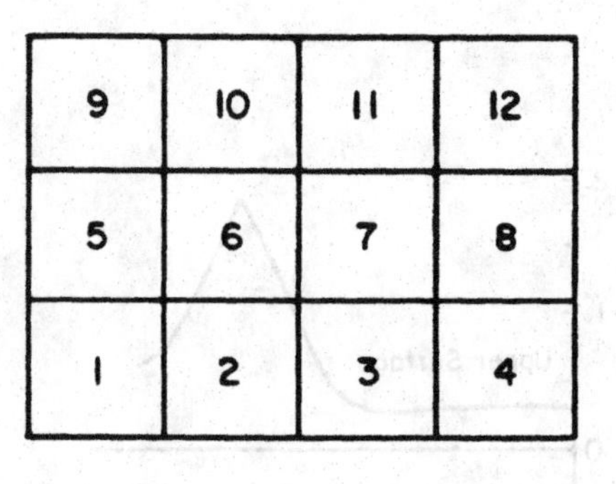

Figure 3(a) Order in which blocks might be updated.

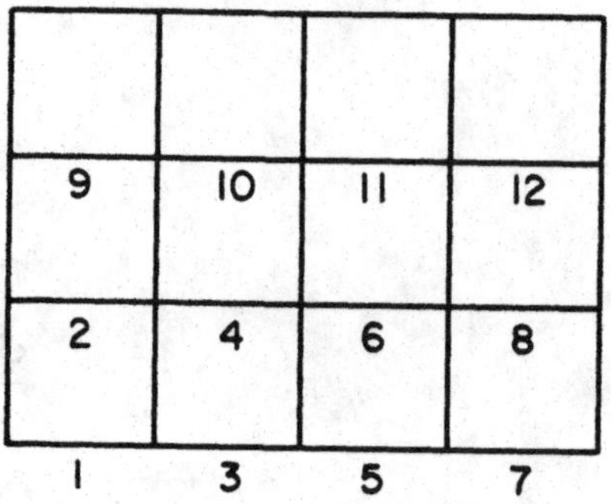

Figure 3(b) Corresponding order in which buffer file MJ1 would be updated.

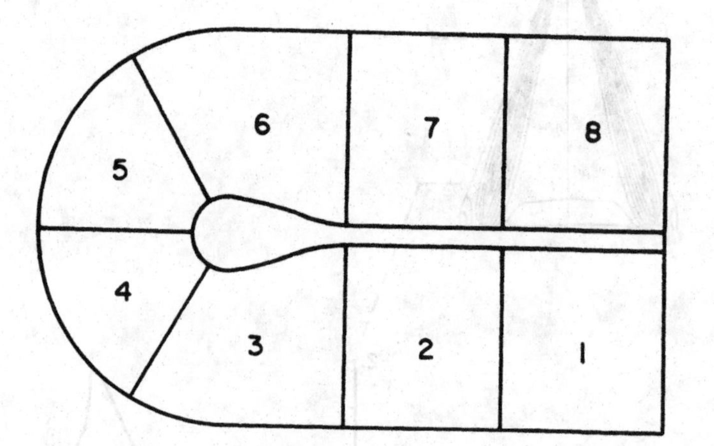

Figure 4 Interfaces introduced between non-contiguous blocks by C-mesh topology

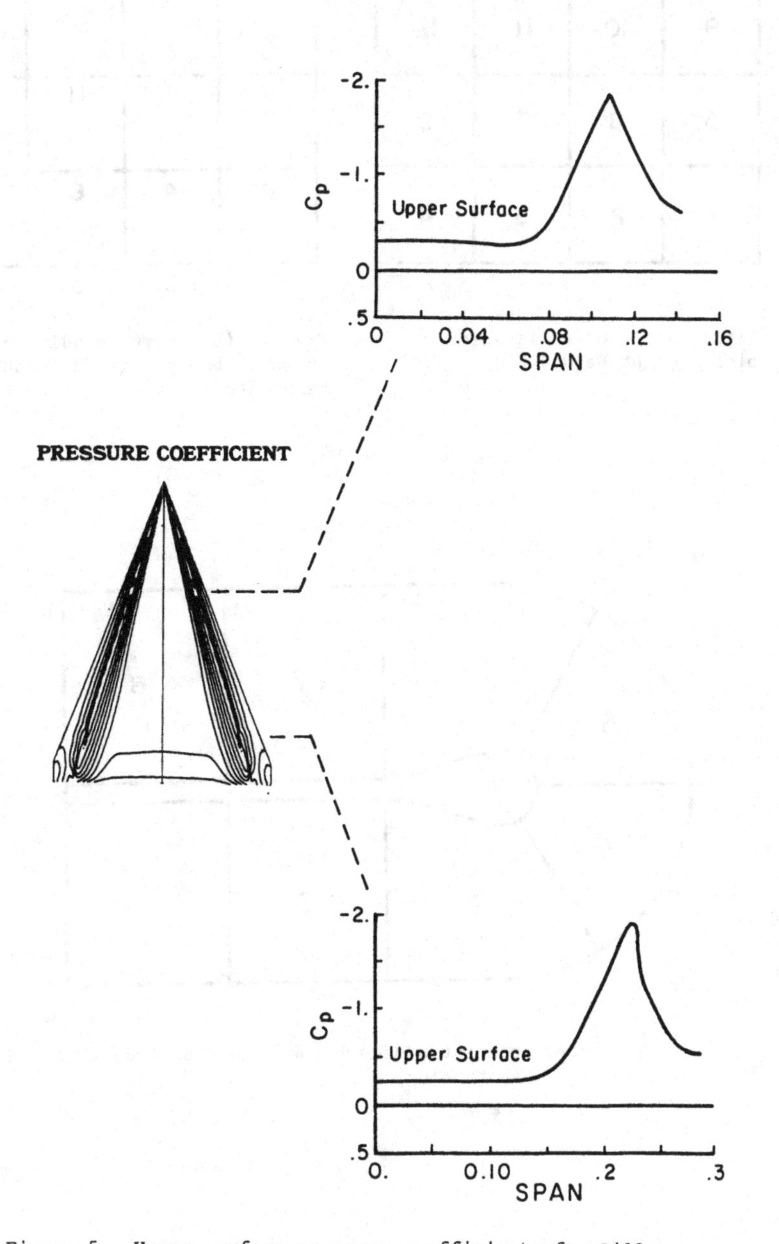

Figure 5. Upper surface pressure coefficients for Dillner wing for M = 1.5, 15 degree angle of attack

Progress in Scientific Computing, Vol. 6
Proceedings of U.S.-Israel Workshop, 1984

Ames Research Center, NASA, Moffett Field, Calif.

DEVELOPMENTS IN THE SIMULATION OF COMPRESSIBLE INVISCID AND VISCOUS FLOW ON SUPERCOMPUTERS

JOSEPH L. STEGER AND PIETER G. BUNING

INTRODUCTION

The near term availability of scientific supercomputers will soon permit routine simulation of three dimensional compressible flow about relatively complex configurations.

General purpose flow simulation codes have been under development at NASA Ames for application to existing large scale computers (such as the CRAY-XMP and the Cyber 205) and future supercomputers. These codes have primarily used the finite difference or finite volume approach, but because of storage and speed limitations, the codes have previously been restricted to simulation of fairly simple geometric configurations. However, work towards generalizing the current codes to treat complex geometries and larger sized meshes has been underway for the past several years in anticipation of supercomputers.

While there is no agreement within NASA Ames as to what is the best general purpose simulation procedure, considerable experience has been obtained with a class of implicit finite difference algorithms written is terms of generalized coordinates. These implicit procedures are widely used in various CFD simulation codes, and they are rapidly being extended to treat quite complex geometries using current and future large scale computers. This paper will review these procedures and discuss how they are being deployed on supercomputers for simulation of complex three dimensional flow fields by using various mesh interface schemes. The importance of flow visualization and diagnostic methods to three dimensional flow simulation will be discussed at the end of this paper.

FINITE DIFFERENCE ALGORITHMS

A general purpose simulation code should be able to solve either steady or unsteady, viscous or inviscid flow with only minor input changes from the user. While it is unlikely to be an optimum code for all tasks, it should be good at its mainstay tasks, and have acceptable efficiency for other tasks. Ideally it will be a modular code whose basic engine can be stripped out and readily applied to other problems. Our mainstay general purpose flow simulation code has been a centrally differenced implicit approximate factorization code [1-6] which solves a conservative form of either the Euler or thin layer Navier Stokes equations cast in generalized coordinates.

1. Fluid Conservation Equations and Transformations

The conservation equations of mass momentum and energy referenced to a Cartesian coordinate system can be represented in the flux vector form (c.f. [7]):

$$\partial_t Q + \partial_x(F + F_v) + \partial_y(G + G_v) + \partial_z(H + H_v) = 0 \tag{1}$$

The viscous flux terms F_v, G_v and H_v contain derivatives and throughout a nondimensional form of the equations will be used. The conservative form of the equations is maintained chiefly to capture the Rankine Hugoniot shock jump relations as accurately as possible.

New independent variables $\tau, \xi, \eta, \varsigma$ are generally chosen to map a curvilinear body conforming discretization into a uniform computational space as shown in Fig. 1. Body conforming curvilinear meshes are generally used in finite difference and finite volume computations for a variety of reasons: to simplify the application of boundary conditions, to allow clustering of grid points to flow field action regions, to help maintain the well-orderliness that is useful for vector processing and for various implicit methods that employ approximate factorization techniques such as ADI, etc. Indeed, the calculation of high Reynolds number viscous flow is simply impractical without the use of a body conforming curvilinear mesh that employs clustering in the direction normal to the body surface.

In the new independent variables the transformed equations can be represented as (c.f. [7] for the detailed terms):

$$\partial_\tau \hat{Q} + \partial_\xi(\hat{F} + \hat{F}_v) + \partial_\eta(\hat{G} + \hat{G}_v) + \partial_\varsigma(\hat{H} + \hat{H}_v) = 0 \tag{2}$$

where the original dependent variables are maintained. (The flux vectors of the transformed equations can be made to resemble their Cartesian counterparts by combining terms into contravarient velocity components [3-4].) If a body conforming coordinate is used, then for high Reynolds number flow it is generally permissable to make a thin layer assumption and to discard viscous terms except for those in the normal-like direction. If ς is the coordinate away from the surface, the thin layer equations can be represented as [3,4]

$$\partial_\tau \hat{Q} + \partial_\xi \hat{F} + \partial_\eta \hat{G} + \partial_\varsigma \hat{H} = Re^{-1} \partial_\varsigma \hat{S} \tag{3}$$

where the viscous terms in ς have been collected into the vector $\hat{S}$ and the nondimensional reciprocal Reynolds number is extracted to indicate a viscous flux term.

In differencing these equations it is often advantageous to difference about a base solution denoted by subscript $_0$ as

$$\begin{aligned} \delta_\tau(\hat{Q} - \hat{Q}_0) + \delta_\xi(\hat{E} - \hat{E}_0) + \delta_\eta(\hat{F} - \hat{F}_0) + \delta_\varsigma(\hat{G} - \hat{G}_0) - Re^{-1}\delta_\varsigma(\hat{S} - \hat{S}_0) \\ = -\partial_\tau \hat{Q}_0 - \partial_\xi \hat{E}_0 - \partial_\eta \hat{F}_0 - \partial_\varsigma \hat{G}_0 + Re^{-1}\partial_\varsigma \hat{S}_0 \end{aligned} \tag{4}$$

where δ indicates a general difference operator, and ∂ is the differential operator. If the base state is properly chosen, the differenced quantities can have smaller and smoother variation and therefore less differencing error. If a the base solution exactly satisfies the partial differential equation, then the right hand side of Eq.(4)

is identically zero. A uniform free stream satisfies the governing equations and is frequently taken as the base solution so as to minimize far field differencing errors. Such an error can occur if the coordinate transformation terms that have been embedded into the transformed fluxes are not consistently differenced [3,4,8]. There are also anologies between the finite volume method and the transformed difference equations, and the actual discretized equations that result from the two schemes can be made similar.

2. Implicit Central Difference Algorithm

Equations (3) or (4) have been solved using a Beam-Warming noniterative approximate factorization implicit scheme of the form [3-6]

$$\begin{aligned}&\left[I+h\delta_\xi\widehat{A}^n-D_i|_\xi\right]\left[I+h\delta_\eta\widehat{B}^n-D_i|_\eta\right]\left[I+h\delta_\zeta\widehat{C}^n-hRe^{-1}\bar{\delta}_\zeta J^{-1}\widehat{M}^nJ-D_i|_\zeta\right]\Delta\widehat{Q}^n\\&=-\Delta t\left[\delta_\xi(\widehat{E}^n-\widehat{E}_\infty)+\delta_\eta(\widehat{F}^n-\widehat{F}_\infty)+\delta_\zeta(\widehat{G}^n-\widehat{G}_\infty)-Re^{-1}\bar{\delta}_\zeta(\widehat{S}^n-\widehat{S}_\infty)\right]\\&-D_e(\widehat{Q}^n-\widehat{Q}_\infty)\end{aligned} \tag{5}$$

where $h = \Delta t$ or $(\Delta t)/2$ and the free stream base solution is used. Here δ is typically a three point second order accurate central difference operator, while $\bar{\delta}$ is a midpoint operator used with the viscous terms. The matrices $\widehat{A}, \widehat{B}, \widehat{C}$, and $\widehat{M}$ result from local linearization about the previous time level and J is the Jacobian of the coordinate transformation. The factored left hand side operators form block tridiagonal matrices.

Because central space difference operators are used, numerical dissipation terms denoted as D_i and D_e have been inserted into Eq.(5). In their simplest form these have been given as combinations of fourth differences

$$D_e = \epsilon_e \Delta t J^{-1}[(\nabla\Delta)_\xi^2 + (\nabla\Delta)_\eta^2 + (\nabla\Delta)_\zeta^2]J \tag{6a}$$

and second differences

$$D_i|_\xi = \epsilon_i \Delta t J^{-1}(\nabla\Delta)_\xi J, \quad D_i|_\eta = \epsilon_i \Delta t J^{-1}(\nabla\Delta)_\eta J, \quad D_i|_\zeta = \epsilon_i \Delta t J^{-1}(\nabla\Delta)_\zeta J \tag{6b}$$

where ∇ and Δ are two-point backward and forward difference operators and where $\epsilon_i > 2\epsilon_e$, and $\epsilon_e = O(1)$. The implicit second difference numerical dissipation operators were chosen to keep the left hand side factors block tridiagonal, and as D_i works on ΔQ, accuracy is not impaired. More robust dissipation terms are often used in which the ϵ coefficients are scaled with some approximate modulus of the A, B, and C Jacobian matrices and gradients of an appropriate variable. For example, explicit and implicit ξ-dissipation terms such as

$$D_e = (\Delta t)J^{-1}(|\xi_x| + |\xi_y| + |\xi_z|)[\epsilon_2 \delta \frac{|\bar{\delta}^2 p|}{|(1+\bar{\delta}^2)p|}\bar{\delta} + \epsilon_4 \delta^4]J$$

$$D_i = (\Delta t)J^{-1}(|\xi_x| + |\xi_y| + |\xi_z|)[\epsilon_2 \delta \frac{|\bar{\delta}^2 p|}{|(1+\bar{\delta}^2)p|}\bar{\delta} + 3\epsilon_4 \bar{\delta}^2]J$$

have been used where p is the nondimensionalized fluid pressure and ϵ_2 is $O(1)$ while ϵ_4 is $O(0.1)$. In transonic flow the term $(|\xi_x| + |\xi_y| + |\xi_z|)$ is an estimate to the spectral radius of A for variables that have been nondimensionalized with respect to the sound speed. Since $(|\xi_x| + |\xi_y| + |\xi_z|)$ does not contain a fluid variable, it can be left as an outside coefficient to the dissipation operator without effecting the weak conservation form of the difference equations.

Body surface boundary conditions have usually been supplied by using a combination of normal-momentum, tangency or no slip, and extrapolation [3,4]. Various far field conditions have been used including characteristic-like conditions [9-11]. Because of their simplicity, in most of the application codes the boundary conditions have been imposed explicitly, or a combination of simplified implicit-explicit conditions [12] have been used. However, fully implicit boundary conditions have also been used (c.f. [1,11,13]) and an elegant implicit characteristic-like formulation has been given and tested by Chakravarthy[11].

3. Vectorization and Multi-Tasking

The structure of the above algorithm lends itself to vectorized computer coding. Vectorization has been implemented by inverting "pencils" of data [14], but FORTRAN based codes generally follow an approach coded by Benek (unpublished, circa 1980). In this approach strings of block tridiagonal matrices are inverted simultaneously so as to avoid the recursive nature of matrix elimination procedures. For example, if indices j, k, l correspond to ξ, η, ζ, the first factor of Eq.(5) forms a block tridiagonal between, say, points $j = 1$ to $j = jmax$. There is one such ξ-block tridiagonal for each k, l index so by simultaneously inverting ξ-block tridiagonals over, say, $k = 1, kmax$ a vector length of $kmax$ is achieved, see Fig. 2. However, in inverting a ξ-block tridiagonal temporary storage is needed for the backward elimination and this storage requirement is increased by $kmax$. (Means of reducing the block size are discussed in [15-17] as a way to improve efficiency, and these same techniques *reduce* temporary storage as well.) Overall, vectorization tends to complicate the coding, but not unduely so. On the CRAY-XMP the ARC2D vectorized code [5] runs about 5 times faster than the unvectorized code for a two-dimensional grid of order 200×40.

The central differenced implicit algorithm, Eq.(5), has also been experimentally coded for multi-tasking on two minicomputers by assigning each processor a portion of the code [18]. For example, processor 1 can solve the ξ-block tridiagonal in the range $l = 1, l_1$, processor 2 can solve the ξ-block tridiagonal from $l = l_1 + 1, l_2$ and so on. Because of a lack of readily available hardware and software, multi-tasking has not yet worked its way into the codes used for routine applications.

4. Improvements in Efficiency and Accuracy

The introduction of a faster computer (that is still easy to use) drives the development of much more efficient numerical algorithms. This is because as the machines become more powerful, it is easier to experiment will new ideas in numerical algorithms. New algorithms can be quickly verified, problem areas can be isolated, and optimum convergence parameters can be found. As a result, improvements in numerical algorithms have kept pace with improvements in computer hardware. Such has been the experience with the central differenced implicit algorithm. Originally coded for the Control Data 7600, it was transferred to the CRAY-1S and vectorized [5] to run an order of magnitude faster. Increased computer speed has in turn hastened the numerical optimization process indicated below which has lead to another order of magnitude improvement in the code's steady state performance.

Improvements to the basic algorithm in both efficiency and accuracy have been made by a variety of contributors. To improve its overall efficiency, simple changes have been optionally implemented into Eq.(5). These changes have been selected so that they do not unduely complicate the basic algorithm. They include the use of space varying Δt, use of a sequence of coarsened grids to provide a good initial guess, cutting inversion costs by using either diagonalization [15,16] or block reduction [17] methods, implementation of better numerical dissipation terms, and more implicit treatment of the numerical dissipation terms. As described in [5,6], these combined changes can improve steady state efficiency by an order of magnitude. Details and appropriate references are described in [5,6].

Although the implicit algorithm has been presented with three point central differencing, versions of the algorithm that have fourth order accuracy in space have been available [1-4] and are preferred unless strong shock waves are present. Reddy [19] has also demonstrated a version in which a pseudo-spectral operator is used in place of the right hand side convection operators. Improved dissipation models [5,6] and total variation diminishing (TVD) implementations [20] have also been carried out to better capture shocks, and perturbation about approximate base solutions has been used to reduce the number of needed grid points [21].

5. Applications

The factored implicit scheme described above has been ultilized in a variety of flow applications. Two-dimensional simulations include steady and unsteady flow about airfoils [c.f. 3,5,6,22-25], cascades [9], projectiles [26-27], and inlets[16]. Three dimensional calculations have been carried out about simple configurations using relatively coarse meshes. Simulations include the supersonic blunt body problem[28], simple bodies [4-6], wings [29], afterbodies [30], and the space shuttle [31]. Figures 3 and 4 are reproduced from the publications of Diewert [30] and Chaussee et al [31] and show representative three dimensional viscous flow solutions that have been obtained on a CRAY-1S. To save computer time and storage, both of these calculations where computed in stages. In the boattail case a complete solution was first obtained using a grid that inadequately resolved the base flow region. A refined grid was then introduced about the afterbody section and the computation for only this region was carried out using inflow conditions taken from the coarsened grid result. In the shuttle calculation a parabolized Navier Stokes code was used in those flow regions in which the main outer flow remains supersonic and the inner viscous layer remains attached with respect to the mainstream direction. These calculations and a majority of the viscous calculations carried out using the factored implicit algorithm at high Reynolds numbers have ultilized a simple algebraic turbulence model [32].

6. Upwind Schemes

As an alternative to using central space differencing for the convection terms, upwind (i.e. backward and forward) space differencing can be used if the fluxes are properly split according to their characteristic properties. As discussed in [33] upwind schemes can have several advantages over central difference schemes, including natural numerical dissipation, better explicit stability, and more readily inverted implicit schemes. Conversely, upwind schemes for systems of equations have generally been more complicated and computationally expensive than central difference schemes and are not very suitable for treating viscous terms.

Although upwind schemes are not as extensively ultilized in aerodynamics simulations as central difference schemes, upwind schemes have been used on curvilinear grids in both finite difference and finite volume formulations and good results have been obtained. Figures 5a and 5b, for example, show an inviscid flow result for a NACA 0012 airfoil at $M_\infty = 0.8$ and an angle of attack of 1.25° using a flux split class of upwind scheme [34] (see [35] for similar results). Various upwind schemes have also proved to be quite effective for capturing strong shock waves, c.f. [36] and [37]. However, because of their complexity and cost, upwind based schemes have not yet been widely distributed as general purpose simulation codes for either inviscid or viscous flow.

Upwind and partially upwind based schemes will likely first emerge as important flow simulation schemes in two areas: 1) supersonic and hypersonic flows in which there are very strong embedded shocks, and, 2) transonic and supersonic flows in which one coordinate direction can be mostly aligned with the flow streamlines. In this latter case, a mixed upwind-central difference scheme can be very effective if the upwind differencing is used in the streamline direction. For example, an implicit algorithm for the thin layer Navier Stokes equations could have the form

$$\begin{aligned}
&\left[I + h\delta_\xi^b(\hat{A}^+)^n + h\delta_\varsigma\hat{C}^n - hRe^{-1}\bar{\delta}_\varsigma J^{-1}\hat{M}^n J - D_i|_\varsigma\right] \\
&\quad \times \left[I + h\delta_\xi^f(\hat{A}^-)^n + h\delta_\eta\hat{B}^n - D_i|_\eta\right]\Delta\hat{Q}^n = \\
&\quad -\Delta t\{\delta_\xi^b[(\hat{E}^+)^n - \hat{E}_\infty^+] + \delta_\xi^f[(\hat{E}^-)^n - \hat{E}_\infty^-] + \delta_\eta(\hat{F}^n - \hat{F}_\infty) \\
&\quad + \delta_\varsigma(\hat{G}^n - \hat{G}_\infty) - Re^{-1}\bar{\delta}_\varsigma(\hat{S}^n - \hat{S}_\infty)\} - D_e(\hat{Q}^n - \hat{Q}_\infty) \qquad (7)
\end{aligned}$$

where δ_ξ^b and δ_ξ^f are backward and forward three-point difference operators and D_e contains only η and ς numerical dissipation operators. This two factor implicit scheme is readily vectorized or multi-tasked in planes of ξ = constant. A semi-implicit scheme is obtained by neglecting the calculation of $h\delta_\xi^f\hat{A}^-$ in the implicit backsweep operating on $\Delta\hat{Q}^n$.

COMPOSITE GRIDS

Use of a single well-ordered body conforming curvilinear mesh simplifies the application of boundary conditions and can lead to use of efficient solution procedures. However, the generation of body conforming well-clustered curvilinear grids that are not overly skewed and have smooth variation can often be quite difficult. In particular, it is generally impractical to build a single grid of this type for complex three dimensional configurations. Of course, by judiciously introducing cuts in the grid, some fairly complex bodies can be meshed with a single grid, Fig. 6 illustrates this possiblility. But the trend has been to introduce more than one grid and to patch or overset the grid systems together. The sketches shown in Fig. 7 illustrate simple patch and overset grid configurations in two dimensions. (There has also been some activity in reverting back to the use of pure Cartesian grids that intersect the body in a random way and so require special logic at the body interface [38]. Such a procedure appears to be impractical, however, for high Reynolds number viscous flow problems unless the Cartesian scheme is patched-to or overset-with a body conforming grid that is used only near the body. In this case the Cartesian mesh approach would be a special case of either the patched or overset grid method.)

The use of a set of patched or overset grids to form a larger composite grid carries the discretization process one step further. In a sense such a finite difference process assumes some of the characteristics of a finite element scheme that uses large powerful elements in which each element is itself discretized. In this discretization process each individual grid in the system is well ordered and is thus suitable for efficient finite difference solution using any available single grid scheme. The problem with such a composite grid scheme is the difficulty of interfacing each mesh without degrading numerical accuracy or convergence. Moreover, as few such meshes should be generated as necessary to achieve grid efficiency.

Limited experience with both patched [39-41] and overset grids [42-46] has not shown which method is preferable - an optimum method will perhaps combine both patched and overset grids. Both schemes necessitate extensive bookkeeping procedures. The patched grid method has as its chief drawback a grid generation problem that is still relatively difficult because various interfaces have to be defined and grids have to be generated with both inner and outer defined boundary surfaces. Drawbacks to using overset grids include having to interpolate data points along an irregular boundary and the bookkeeping can be especially complex if more than two levels of overset grids intersect each other.

Figures 8 - 9 show recently published two dimensional inviscid flow results obtained using patched and overset meshes. Both grids and flow contour levels are shown. The supersonic biplane results of Hessenius and Rai [41] were computed using a patched grid method in which a flux balance interface scheme is used at grid boundaries to ensure fluid conservation. In this method [40], meshes must share a common boundary, but grid lines need not have a common slope or even join together. These results were obtained using a first order accurate Osher scheme. The overset grid results of Dougherty [46] for an airfoil and flap were computed using the factored implicit algorithm. Only simple interpolation is used to interface the mesh boundaries using data from nearby points from the underlying grid. Figure 9c shows the nearby interpolation points used to update the outer boundary of the flap grid (open symbols) and grid points which are excluded from the calculation (filled symbols). Simulation of wing-body configurations using the implicit algorithm with composite grids are currently underway by Holst and coworkers and Benek and coworkers and will appear shortly.

POSTPROCESSING

1. Flow Visualization

One of the major problems that is being encountered in three dimensional flow simulation is the difficulty of displaying a limited and proper kind of data that will lead to better understanding. In the future, much of the computational aerodynamicist's time will be devoted to extracting and displaying various features of the solution. In three dimensions, flow phenomenon such as flow-reversals, shocks, shear-layers, vortices, etc. can often be difficult to identify and visualize, especially if the flow is also unsteady. Graphic displays of contour surfaces and particle paths and the like can also be expensive to generate. As a result, more computer resource could be expended on analyzing a solution and displaying it in a meaningful way to a human being than what was needed to generate the solution in the first place. To become convinced of how costly this can become one need merely contemplate generating numerous detailed displays of a three dimensional flow field in which reliable hidden line removal is needed to keep the display from being too confusing.

Two approaches to extracting graphic information and flow visualization are evolving. In one approach the engineer works at a graphics workstation and displays various portions of the flow field, observing the solution from different vantage points, and otherwise interacts with the computed solution, perhaps by seeding particles and observing their behaviour. Because the three dimensional data base can be quite large, this approach will require very powerful graphic work stations and high speed data links between the graphics station and the supercomputer. The advantage of this approach is that one can stumble across information that might not otherwise be anticipated.

The other approach is to program the supercomputer itself to diagnose the data base. Such a computer program is needed in order to throughly search a flow field to find details that a human at a workstation would find too tedious to locate. Algorithms must be developed to search out and display special features such as shock waves, vortices, and separation lines. Because the data bases are so large, only the supercomputer itself will be able to perform many of these calculations.

Reliable software to locate interesting flow regions is not available and will be difficult to generate. For the CFD algorithm developer, developing flow visualization procedures offers interesting opportunities. Consider, for example, the difficulty of building an algorithm that automatically identifies embedded shock waves in a complex three dimensional flow, especially weak oblique shocks. Or consider the apparently straightforward task of building an efficient and accurate particle trace scheme. To trace particles given the velocity field, one need only numerically integrate simple ordinary differential equations. But if too many particles are displayed, the picture will be confusing. Interpolation of the velocity components for each particle location can also be a costly process, so efficient algorithms are needed to minimize this expense. Finally, unless the particles are seeded in the right locations, the most interesting features of the flow will not be observed. For example, particles seeded outside of a small vortex core will not be entrained inside the core.

2. Diagnostics

The accuracy of a finite difference solution is generally appraised by successively refining the grid and comparing solutions from one grid to another. If the solution is unchanged, the flow result is likely resolved. (Here we assume that a Reynolds averaged Navier Stokes solution can become invariant with grid refinement.) Because the next generation of supercomputers will not be 'super enough' to carry this process too far, the problem of determing solution accuracy may stimulate additional research in approximate methods such as potential and boundary layer schemes. We might compute, for example, a boundary layer solution to verify the correctness of the skin friction and heat transfer found from a Navier Stokes procedure. By taking the Navier Stokes computed pressure gradient, the boundary layer equations can be solved using a very refined grid near the wall. If the boundary layer solution returns the same wall values computed from the Navier Stokes equations, then the Navier Stokes solution is adequately resolved. (Because the boundary layer will likely be run in an inverse mode if the flow is separated, the check might be whether or not the boundary layer solution returns the same surface pressure distribution.)

For complex three dimensional flows it will also be difficult to provide code verification. Comparison with linear theory may be inadequate and experimental data may not be available or reliable. One possiblility is to compare the Navier Stokes computed results to those obtained from the solution of a simplified set of equations that are solved over the entire field, but in which rotational and perhaps even compressibility effects are directly taken from the Navier Stokes solution. As a quite simple illustration, the Poisson equation for the incompressible streamfunction ψ

$$\psi_{xx} + \psi_{yy} = -\omega$$

should reproduce an incompressible two-dimensional Navier Stokes result if the vorticity, ω, is taken directly from the Navier Stokes solution. Because this is a simple scalar equation, it should be possible to solve for it very quickly and using a more refined grid than what was used for the Navier Stokes simulation. If the refined grid Poissson calculation can return the Navier Stokes solution then considerable validation of the code is obtained. Carrying out such a process for complex configurations in three dimensions will be difficult but useful.

CONCLUDING REMARKS

Evolution of existing finite difference schemes should provide reliable and efficient single, mesh codes for nonlinear three dimensional flow field simulation. To treat viscous flow about highly complex body configurations, however, composite meshes will be needed and considerable work in interfacing these schemes into multiple grid codes must yet be carried out. Just as the task of optimizing the flow solvers has been accelerated by the availability of large fast computers, so too the task of developing interface schemes will be accelerated with the availablity of forthcoming supercomputers, provided these machines are easy to use.

Cost considerations are leading to supercomputers that achieve more and more of their performance advantage by use of repetitive processors. Many of these machines are more difficult to use than machines that achieve higher performance by having a faster single instruction time. Hopefully these future supercomputers will not be a machines of 'insurmountable opportunity'.

It is generally accepted within the CFD community that three dimensional data bases are so large that new ways of displaying and interacting with the data is necessary. It must be emphasized that hardware and graphics software provides only part of the solution. If the computed solutions are to be properly analyzed and understood, extensive new algorithm developments in flow visualization and diagnostics will be needed, and these algorithms will require the same kind of creativity that went into generating the flow simulation procedures themselves.

REFERENCES

1. Beam, R. M. and Warming, R. F., *An Implicit Finite - Difference Algorithm for Hyperbolic Systems in Conservation - Law - Form*, J. Comp. Phys. **22** (Sept. 1976,), p. 87-110.
2. Warming, R. F. and Beam, R. M., *On the Construction and Application of Implicit Factored Schemes for Conservation Laws*, Symposium of CFD, New York, **11** (April 16-17, 1977, SIAM-AMS Proceedings).
3. Steger, J. L., *Implicit Finite Difference Simulation of Flow About Arbitrary Two Dimensional Geometries*, AIAA J. **16, No. 7** (July 1978).
4. Pulliam, T. H. and Steger, J. L., *On Implicit Finite Difference Simulations of Three Dimensional Flows*, AIAA J. **18** (Feb. 1980).
5. Pulliam, T. H., *Euler and Thin Layer Navier Stokes Codes : ARC2D, ARC3D, Notes for Computational Fluid Dynamics User's Workshop*, The University of Tennessee Space Institute, Tullahoma, Tenn. (March 12-16, 1984).
6. Pulliam, T. H. and Steger, J. L., *Recent Inprovements in Efficiency, Accuracy, and Convergence of an Implicit Approximate Factorization Algorithm*, AIAA Paper 85-0360 (Jan. 1985).
7. Viviand, H., *Conservative Forms of Gas Dynamics Equations*, La Recherche Aerospatiale **1** (1978).
8. Thomas, P. D. and Lombard, C. K., *Geometric Conservation Law and Its Application to Flow Computations on Moving Grids*, AIAA Journal **17** (1979).
9. Steger, J. L., Pulliam, T. H. and Chima, R. V., *An Implicit Finite Difference Code for Inviscid and Viscous Cascade Flow*, AIAA paper 80-1427 (1980).
10. Pulliam, T. H., *Characteristic Boundary Conditions for the Euler Equations*, NASA Publication 2201, Numerical Boundary Condition Procedures (Oct. 1980).
11. Chakravarthy, *Euler Equations - Implicit Schemes and Implicit Boundary Conditions*, AIAA 82-0228 (Jan. 11-14, 1982).
12. Schiff, L. B. and Steger, J. L., *Numerical Simulation of Steady Supersonic Viscous Flow*, AIAA Journal **18** (Dec. 1980).
13. Nicolet, W. E., Shanks, S., Srinivasan, G. and Steger, J. L., *Flowfield Predictions About Lifting Entry Vehicles*, AIAA Paper 82-0026 (Jan. 11-14, 1982).
14. Lomax, H. and Pulliam, T. H., *A Fully Implicit Factored Code for Computing Three - Dimensional Flows on the ILLIAC IV*, Parallel Computations, G. Rodrigue, Ed. Academic Press, New York (1982).
15. Pulliam, T. H. and Chaussee, D. S., *A Diagonal Form of an Implicit Approximate - Factorization Algorithm*, J. Comp. Phys. **39, no. 2** (Feb. 1981), p. 347.
16. Chaussee, D. S. and Pulliam, T. H., *A Diagonal Form of an Implicit Approximate Factorization Algorithm with Application to a Two Dimensional Inlet*, AIAA J. **19, no. 2** (Feb. 1981), p. 153.
17. Barth, T. J. and Steger, J. L., *An Efficient Approximate Factorization Implicit Scheme for the Equations of Gasdynamics*, NASA TM 85957 (June 1984).
18. Eberhardt, D. S., *A Study of Multiple Grid Problems on Concurrent Processing Computers*, Stanford University Dissertation, Department of Aeronautics and Astronautics, Stanford, CA. (Sept., 1984).

19. Reddy, K. C., *Pseudospectral Approximation in a Three - Dimensional Navier - Stokes Code*, AIAA J. **21, no. 8** (Aug. 1983).
20. Yee, H. C., Warming, R. F. and Harten, A., *Implicit Total Variation Diminishing (TVD) Schemes for Steady - State Calculations*, NASA TM 84342 (March 1983).

21. Chow, L. J., Pulliam, T. H. and Steger J. L., *A General Perturbation Approach for the Equations of Fluid Dynamics*, AIAA Paper 83-1903 (July 13-15, 1983).
22. Anderson, W. K., Thomas, J. L. and Rumsey, C. L., *Application of Thin Layer Navier - Stokes Equations Near Maximum Lift*, AIAA Paper 84-0049 (Jan. 1984).
23. Barth, T. J., Pulliam, T. H. and Buning, P., *Navier - Stokes Computations for Exotic Airfoils*, AIAA Paper 85-0109 (Jan. 1985).
24. Barton, J. T. and Pulliam, T. H., *Airfoil Computations at High Angles of Attack, Inviscid and Viscous Phenomena*, AIAA Paper 84-0524 (1984).
25. Srinivasan , G. R., McCroskey, W. J. and Kutler, P., *Numerical Simulation of the Interaction of a Vortex with Stationary Airfoil in Transonic Flow*, AIAA Paper 84-0254 (Jan. 1984).
26. Nietubicz, C. J., *Navier Stokes Computations for Conventional and Hollow Projectile Shapes at Transonic Velocities*, AIAA Paper 81-1262 (June 1981).
27. Sahu, J., Nietubicz, C. J. and Steger, J. L., *Numerical Computation of Base Flow for a Projectile at Transonic Speeds*, AIAA 82-1358 (Aug. 9-11, 1982).
28. Kutler, P., Pedelty, J. A. and Pulliam, T. H., *Supersonic Flow over Three - Dimensional Ablated Nosetips Using an Unsteady Implicit Numerical Procedure*, AIAA Paper 80-0063 (1980).
29. Mansour, N. N., *Numerical Simulation of the Tip Vortex Off a Low - Aspect Ratio Wing at Transonic Speed*, AIAA Paper 84-0522 (Jan. 1984).
30. Diewert, G. S., *Topological Analysis of Computed Three -Dimensional Viscous Flow Fields*, In 'Recent Contributions to Fluid Mechanics', W. Haase, ed. Springer-Verlag, Berlin (1982).
31. Chaussee, D. S., Rizk, Y. M. and Buning, P. G., *Viscous Computation of a Space Shuttle Flow Field*, NASA TM 85977 (June 1984).
32. Baldwin, B. S. and Lomax, H.,, *Thin Layer Approximation and Algebraic Model for Separated Turbulent Flows*, AIAA Paper 78-257 (Jan. 1978).
33. Steger, J. L. and Warming, R. F., *Flux Vector Splitting of the Inviscid Gasdynamics Equations with Applications to Finite Difference Methods*, J. of Comp. Physics 40 (1981).
34. Buning, P. G., *Computation of Inviscid Transonic Flow Using Flux Vector Splitting in Generalized Coordinates*, Stanford University Dissertation, Department of Aeronautics and Astronautics, Stanford, CA (1983).
35. Anderson, W. K., Thomas, J. L. and Van Leer, B., *A Comparison of Finite Volume Flux Vector Splittings for the Euler Equations*, AIAA Paper 85-0122 (Jan. 1985).
36. Chakravarthy, S. R. and Osher, S., *Numerical Experiments with the Osher Upwind Scheme for the Euler Equations*, AIAA Paper 82-0975 (June 1982).
37. Chakravarthy, S. R. and Osher, S., *High Resolution Applications of the Osher Upwind Scheme for the Euler Equations*, AIAA Paper 83-1943 (1983).

38. Clarke, D. K., Hassan, H. A. and Salas, M. D., *Euler Calculations for Multi-element Airfoils Using Cartesian Grids*, AIAA Paper 85-0291 (Jan. 1985).
39. Lasinski, T. A., et. al., *Computation of the Steady Viscous Flow Over a Tri-Element Augmenter Wing Airfoil*,, AIAA Paper 82-0021 (Jan. 1982).
40. Rai, M. M., *A Conservative Treatment of Zonal Boundaries for Euler Equation Calculations*, AIAA Paper 84-0164 (1984).
41. Hessenius, K. A. and Rai, M. M., *Applications of a Conservative Zonal Scheme to Tansient and Geometrically Complex Problems*, AIAA Paper 84-1532 (June 1984).
42. Atta, E. H. and Vadyak, T., *A Grid Interfacing Zonal Algorithm for Three Dimensional Transonic Flows About Aircraft Configurations*, AIAA Paper 82-1017 (June 1982).
43. Steger, J. L., Dougherty, F. C. and Benek, J. A., *A Chimera Grid Scheme*, ASME Mini-Symposium on Advances in Grid Generation, Houston, Texas, (June 1983).
44. Benek, J. A., Steger, J. L. and Dougherty, F. C., *A Flexible Grid Embedding Technique with Application to the Euler Equations*, AIAA 83-1944 (July 13-15, 1983).
45. Berger, M. J., *Adaptive Mesh Refinement for Hyperbolic Partial Differential Equations*, STAN-CS-82-924, Stanford University (Aug. 1982).
46. Dougherty, F. C., *Development of a Chimera Grid Scheme with Applications to Unsteady Problems*, Stanford University Dissertation, Department of Aeronautics and Astronautics, Stanford, CA (Oct. 1984).

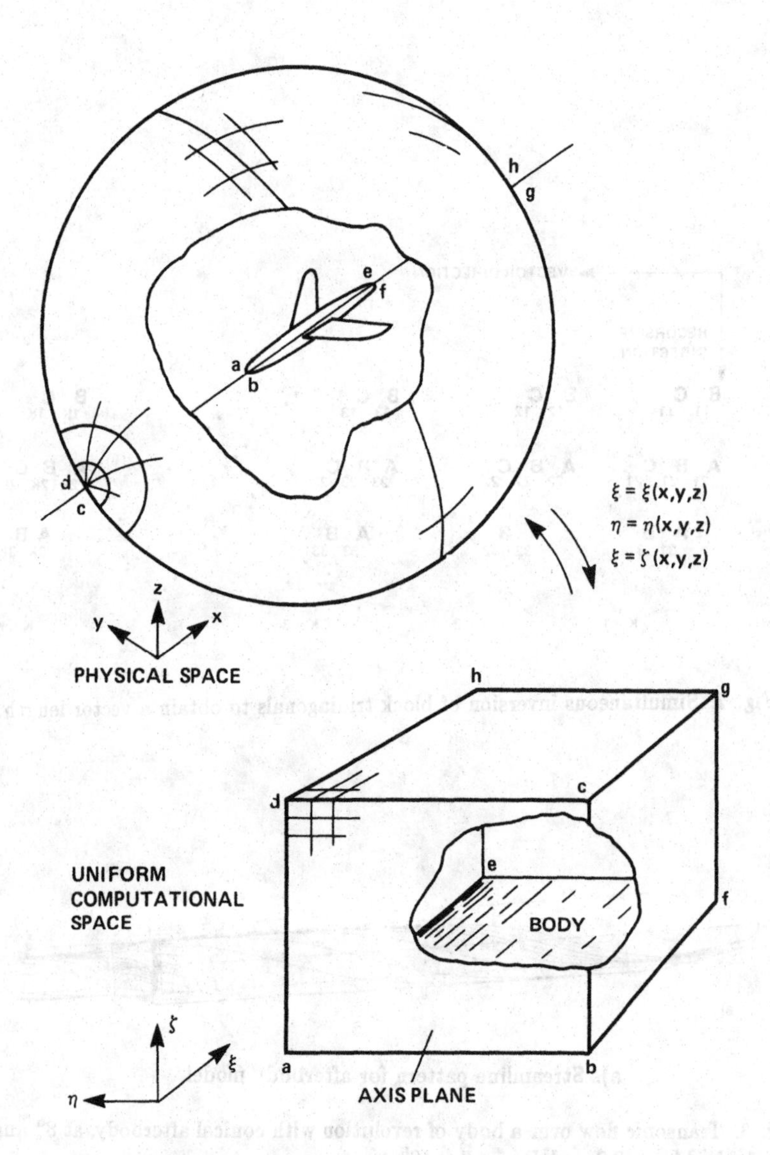

Fig. 1. Sketch showing mapping of physical space to computational space for a well-ordered warped spherical grid.

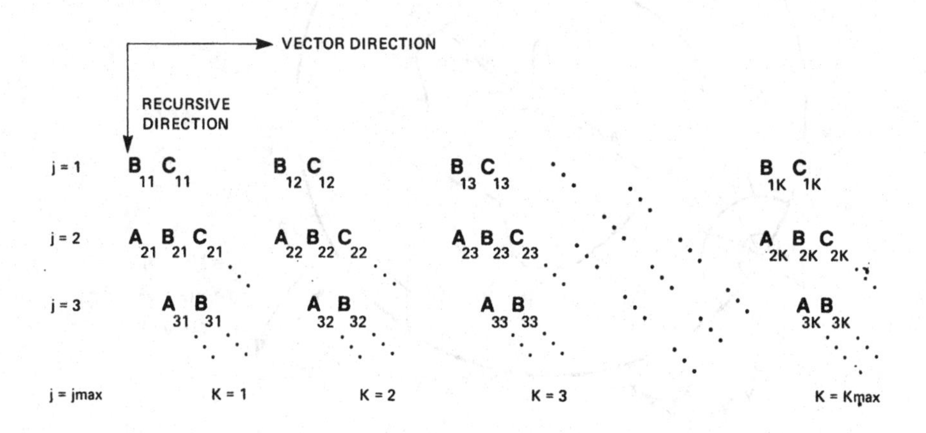

Fig. 2. Simultaneous inversion of block tridiagonals to obtain a vector length.

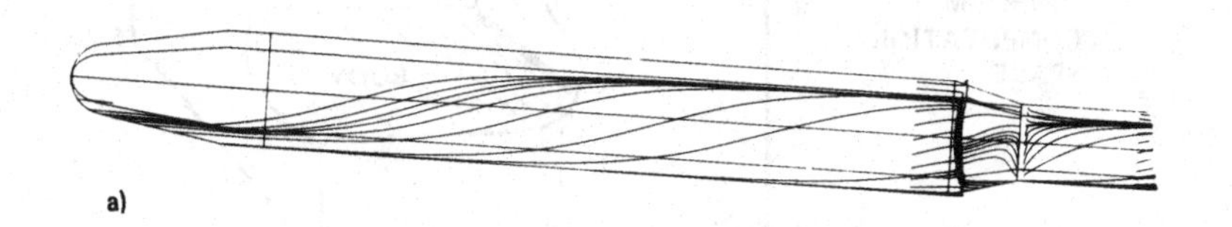

a). Streamline pattern for afterbody model

Fig. 3. Transonic flow over a body of revolution with conical afterbody, at 8° angle of attack, $M_\infty = 0.9$ and $Re_d = 3 \times 10^6$.

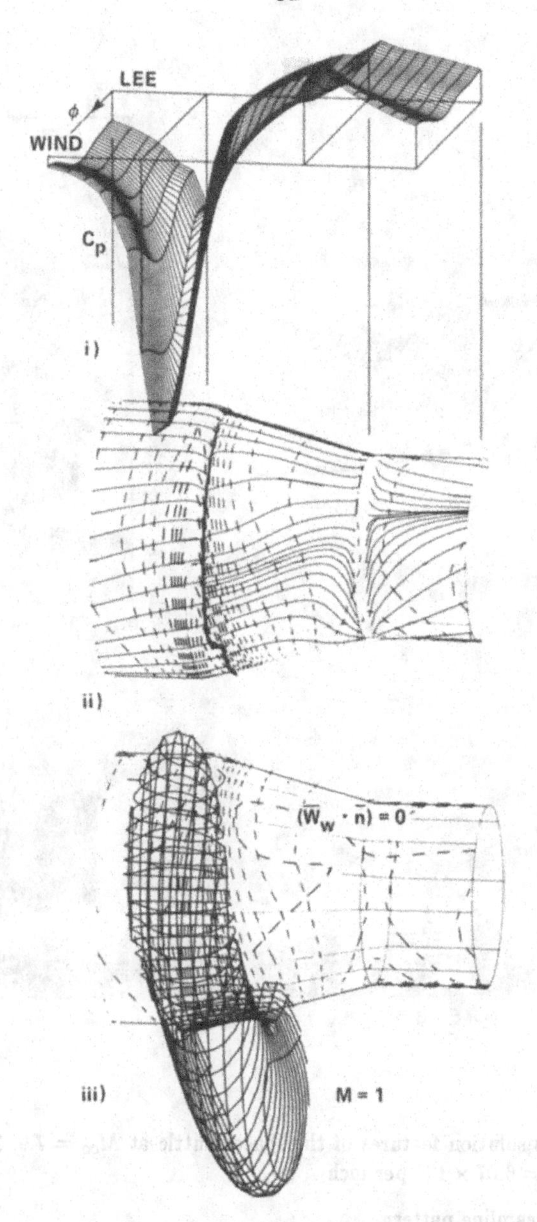

b). Lateral perspective of flow features for conical afterbody showing (i) surface pressures, (ii) isobars (dashed line) and surface shear (solid line), and (iii) sonic surface.

Fig. 3. Concluded.

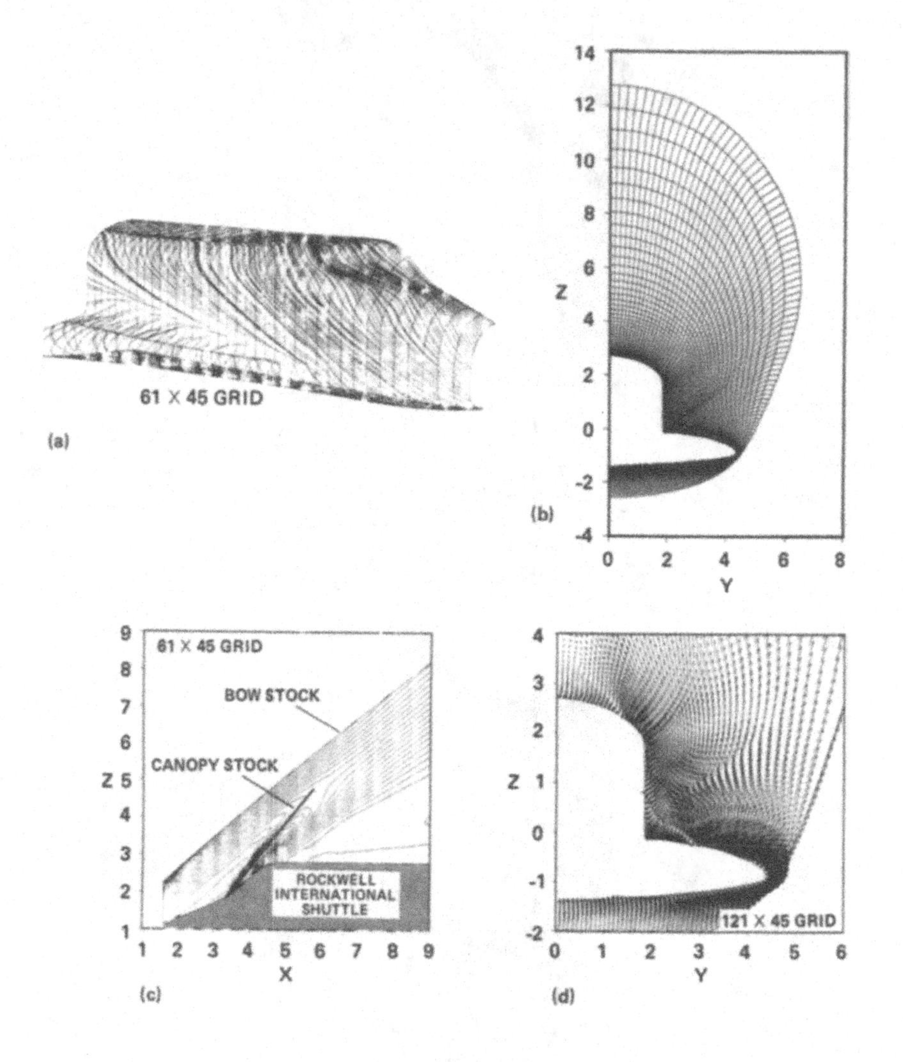

Fig. 4. Various solution features of the space shuttle at $M_\infty = 7.9$, 25° angle of attack and $Re_l = 6.07 \times 10^5$ per inch .

a) Surface streamline pattern.

b) Cross sectional view showing grid slice conforming to outer bow shock.

c) Pressure contours in vicinity of canopy region showing embedded bow shock.

d) Cross sectional flow vectors.

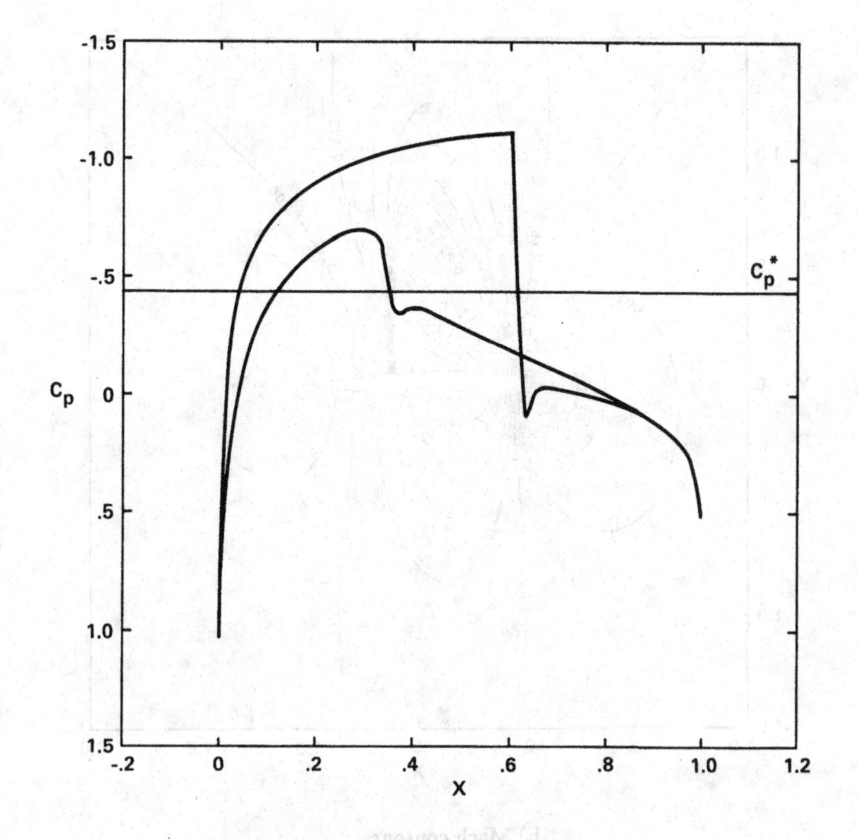

a) Surface pressure distribution, C_p.

Fig. 5. Calculation of inviscid transonic flow using flux vector splitting class of finite difference scheme for NACA0012 airfoil at $M_\infty = 0.8$ and 1.25° angle of attack.

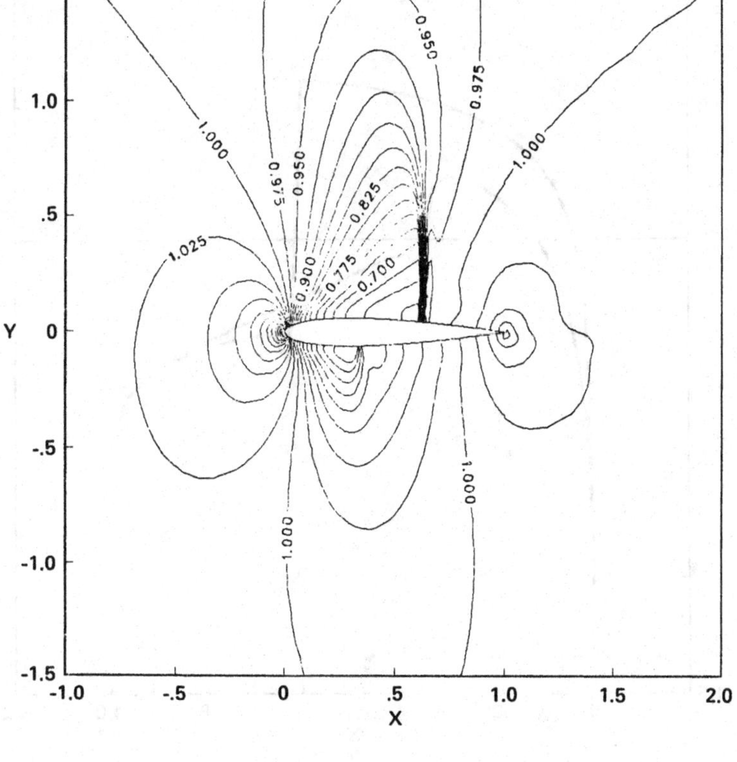

b) Mach contours.

Fig. 5. Concluded.

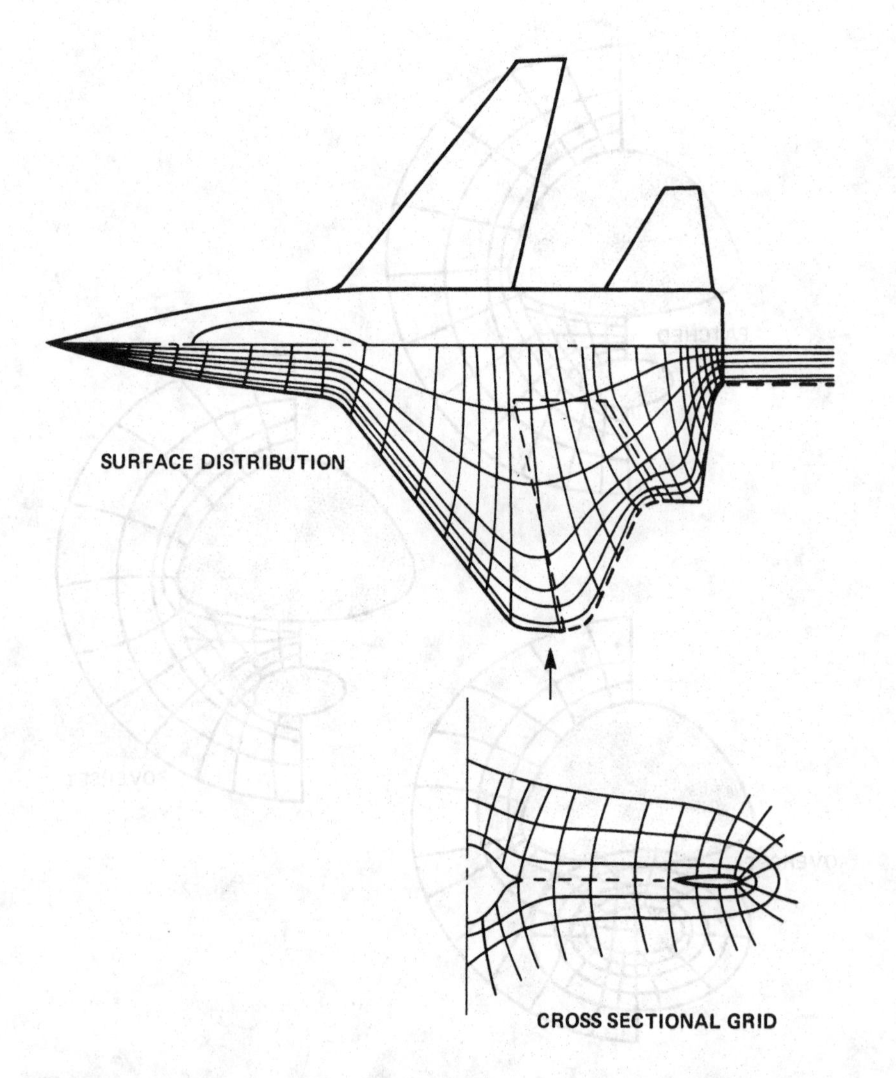

Fig. 6 Sketch illustrating the use of a single body-conforming warped-spherical grid with cuts.

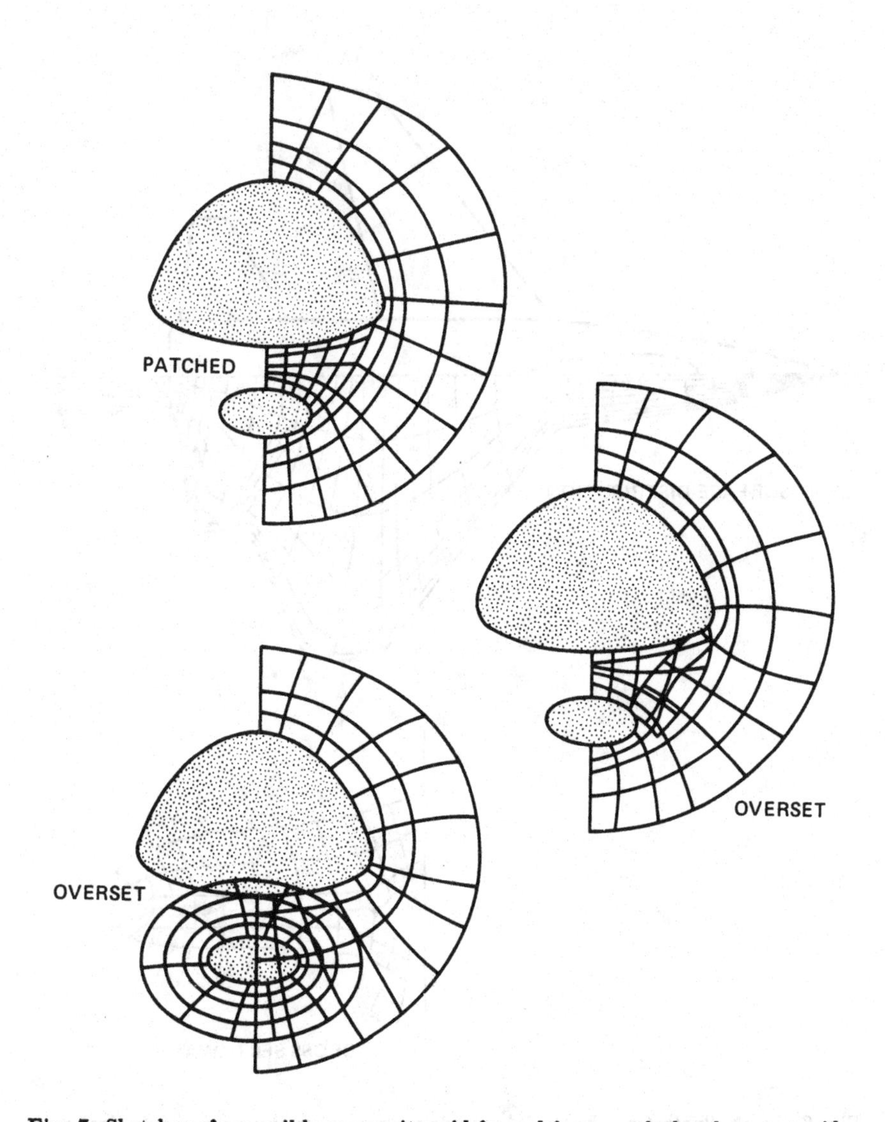

Fig. 7. Sketches of a possible composite grid formed from patched and overset grids about a typical multiple body combination.

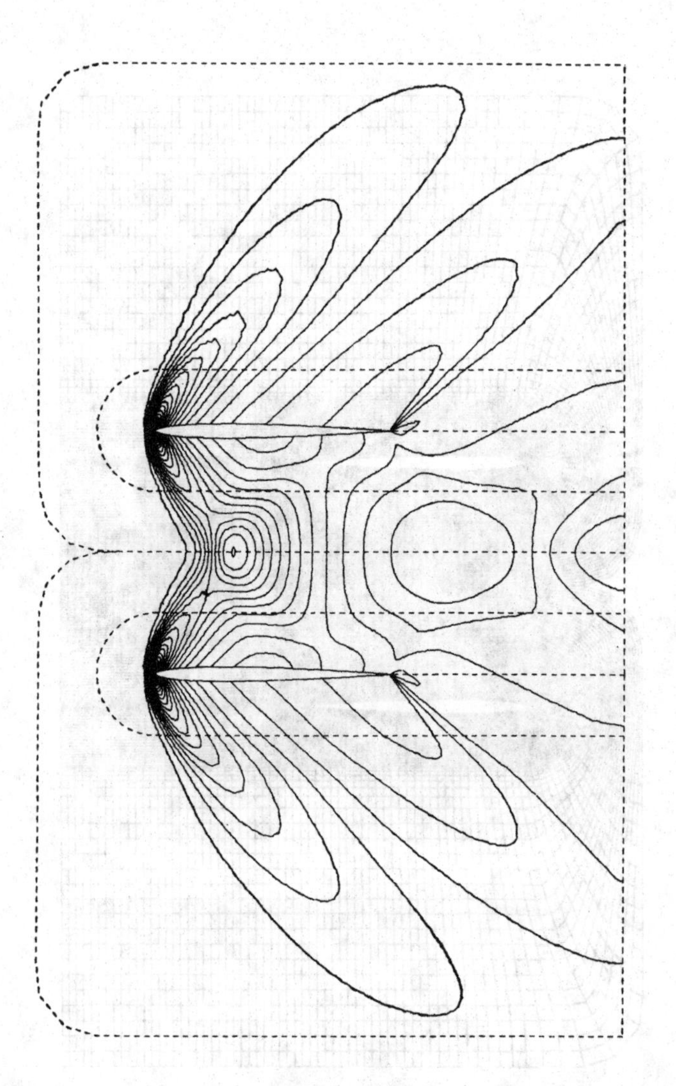

a) Mach contours

Fig. 8. Inviscid supersonic flow about a biplane computed using a first order Osher scheme with patched grids, $M_\infty = 1.5$.

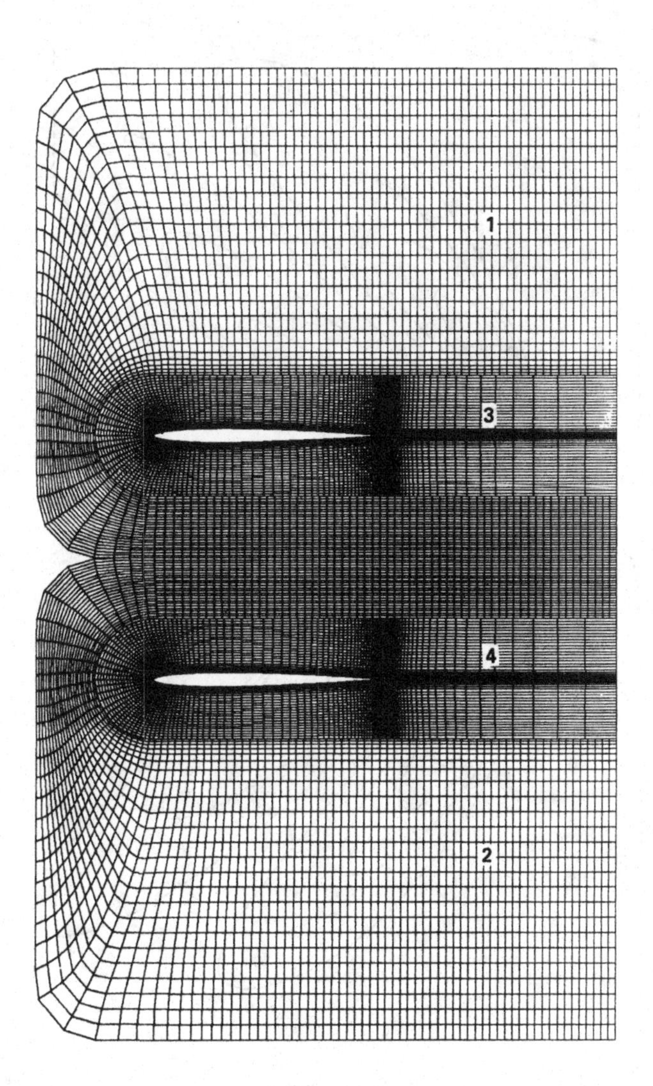

b) Composite grid formed using four patches.

Fig. 8. Concluded.

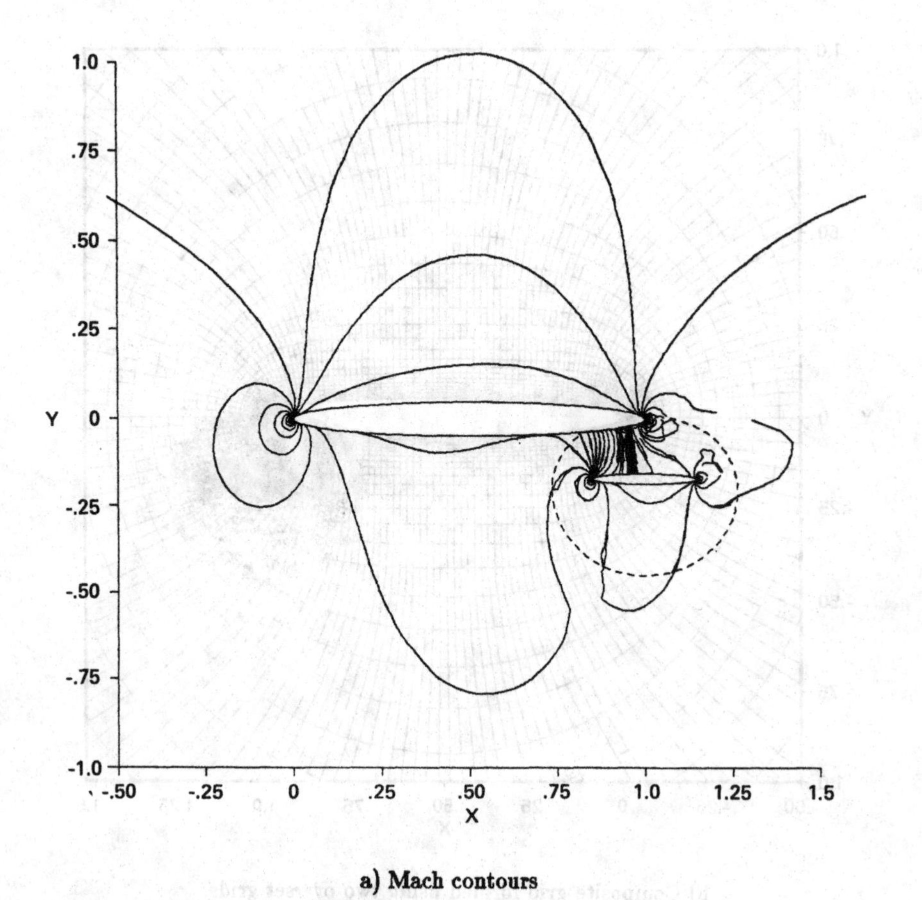

a) Mach contours

Fig. 9. Inviscid transonic flow about a generic airfoil detached flap combination using the implicit algorithm and overset grids, $M_\infty = 0.7$.

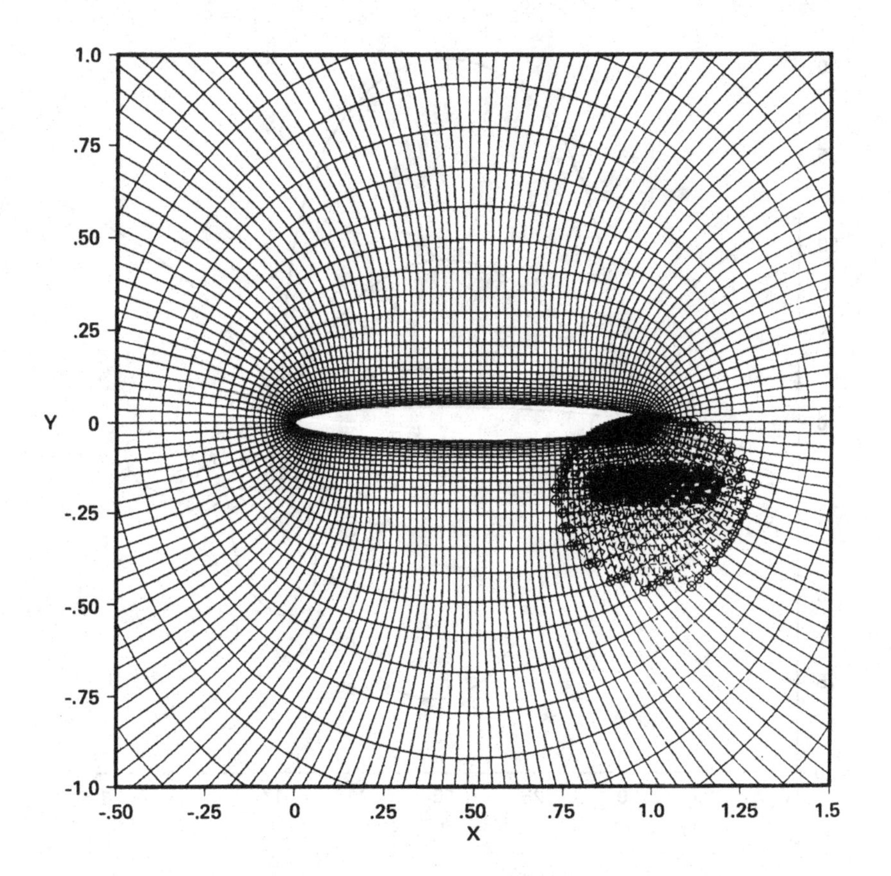

b) Composite grid formed using two overset grids.

Fig. 9. Continued.

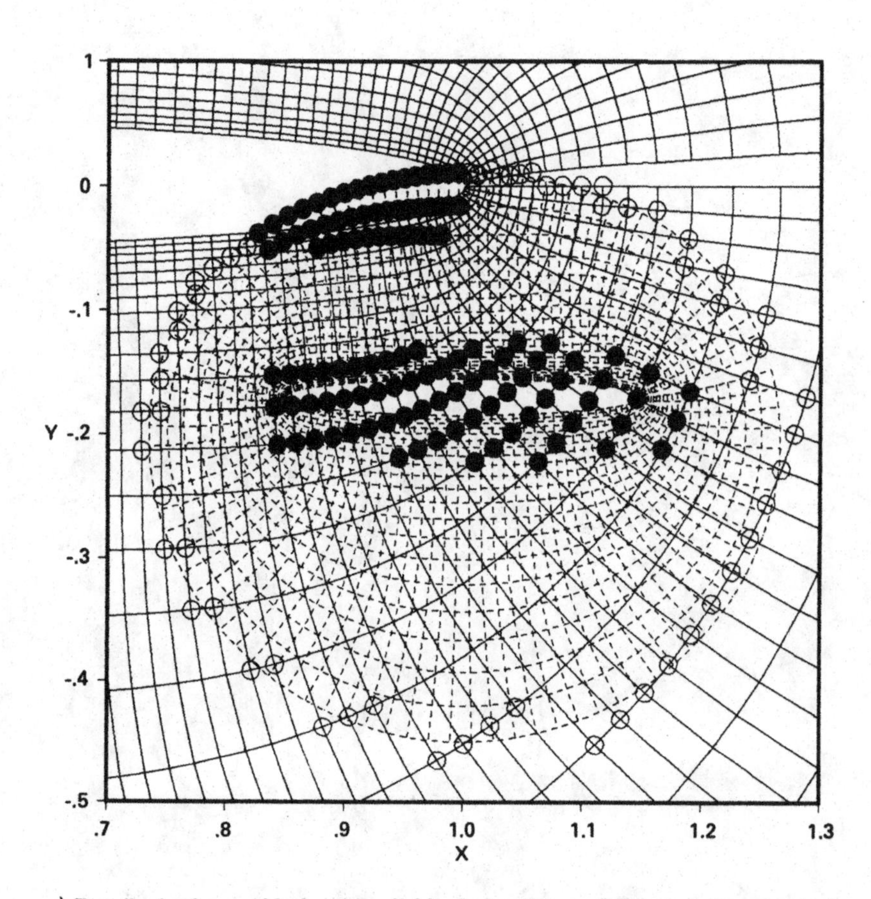

c) Detail of minor grid showing all blanked points and interpolation points for the outer boundary of the minor grid. (Hole interpolation points not shown).

Fig. 9. Concluded.

Progress in Scientific Computing, Vol. 6
Proceedings of U.S.-Israel Workshop, 1984

HIGH RESOLUTION SOLUTIONS OF THE EULER EQUATIONS FOR VORTEX FLOWS

Earll M. Murman
Arthur Rizzi
Kenneth G. Powell

1. ABSTRACT

Solutions of the Euler equations are presented for $M = 1.5$ flow past a 70 degree swept delta wing. At an angle of attack of 10 degrees, strong leading edge vortices are produced. Two computational approaches are taken based upon 1) fully three-dimensional and 2) conical flow theory. Both methods utilize a finite-volume discretization solved by a pseudo-unsteady multistage scheme. Results from the two approaches are in good agreement. Computations have been done on a 16 million word CYBER 205 using $196 \times 56 \times 96$ and 128×128 cells for the two methods. A sizable data base is generated, and some of the practical aspects of manipulating it are mentioned. The results reveal many interesting physical features of the compressible vortical flow field, and also suggest new areas needing research.

2, INTRODUCTION

This paper considers the aerodynamic problem of compressible flow past a sharp-edge delta wing at moderate to high angle of attack. It is well known that at angles of attack above a few degrees, a strong primary vortex appears on the leeside of the wing which is generated at the sharp leading edge (Fig. 1). The leading edge vortices dramatically affect the aerodynamics of high performance wings, and their prediction is of great importance to designers. Recent experimental evidence [1,2] indicates that a variety of flow structures involving vortices and shock waves can exist. However, very little of the detail of these flows is understood.

A number of publications have appeared for solutions to the Euler equations for vortical flows past wings [3-10]. For a sharp-edge geometry, the separation point for the primary vortex is fixed at the wing leading edge, independent of the Reynolds number. It appears that the dissipation in the discrete numerical solution provides enough "viscosity" to mimic the physical viscosity and cause separation. Rizetta and Shang's calculations [8] illustrate that the difference in the primary vortex features predicted by the Euler and Navier-Stokes calculations is slight. However, a viscous model is needed to adequately model the boundary layer and secondary vortex. For a rounded leading edge geometry, the physical separation point is sensitive to the Reynolds number, and an Euler equation model can lead to an algorithm dependent separation point [9]. Rizzi's studies [7] show that quite dense meshes are needed to adequately resolve the details of the flow fields predicted by the Euler equations.

In this paper two separate Euler calculations are presented for supersonic flows past a flat plate delta wing with a 70 degree leading edge sweep. One calculation is done using a fully three-dimensional model, while the other uses a conical flow approximation. The motivation in performing the separate calculations is to compare results from two approaches, which should agree, in order to better understand the accuracy of the 3D model and the resolution required to model the flow physics.

3. GOVERNING EQUATIONS

The three-dimensional Euler equations representing the conservation laws for mass, momentum and energy for a perfect gas are written in integral form for an arbitrary volume V bounded by a closed surface S

$$\int_V \frac{\partial U}{\partial t}\, dV + \int_S (F,G,H) \cdot (n_x,n_y,n_z)\, dS = 0 \tag{1}$$

$$h_0 = E + \frac{p}{\rho} \tag{2}$$

$$p = (\gamma-1)\rho\left(E - \frac{1}{2}\left\{u^2+v^2+w^2\right\}\right) . \tag{3}$$

The state vector $U = (\rho,\rho u,\rho v,\rho w,\rho E)$, and flux vectors F,G,H are the usual convective terms for the continuity, three momentum, and energy equations. In these expressions, ρ is the density, u,v,w are the Cartesian components of velocity, p is the static pressure, E and h_0 are the total energy and enthalpy per unit mass, respectively, and γ is the ratio of specific heats of the gas. The variables x,y,z are the directions in a Cartesian coordinate system fixed in the wing frame of reference (Fig. 2). Since supersonic flow is being considered, the boundary conditions are 1) the freestream state is undisturbed ahead of the bow shock, and 2) there is no flow through the wing surface. Equations (1) are in strong conservation form so that the shock jump relations (Rankine-Hugoniot equations) are valid weak solutions of them.

In the three-dimensional calculations (3D) to be presented, equations (1) will be integrated numerically from an initial state to some final state which is assumed to be a steady flow. Since the total enthalpy is constant in a steady adiabatic inviscid flow, h_0 is taken as a constant throughout the calculations, permitting the elimination of the energy equation. Thus, a pseudo-time path will be followed. It should be noted that the 3D model is not limited to supersonic flows or the simple wing geometry considered in this paper.

The second model to be presented uses the conical flow approximation. This restricts the class of three-dimensional problems to those which are steady supersonic flows with attached shock waves past bodies whose shapes are generated by rays through the apex of the wing. The conical coordinates may be introduced (Fig. 2)

$$r = \sqrt{x^2 + y^2 + z^2} \qquad \eta = \frac{y}{x} \qquad \xi = \frac{z}{x} . \tag{4}$$

With the observation that the dependent variables are independent of r, the Euler equations become

$$\int_A \frac{\partial \hat{U}}{\partial t} \, dA + \int_S \left(\hat{F},\hat{G}\right) \cdot \left(n_\xi, n_\eta\right) dS + \int_A \hat{H} \, dA = 0 . \tag{5}$$

The modified state variable $\hat{U}$ and flux functions $\hat{F},\hat{G},\hat{H}$ are given as

$$\hat{U} = \frac{r^2 U}{\left(1+\eta^2+\xi^2\right)^{3/2}} \qquad \hat{F} = \frac{r(G-\eta F)}{\left(1+\eta^2+\xi^2\right)}$$
$$\hat{G} = \frac{r(H-\xi F)}{\left(1+\eta^2+\xi^2\right)} \qquad \hat{H} = \frac{2r(F+\eta G+\xi H)}{\left(1+\eta^2+\xi^2\right)^2} . \tag{6}$$

Since the conical equations are also in divergence form, their weak solutions are the Rankine-Hugoniot relations [11]. The boundary conditions are the same as the 3D problem, except that the wing must be of infinite extent, thereby excluding any effect of a trailing edge. It should be noted that although the fluid dynamic variables (ρ,u,v,w, p,etc) are independent of direction r, either derivatives or integrals of them, such as vorticity and circulation, involve a length and therefore depend upon r. It is usual to solve the conical equations on the unit sphere ($r = 1$) and that is done herein. The present conical flow approach has been developed by Powell [12] and Perez [13].

4. NUMERICAL METHOD

The solution of equations (1) or (5) is accomplished by a pseudo-unsteady finite volume method. The basic approach is completely discussed by Rizzi and Eriksson [14], and will only be outlined here. Some differences exist between the 3D and the conical implementation which will be noted. A suitable grid system composed of a large number of cells is constructed to fit the body shape, as shown in Figure 3 for the conical wing and Figure 4 for the 3D wing. Note that the grid shown in Figure 4 is for a wing with thickness, and in the present calculations the wing has no thickness. Each cell has associated with it a value of the state vector U or $\hat{U}$. The flux vectors F,G,H or $\hat{F},\hat{G},\hat{H}$, are evaluated at the sides of each cell by simple averages of the flux vectors for the two adjacent cells, computed using values of U or $\hat{U}$. Equation (1) or (5), respectively, is then applied to an arbitrary cell as

$$\frac{dU_{ijk}}{dt} V_{ijk} + \sum_{\ell=1}^{6} \left(F_\ell, G_\ell, H_\ell\right) \cdot \left(n_{x\ell}, n_{y\ell}, n_{z\ell}\right) \Delta S_\ell = 0 \tag{7}$$

$$\frac{dU_{ij}}{dt} A_{ij} + \sum_{\ell=1}^{4} \left(\hat{F}_\ell, \hat{G}_\ell\right) \cdot \left(n_{\xi\ell}, n_{\eta\ell}\right) \Delta S_\ell + \hat{H}_{ij} A_{ij} = 0 \ .$$

The areas of the cell sides projected in the three directions, as well as the cell volume, are evaluated using simple formulas involving the corner points of the cells (node or grid points).

This semi-discrete approximation produces a large system of coupled ordinary differential equations for the state vector. The ODE's are integrated using a multistage method. In the 3D model a three stage method is used [14]. For the conical model, the four stage method of Jameson, Schmidt and Turkel [13] is adopted. In order to avoid the slow convergence to steady state which would result from using a constant $\Delta t = \min \Delta t$ over all cells, the usual practice of setting Δt to its local stability limit is adopted.

The above discrete approximation to the governing equations contains modes with wavelengths on the order of the mesh spacing which must be damped. In addition, the possibility of shock waves requires that suitable artificial viscosity be added. This alters the true discontinuous solutions into smooth but steep gradients over a few mesh cells. The damping terms added to the right hand side of Eq. (7) form

$$\nu_2 \delta_\ell (S \delta_\ell U) - \nu_4 \delta_\ell^4 U \qquad \ell = i,j,k \tag{8}$$

$$S = \frac{|\delta_\ell^2 p|}{\max|\delta_\ell^2 p|} \qquad \delta_\ell U = U_{\ell+\frac{1}{2}} - U_{\ell-\frac{1}{2}} . \tag{9}$$

The first component in equation (8) is a second difference term with a nonlinear coefficient normalized to vary between 0 and 1 in the domain. It is designed to be large at shocks and small elsewhere. The second component is a fourth difference term with a constant coefficient designed to damp the unwanted high frequency modes. In the current calculations $\nu_2 = 0.05$ and $\nu_4 = 0.01$. This is the model described in detail in Ref. [14], wherein the treatment of the damping at the boundary is also described. For the conical calculations it was found that the second difference damping at the wing tip, as controlled by the nonlinear switch, was not sufficient to prevent negative pressures from appearing. Therefore, in a few cells around the tip, the switch was set to unity. This adjustment of the switch was not used in the 3D calculations. Also, in the conical calculation, the energy equation is solved, and the damping is applied to ρh_0 rather than ρE. With this, the maximum error in total enthalpy for the results presented is 0.4%. Damping applied to ρh_0 results in the same damping being added to both the continuity and energy equations for the converged results.

The no flow boundary condition at the body requires that only the pressure on the cell side adjacent to the body be known. For the flat plate wing geometry under study, both approaches used the approximation that the derivative of pressure normal to the body is zero. For the 3D method, the outer boundary is placed far enough away that the bow shock was captured. A characteristics treatment was used for the boundary conditions [14]. For the conical problem, the bow shock wave was fitted [12,13]. In the grid system shown in Figure 3, the outer boundary corresponds to the converged shock shape.

In both calculations, the computations are started on a coarse grid and continued for several hundred iterations. The grid spacing is then halved, the state vectors interpolated to the new points, and the iterations continued. For the 3D model, two coarse grids are used along with the final grid. One thousand iterations were done on the fine grid with a final rms residual level of approximately 10^{-4}. For the conical calculations, the solution is started on a grid with 16 x 16 cells and continued until the 128 x 128 grid is reached. Two thousand iterations were done on the fine grid with a final rms residual level of 10^{-8}. A posteriori analysis indicated the calculations were essentially converged after 500 iterations.

5. COMPUTATIONAL CONSIDERATIONS

Both the 3D and the conical computations are representative of methods for analyzing compressible flows which require supercomputer capability to produce useful results. For both methods, approximately 25 words of memory are required per cell, and on the order of 1000 floating point operations per cell per time step are needed. About 1000 iterations are needed on a fine grid for convergence. Coarser grid calculations add a little extra CPU time over the above, but greatly reduce the number of iterations needed on the fine grid.

The three-dimensional computations were done on a 16 million word CYBER 205 which, at the time, was at CDC. The program has been highly vectorized for the 205 [15] and achieves a sustained performance rate of 125 MFLOPS (million floating point operations per second). The calculations are done in half precision (32 bit words) except for the grid generation and metric calculations, which are done in full precision (64 bit words) and then truncated. The final grid was 196 x 56 x 96 cells, which translate into a 25 million, 32 bit word main memory requirement and about one trillion operations to reach convergence.

The conical flow model with one less space dimension has correspondingly less computer requirements. With 128 x 128 cells and all variables using full precision, about 600K of main memory are needed with about 100 billion floating point operations to reach convergence. The calculations were done on the same CYBER 205 after it was installed as the VPS 32 at NASA Langley Research Center. The program has not yet been vectorized and runs at about 5 MFLOPS in the scalar mode.

An important consideration in the use of supercomputers like the VPS 32 is the transmission, manipulation, and digestion of the resulting data base representing the solution. With communications and graphics processing technology which is available to the average researcher, it is not feasible to work with the entire data bases of the 3D results. The present researchers are located remotely from the supercomputers, as will be many of the users of large machines during the next decade. Data transmission and manipulation will become bottlenecks unless considerations are given on how to handle the generated numbers. It is clear that graphical output is the only meaningful way to evaluate the results and learn what they contain. One way to alleviate the bottleneck is to first work with data bases which have been thinned of many state vectors for display purposes. Figure 6 is an example of this. These results were computed on a 128 x 128 grid, but thinned to a 64 x 64 grid for plotting. After viewing the solution on the thinned grid, regions of interest can be determined and cuts in the solution grid can then be transmitted and displayed more effectively. Most of the flow field plots shown were constructed this way.

6. RESULTS

Computations for the two methods discussed above have been done for 5, 10 and 15 degrees angle of attack. Only the 10 degree case is reported in detail as analysis of the other two cases is still in progress. Figure 5 shows the isobar lines on the surface of the wing for the three-dimensional calculation. This and similar off-surface plots demonstrate that the fully 3D calculation is reasonably conical in structure except near the apex and trailing edge region. With this observation, the 3D results have been processed by the same graphics reduction program as the conical solutions to permit a direct comparison. A computational surface of data which intersects the wing at about 80% chord station has been used. The approximate location is marked on Figure 5. The topology of the surface is like that in the upper left hand corner of Figure 4. X,Y,Z coordinates were converted to r,η,ξ coordinates using Equation (4). With the assumption that the solution is independent of r (conical assumption), the values of the state vectors at these points should be directly comparable to the conical solution when presented in η,ξ coordinates.

Contours of static pressure coefficient from the conical

computation are shown in Figure 6. The leading edge vortex structure is the dominant feature in this plot as it is for the other flow variables. The remaining contour plots will be confined to the region near the wing. Figure 7 displays the surface values of static pressure coefficient for both the conical and the 3D calculations. The strong suction region under the vortex is shown in both Figures 6 and 7. Other features include a cross flow shock under the vortex and a rapid expansion around the wing tip. The agreement between the two calculations is surprisingly good. There is some difference in the "spikeness" at the wing tip due to differences in mesh skewness or damping in these cells.

Figure 8 compares other surface variables from the two calculations. With the exception of the total pressure distributions, the agreement is as good as for the static pressure values. Total pressure results show qualitative agreement only. The physical interpretation of these results will be discussed in the following paragraphs with reference to the contour plots in the field.

Figure 9 compares contours of the cross flow Mach number computed by subtracting the r-component from the total velocity magnitude and dividing by the speed of sound. Comparison of the two results shows quitw good qualitative agreement. In both models, the cross flow Mach number reaches 1.5 above the vortex, and 1.7 (in the opposite) direction underneath the vortex. Most features appear approximately the same in the two calculations. However, the contours for the 3D calculations are displaced further from the wing surface than are the conical ones. This observation holds for all of the contour plots, which is somewhat surprising since the surface values are in excellent agreement. Perhaps, since the 3D calculations are not perfectly conical, a distortion arises in the off surface values when reducing the data to conical form.

Contours of radial velocity are given in Figure 10, which demonstrates the level of agreement discussed above. This parameter is the velocity normal to the unit sphere normalized by the projection of the freestream velocity along the wing leading edge. A rather strong "shear layer" is seen to emanate from the wing tip. In this and the other contour plots, the 3D calculations have a more diffused structure in this region than do the conical calculations. The velocity on the windward side is nearly constant, as it is on the leeward side away from the vortex. Underneath the vortex, the radial velocity increases towards the wing tip as the cross flow component decreases and the flow turns to the downstream direction. The velocity in the vortex core is significantly greater than the freestream value, and the core is pushed inboard by the high pressure region behind the cross flow shock.

Magnitudes of the total pressure loss parameter $(1 - Pt/Pt_{\infty})$ are shown in Figure 11. The vortical region is very evident as are meandering contours of zero total pressure loss. For reference purposes, the maximum total pressure loss across the bow shock is 0.3×10^{-4}, which is insignificant on this plot. Total pressure loss is a sensitive measure of accuracy of the Euler equation solutions, and the wandering freestream level curves are typical of such solutions. This is also the reason that surface values of total pressure do not overlay each other for the two calculations.

The large structure indicating the extent of the vortex core represents a different phenomenon. It is important to note that the two computations yield approximately the same magnitude of total pressure loss in the core. The appearance of large total pressure losses in the primary vortex cores is evident in all Euler and Navier-Stokes finite volume calculations examined by these authors. The magnitude of the loss is insensitive to computational parameters as well as inclusion of viscous terms of the Navier-Stokes equations [8]. Rizzi [7] has discussed this issue, and further numerical results and explanations will be given by Powell et al [16].

The final results to be presented are the cross flow streamlines shown in Figure 12. These are the lines which are everywhere parallel to the cross flow velocity vectors on the unit sphere. Since the radial velocity component is not considered, these trajectories represent the projection onto the unit sphere of helical streamlines which start at the bow shock and spiral around as they move downstream. In the numerical integration of the trajectory equations, the calculations are stopped after a preset number of cells have been traversed. This is the cause of the void of streamlines in the vortex core. Several features can be observed. First, there is an attachment line about mid span on the leeward surface dividing the flow which gets wrapped up in the vortex from that which ends up at the symmetry plane. Second, there is an attachment line just inboard of the tip on the windward surface which similarly divides the flow. Third, the streamlines under the vortex have a minimum area near where the cross flow goes sonic, followed by a mildly diverging area region up to where the cross flow shock appears. It resembles a quasi one-dimensional nozzle behavior.

Overall the agreement between the two independently obtained solutions is quite good. One question which had been of interest to the authors regarded the effect of the wing apex modeling on the fidelity of the 3D calculations. Since the vortex starts at this location, any difficulties in resolving this region might have lingering effects over the entire wing. Certainly it can be said that the surface values are not affected adversely in this regard.

Results have been computed for 5 and 15 degree angles of attack, but have not been fully analyzed and compared. A brief description will be given herein of the findings. For 5 degrees, a small vortex is formed which lies on the surface of the wing. Although the cross flow under the vortex goes slightly supersonic, there is no indication of a shock. For 15 degrees, a very strong vortex is present and the attachment line on the leeward surface has moved to the symmetry plane. The calculations indicate unsteadiness in the flow and are being further examined before being reported.

7. DISCUSSION

The quantitative agreement between the two computational approaches and qualitative agreement with experimentally observed features lends support to the usefulness of Euler equation solutions in analyzing vortex flows. However, the problem studied serves to illustrate future trends over the next decade in CFD. The paper is concluded with our observations in this area.

The ability of CFD to analyze new classes of problems opens up new questions and issues for fluid dynamicists to resolve. Both analysis and experiment need to be done to investigate these questions. For example, there is not yet a completely satisfactory description of the mechanism of total pressure loss in the vortex core. There is the probability of unsteadiness being found in the compressible rotational flows modeled here. Fluid mechanics has always enjoyed the benefit of theory and experiment done side by side. Now, with CFD providing a new approach, there will be plenty of new problems for the curious researcher.

The development of better algorithms must continue. It is clear that zonal or embedded methods for grids and/or equations are very important to develop. For example, the bow shock position has converged on the 32 x 32 grid, and only a relatively small zone in the flow field needs to be resolved to the extent of resolution done here. Another example is that the requirements of the 3D method to resolve the flow structure at the apex of the wing can be greatly reduced if a conical solution can be embedded as a local solution to the equations.

Finally there are the issues of the advancing computer technology. From the experience of the second author, the algorithms being used here can be vectorized to get good processing rates. They most probably can be made to work efficiently on multiprocessor architectures also. However, it takes some time to understand the requirements for each computer. The other important aspect is the manipulation of the data base for post-processing. Graphics are crucial, but working with the large data bases with available telecommunications and post-processing technology requires many hours.

8. ACKNOWLEDGMENTS

The authors would like to acknowledge the financial support of FFA, The Aeronautical Research Institute of Sweden, and NASA Langley Research Center under Grant NAG-1-358. Computer time for the 3D calculations was provided by Control Data Corporation and valuable support given by Chuck Purcell. Computer time for the conical calculations was provided by NASA Langley and valuable support given by Kara Haigler. The first author is a Professor of Aeronautics and Astronautics at MIT, the second author is a research scientist at FFA, and the third author is a research assistant at MIT.

9. REFERENCES

[1] Miller, D.S. and Wood, R.M., "An Investigation of Wing Leading-Edge Vortices at Supersonic Speeds," AIAA Paper 83-1816, July 1983.

[2] Vorropoulos, G. and Wendt, J.F., "Laser Velocimetry Study of Compressibility Effects on the Flow Field of a Delta Wing," AGARD-CP-342, Paper 9, 1983.

[3] Rizzi, A., "Damped Euler Equation Method to Compute Transonic Flow Around Wing-Body Computations," AIAA J. Vol. 20, pp 1321-1328, October 1982.

[4] Rizzi, A., Eriksson, L.E., Schmidt, W. and Hitzel, S., "Numerical Solutions of the Euler Equations Simulation of Vortex Flows Around Wings," AGARD-CP-342, Paper 21, 1983.

[5] Weiland, C., "Vortex Flow Simulations Past Wings Using the Euler Equations," AGARD-CP-342, Paper 19, 1983.

[6] Raj, P. and Sikora, J.S., "Free-Vortex Flows: Recent Encounters with and Euler Code," AIAA Paper 84-0135, January 1984.

[7] Rizzi, A., "Euler Solutions of Transonic Vortex Flow Around the Dilner Wing - Compared and Analyzed," AIAA Paper 84-2142, August 1984.

[8] Rizetta, D.P. and Shang, J.S., "Numerical Simulation of Leading-Edge Vortex Flows," AIAA Paper 84-1544, June 1984.

[9] Newsome, R.W., "A Comparison of Euler and Navier-Stokes Solutions for Supersonic Flow Over A Conical Delta Wing," AIAA Paper 85-0111, January 1985.

[10] Fujii, K. and Kutler, P., "Numerical Simulation of the Viscous Flow Fields Over Three-Dimensional Complicated Geometries," AIAA Paper 84-1550, June 1984.

[11] Viviand, H., "Formes Conservatives des Equations de la Dynamique des Gaz," Recherche Aerospatiale, No. 1, Fevrier 1974.

[12] Powell, K., "The Effects of Artificial Viscosity Models on Conically Self-similar Solutions to the Euler Equations," S.M. Thesis, MIT, 1984.

[13] Perez, E., "Computation of Conical Flows with Leading Edge Vortices," S.M. Thesis, MIT, 1984.

[14] Rizzi, A. and Eriksson, L.E., "Computation of Flow Around Wings Based on the Euler Equations," Journal of Fluid Mechanics, October 1984.

[15] Rizzi, A., "Vector Coding the Finite-volume Procedure for the CYBER 205," in Lecture Series Notes 1983-84, von Karmen Institute, Brussels, 1983.

[16] Powell, K.G., Perez, E.S., Murman, E.M. and Baron, J.R., "Total Pressure Loss in Vortical Solutions of the Conical Euler Equations," AIAA Paper 85-1701, July 1985.

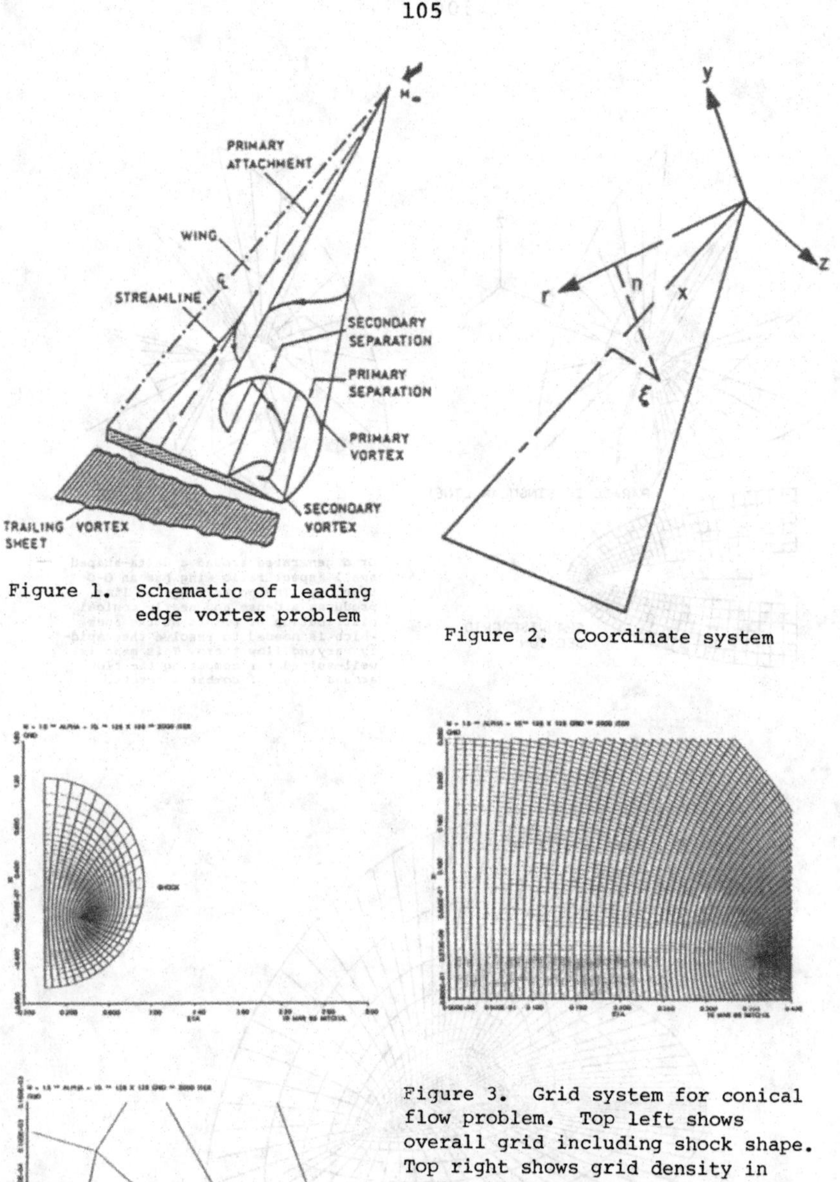

Figure 1. Schematic of leading edge vortex problem

Figure 2. Coordinate system

Figure 3. Grid system for conical flow problem. Top left shows overall grid including shock shape. Top right shows grid density in regions corresponding to Figs. 9-12. Bottom left shows grid topology in wing tip region.

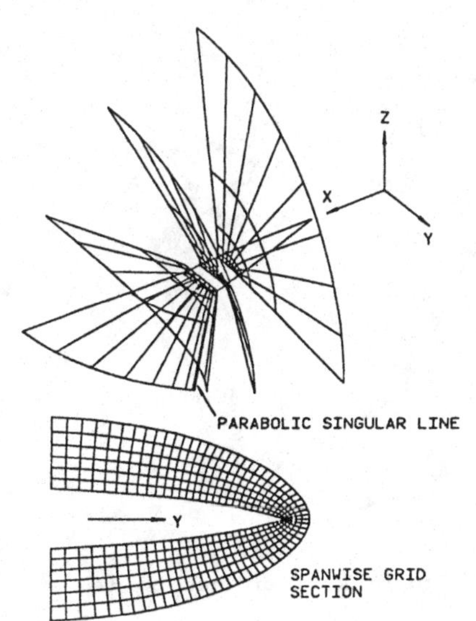

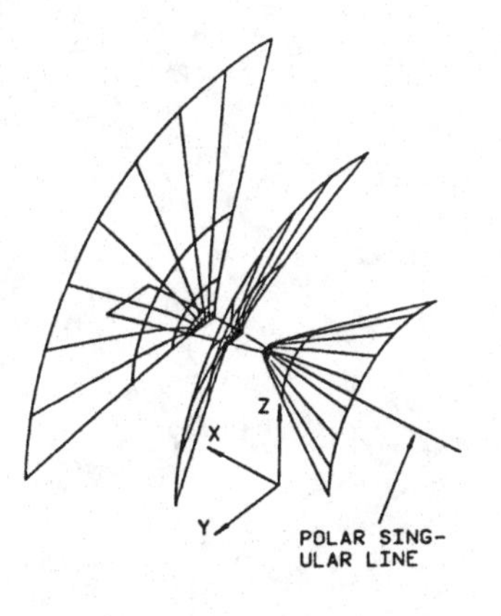

Grid generated around a delta-shaped small aspect ratio wing has an O-O topology. The polar singular line produces a dense and nearly conical distribution of points at the apex which is needed to resolve the rapidly varying flow there. This mesh is well-suited for computing the flow around wings of combat aircraft.

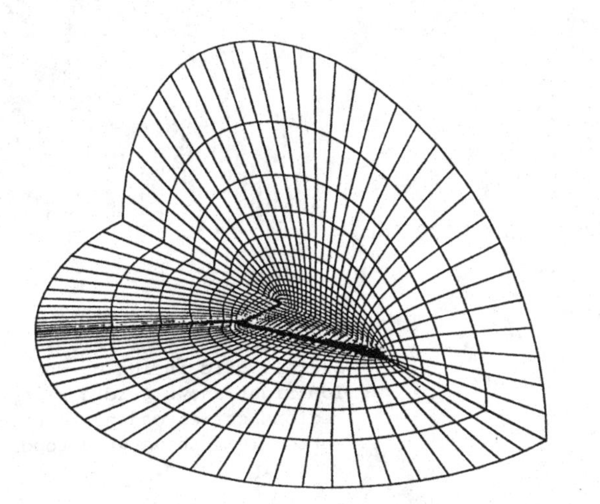

Figure 4. Grid system for three-dimensional flow problem

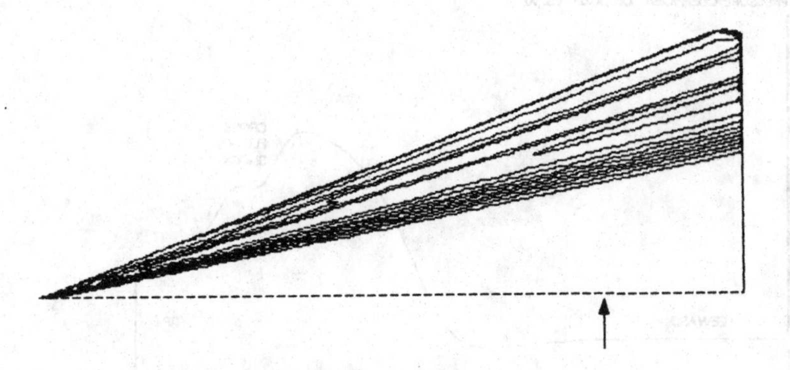

Figure 5. Upper surface isobars for 3D calculation illustrating near conical flow behavior. Arrow indicates station used for other 3D plots.

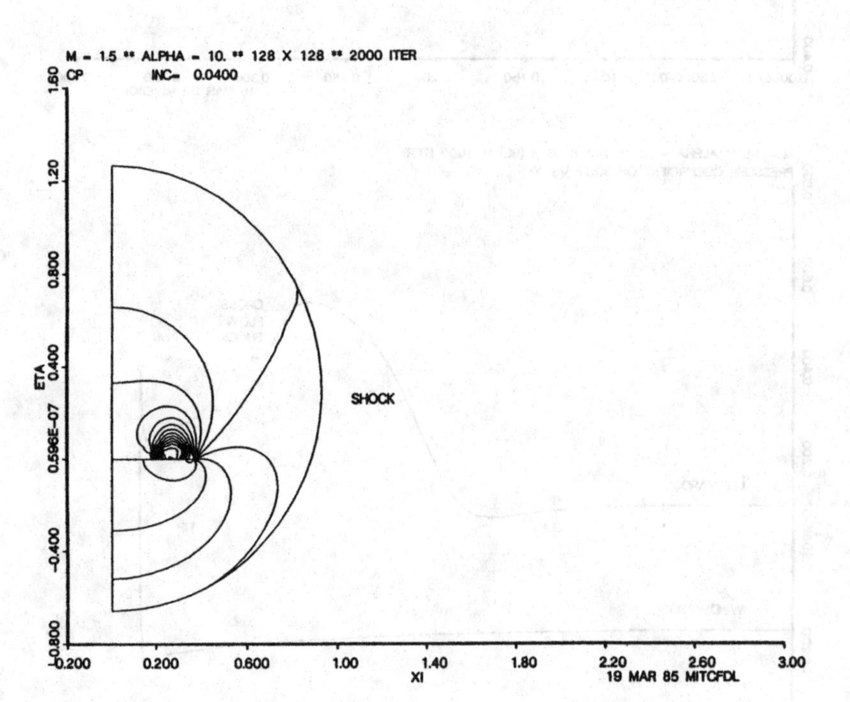

Figure 6. Pressure coefficient contours from conical flow calculation showing shock shape and dominant vortex structure.

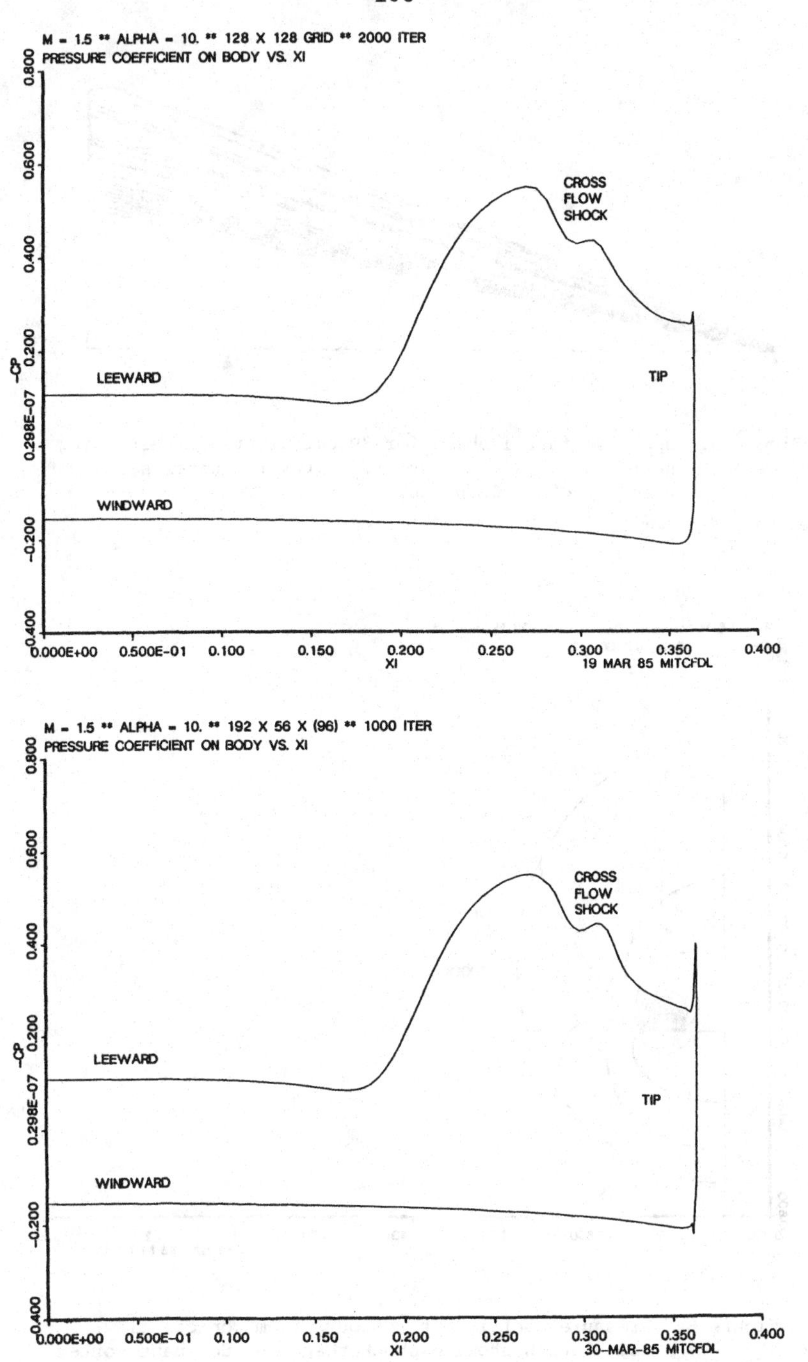

Figure 7. Pressure coefficient contours on wing surface for conical (top) and 3D (bottom) calculations.

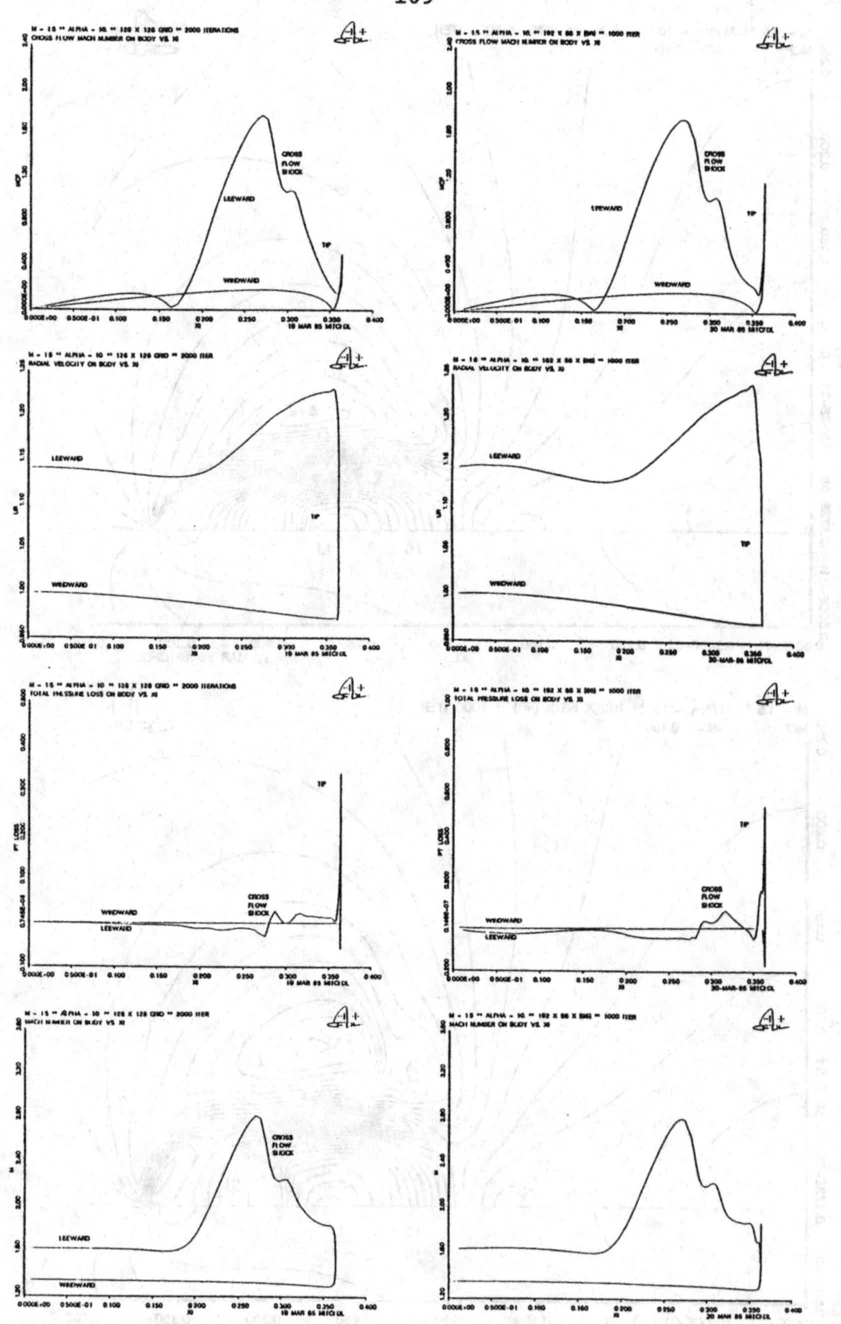

Figure 8. Wing surface values for conical (left) and 3D (right) calculation. From top to bottom, cross flow Mach number, radial velocity, total pressure loss, and total Mach number.

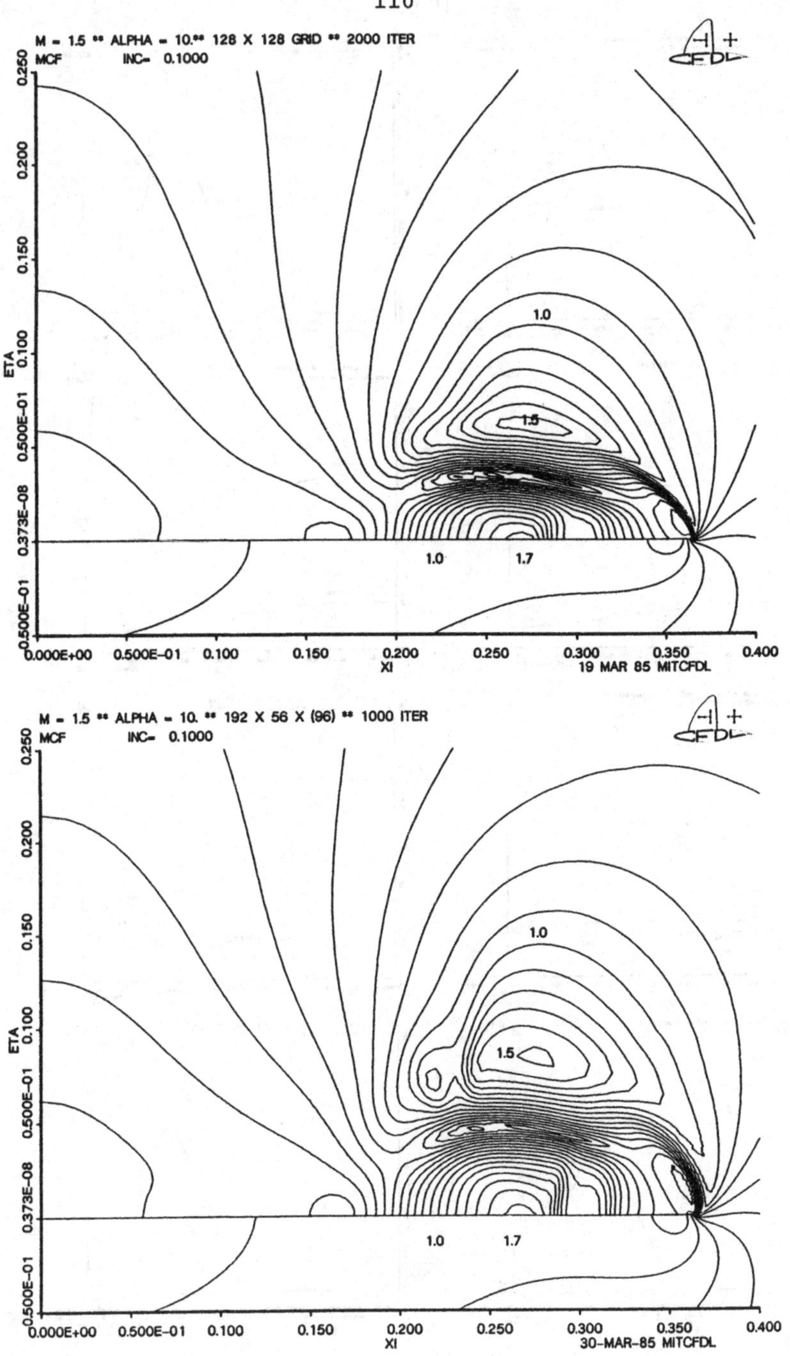

Figure 9. Cross flow Mach number contours for conical (top) and 3D (bottom) calculations.

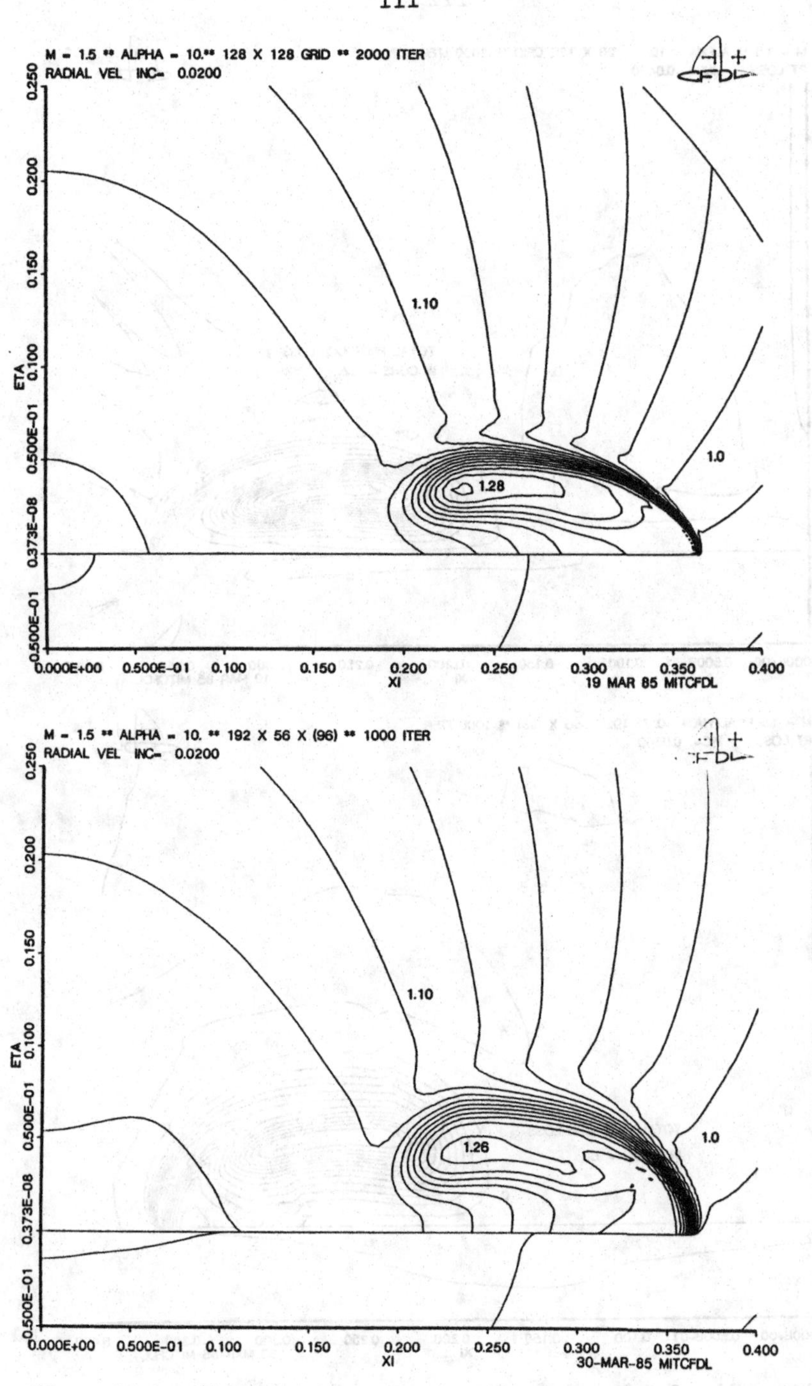

Figure 10. Radial velocity contours for conical (top) and 3D (bottom) calculations.

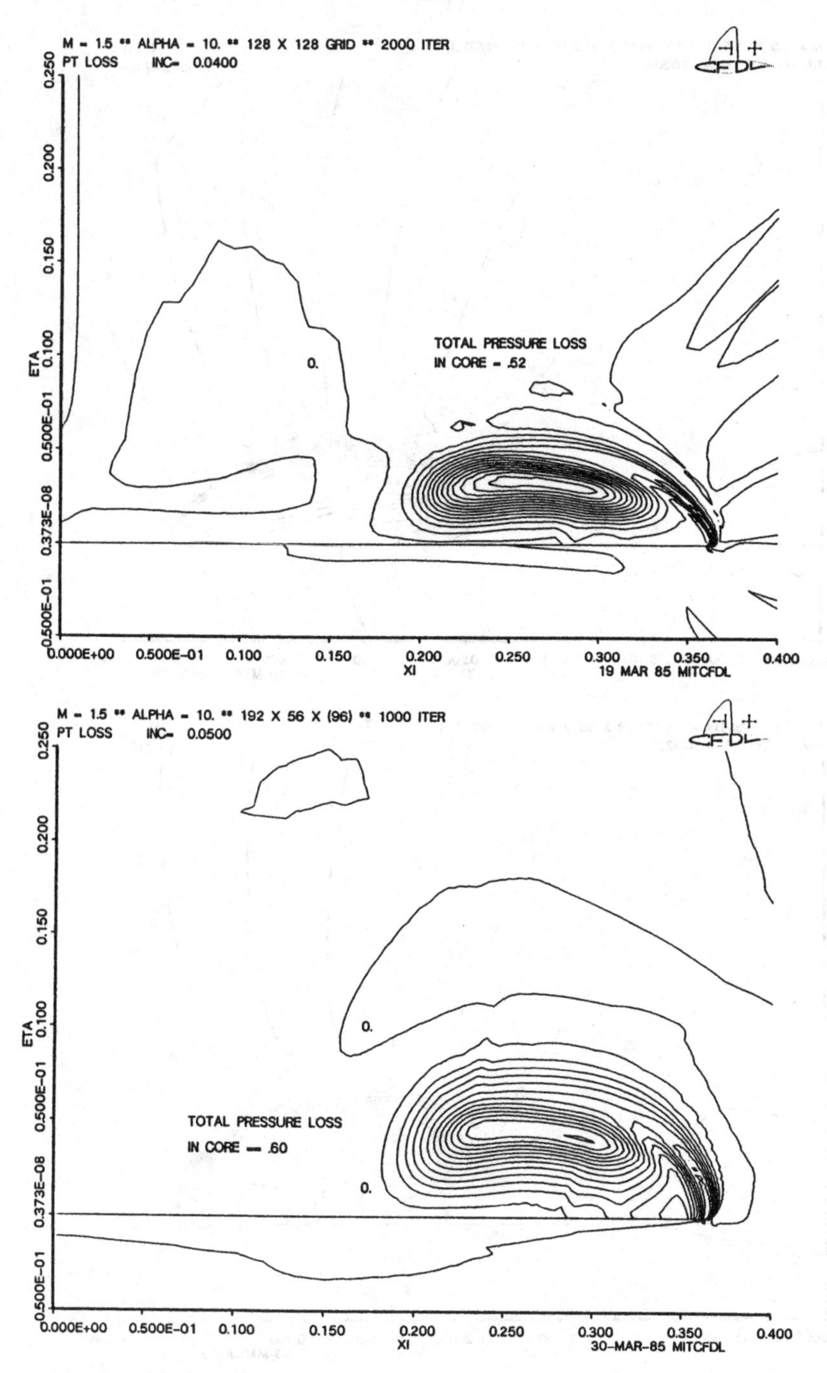

Figure 11. Total pressure loss contours for conical (top) and 3D (bottom) calculations.

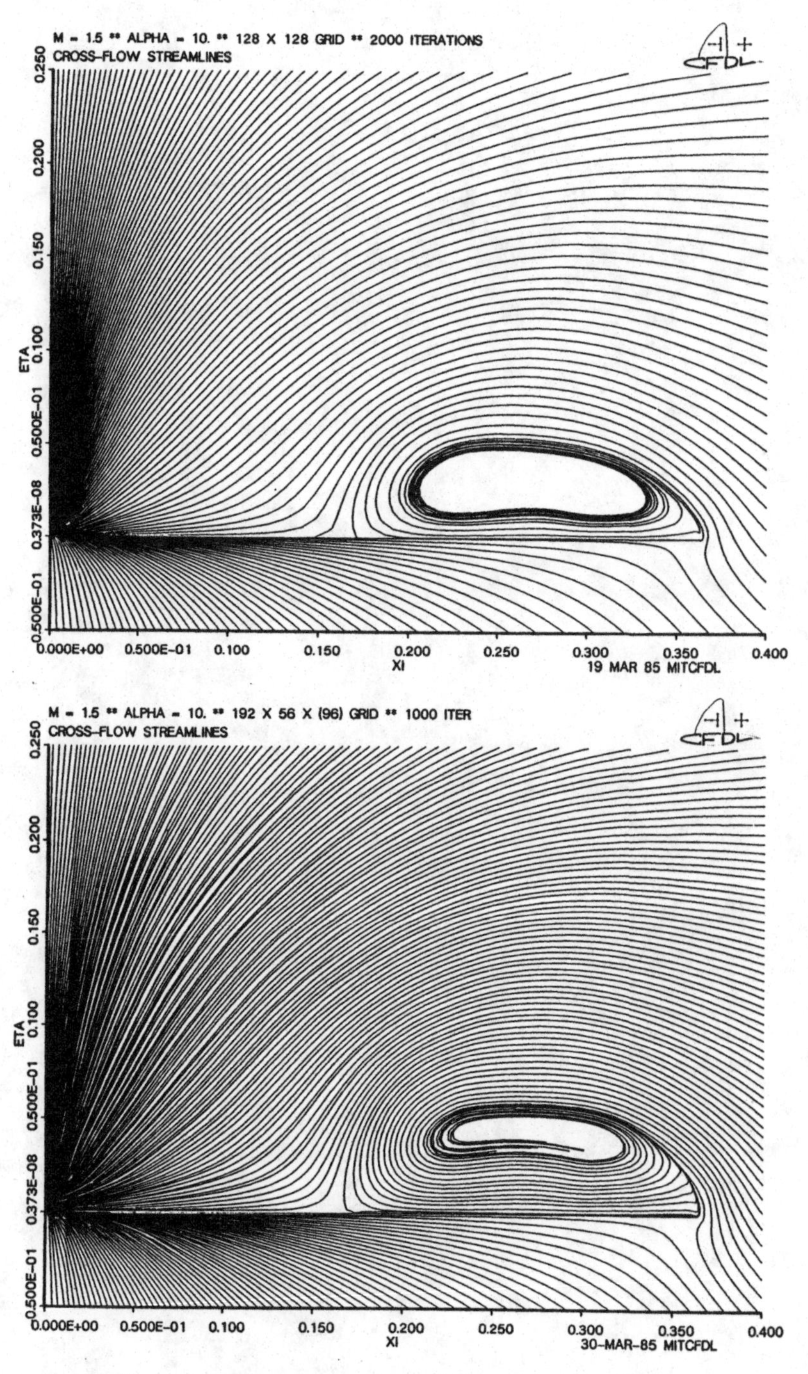

Figure 12. Cross flow streamlines for conical (top) and 3D (bottom) calculations.

Progress in Scientific Computing, Vol. 6
Proceedings of U.S.-Israel Workshop, 1984

AN EFFICIENT ITERATION STRATEGY FOR THE SOLUTION OF THE EULER EQUATIONS

Robert W. Walters
National Research Council
and
Douglas L. Dwoyer
Head, Computational Methods Branch

NASA Langley Research Center
Hampton, Virginia

Abstract

A line Gauss-Seidel (LGS) relaxation algorithm in conjunction with a two-parameter family of upwind discretizations of the Euler equations in two-dimensions is described. The basic algorithm has the property that convergence to the steady-state is quadratic for supersonic flows and linear for subsonic flows. This is in contrast to the central differenced block ADI methods, the upwind biased relaxation schemes, and also to the spatially split fully upwind methods, all of which converge linearly, independent of the flow regime. Moreover, the algorithm presented here is easily enhanced to detect regions of mixed supersonic/subsonic flow; marching by lines in the supersonic regions, converging each line quadratically, and iterating in the subsonic regions, thus yielding a very efficient iteration strategy. Numerical results are presented for two-dimensional supersonic and transonic flows containing both oblique and normal shock waves which confirm the efficiency of the iteration strategy.

Introduction

A large class of problems of fundamental importance to the field of computational aerodynamics is that of simulating the flow of an inviscid, compressible gas in the transonic and supersonic regimes. The most popular approaches used in obtaining steady-state solutions to these problems have been the implicit spatially split methods of the approximate factorization (AF) type such as those due to Briley and McDonald[1] and Beam and Warming[2] and the highly vectorizable explicit schemes such as the multi-stage Runge-Kutta method used by Jameson, et al.[3] and the predictor-corrector scheme of MacCormack.[4] A motivation for algorithm research stems from the fact that the highly vectorizable explicit schemes suffer from stringent stability restrictions and though the implicit AF methods are unconditionally stable in two-dimensions, they require an optimal set of iteration parameters for rapid convergence which are difficult and time consuming to obtain.[5] Moreover, the implicit AF schemes are unstable for the three-dimensional Euler equations.

The recent emergence of upwind differencing technology for the Euler equations has opened the door to a new class of solution strategies aimed at improving computational efficiency. Since the upwind differencing is more costly per iteration to implement than central differencing, overall efficiency gains must be derived from improved convergence rates. This is made possible by the improved conditioning of the system of difference equations afforded by the more physical 'upwind' discretization.

Some preliminary work by Chakravarthy[6] and van Leer and Mulder[7] using relaxation methods and upwind schemes have been promising. In this study, the fact that upwind differencing closely models the propagation of information along characteristics is exploited by choosing a line relaxation/upwind discretization combination such that, for supersonic flow in the streamwise direction, the algorithm becomes a direct solver of the linearized problem. That is to say, the family of algorithms presented here all revert to an efficient implementation of Newton's method in the supersonic regime resulting in quadratic or 'near' quadratic convergence to the steady state. The 'near' quadratic is added because Newton's method is only guaranteed to converge quadratically to a root of a function if the function is sufficiently smooth (continuous). For flows with shock waves, this may degrade the convergence rate, but as is shown in the results section for a two-dimensional flow and also in reference 8 for a one-dimensional problem, this degradation is minor.

Further efficiency gains are realized in supersonic regions by solving the problem in a marching fashion, iterating on each line in order to remove the linearization error before proceeding to the next line. This is done without parabolization but can be accomplished since, in the absence of adverse waves, the line relaxation/upwind discretization combinations presented in this study result in the uncoupling of the discrete algebraic problem by lines. Thus, one can effectively recover the computational efficiency of parabolized methods without suffering from their inherent limitations. For problems with mixed supersonic/subsonic regions, such as inlet/diffuser and dual throat inlet problems, the solution strategy can be extended to detect the flow regime interfaces, march in the supersonic regions and iterate with 'classical' line Gauss-Seidel in the subsonic regions.

A comparison of CPU times with and without the new iteration strategy are provided in the results section.

Spatial Discretization

The individual is free to choose between either the general control volume formulation or a finite-difference formulation in generalized coordinates. With upwind schemes, the authors prefer the control volume formulation since free-stream subtraction[9] is not required. The results presented in this study have been obtained, however, by both formulations.

The integral form of the conservation equations governing the flow of an inviscid fluid can be written as:

$$\frac{\partial}{\partial t} \int_V q \, dV + \int_S \vec{F} \cdot \hat{n} \, ds = 0 \qquad (1)$$

where V is the volume bounded by the surface S and $\hat{n}$ is the outward pointing unit normal to the surface, q is the vector of conserved state variables and $\vec{F}$ is the vector of convective fluxes. For simplicity in describing the upwind method, consider the semi-discrete formulation of equation (1) for a uniform Cartesian grid as shown in figure 1.

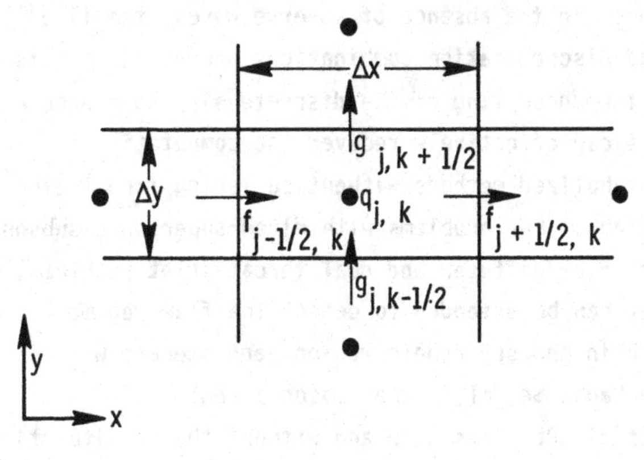

Figure 1.-
Notation for cell j,k. Cell boundaries at $j\pm1/2$ and $j,k\pm1/2$; state variables at j,k.

Dividing by the cell volume ($\Delta x\Delta y$) and rearranging gives,

$$\left(\frac{\partial q}{\partial t}\right)_{j,k} = -\left[\frac{1}{\Delta x}(f_{j+1/2,k} - f_{j-1/2,k}) + \frac{1}{\Delta y}(g_{j,k+1/2} - g_{j,k-1/2})\right] = -R_{jk} \tag{2}$$

In the above,

$$q = \begin{bmatrix} \rho \\ \rho u \\ \rho v \\ e \end{bmatrix} \quad f = \begin{bmatrix} \rho u \\ \rho u^2 + p \\ \rho uv \\ (e + p)u \end{bmatrix} \quad g = \begin{bmatrix} \rho v \\ \rho uv \\ \rho v^2 + p \\ (e + p)v \end{bmatrix} \tag{3}$$

where ρ is the density, u and v are the velocity components in the x and y directions respectively, p is the pressure and e is the total energy per unit volume.

In this study, upwind differencing is implemented by splitting the convective fluxes, f and g, into positive and negative contributions. The particular flux vector splitting (FVS) technique used here is due to van Leer.[10] Although other FVS techniques could be implemented in the general framework to follow, van Leer's was chosen because the individual split flux contributions smoothly transition across sonic points, and the Jacobian matrices associated with the linearization of the positive split flux contributions have eigenvalues $\lambda > 0$ and likewise the Jacobian matrices of the negative split flux contributions have eigenvalues $\lambda < 0$. In addition, an important reason for choosing van Leer's scheme is its simplicity. The splitting of f is given by

$$f(q) = f^{+}(q) + f^{-}(q) \tag{4}$$

where, for $|M_x| < 1$,

$$f^{\pm}(q) = \begin{bmatrix} f^{\pm}_{mass} \\ f^{\pm}_{mass} \; [(\gamma-1)u \pm 2c]/\gamma \\ f^{\pm}_{mass} \cdot \; v \\ f^{\pm}_{mass} \; \{[(\gamma-1)u \pm 2c]^2/[2(\gamma^2-1)] + v^2/2\} \end{bmatrix}$$

and

$$f^{\pm}_{mass} = \pm\rho c[\tfrac{1}{2}(M_x \pm 1)]^2$$

For

$$M_x > 1, \; f^{+} = f, \; f^{-} = 0$$

and for

$$M_x < -1, f^{+} = 0, \; f^{-} = f$$

In the above, γ is the ratio specific heats, c_p/c_v, c is the local sound speed, and M_x is local Mach number based on the velocity component in the x-direction, u. The splitting of g in terms of $M_y = v/c$ follows similarly.

Utilizing the flux vector splitting technique, a pair of state vectors are assigned to a common cell interface and a single numerical flux is derived from this pair. For example, $f_{j+1/2,k}$ can be expressed as

$$f(q)_{j+1/2,k} = f(q_L, q_R)_{j+1/2,k} = f^+(q_L)_{j+1/2,k} + f^-(q_R)_{j+1/2,k} \tag{5}$$

where the subscripts L and R imply that the dependent variables at the cell boundary are evaluated from the 'left' and 'right' respectively. More precisely, a two-parameter family of estimates for the state vector pair at the (j+1/2,k) cell boundary is given by,

$$(q_L)_{j+1/2,k} = q_{j,k} + \frac{\phi}{4}\left[(1-\kappa_x)\nabla + (1 + \kappa_x)\Delta\right]q_{jk} \tag{5}$$

$$(q_R)_{j+1/2,k} = q_{j+1,k} - \frac{\phi}{4}\left[(1-\kappa_x)\Delta + (1+\kappa_x)\nabla\right]q_{j+1,k} \tag{6}$$

where

$$\Delta q_{j,k} = q_{j+1,k} - q_{j,k} \tag{7}$$

$$\nabla q_{j,k} = q_{j,k} - q_{j-1,k} \tag{8}$$

Similar formulas hold for the interpolation in the y-direction with parameter κ_y. In general, no distinction is made between κ_x and κ_y, but as will be shown later, the family of marching algorithms are obtained with $\kappa_x = -1$ and with various values of κ_y. For $\phi = 0$, the method is first-order accurate in space and for $\phi = 1$, the parameters, κ_x and κ_y control the spatial accuracy and hence the truncation error of the method. Some common choices are $\kappa = 1/3$, corresponding to the only third-order accurate member of the family (strictly for a one-dimensional problem) and $\kappa = -1$, the fully upwind second-order scheme. The remaining second-order members are all upwind biased or centered approximations. In the next section, a marching method for supersonic regions will be developed based on a particular choice of κ_x and κ_y.

Relaxation Algorithm

Application of the Euler implicit time integration scheme in delta form and time linearization to the system of semi-discrete conservation laws given by equation (2) yields

$$\left[\frac{I}{\delta t} + \frac{dR}{dq}\right]\delta q_{jk} = - R^n_{jk} \tag{9}$$

where

$$\delta q = q^{n+1} - q^n; \quad q^n = q(n\delta t).$$

In matrix notation, equation (9) can be expressed as:

$$M\delta q = -R \tag{10}$$

where M is a large, banded, block coefficient matrix with block size of four. For δt tending to infinity, the solution of equation (10) by direct inversion is Newton's method. In general, due ultimately to the bandwidth of M, the direct solution of equation (10) is not efficient. For a fixed grid, the bandwidth of M depends on the particular value of κ used in evaluating the state vectors at a cell interface. Even more importantly, however, is the fact that for the upwind methods, the bandwidth is also dependent upon the flow regime of the particular problem under consideration. This is in contrast to standard central difference schemes which maintain the same structure of the coefficient matrix independent of the Mach number.

To be specific, equation (9) can be expanded to read:

$$\begin{aligned}
&\left(\frac{\delta q}{\delta t}\right)_{j,k} + \frac{1}{\Delta x}\{[A^+(q_L)\delta q_L + A^-(q_R)\delta q_R]_{j+1/2,k} \\
&- [A^+(q_L)\delta q_L + A^-(q_R)\delta q_R]_{j-1/2,k}\} \\
&+ \frac{1}{\Delta y}\{[B^+(q_B)\delta q_B + B^-(q_A)\delta q_A]_{j,k+1/2} \\
&- [B^+(q_B)\delta q_B + B^-(q_A)\delta q_A]_{j,k-1/2}\} = - R_{jk}
\end{aligned} \tag{11}$$

where

$$A^{\pm} = \frac{\partial f^{\pm}}{\partial q} \text{ and } B^{\pm} = \frac{\partial g^{\pm}}{\partial q} \tag{11a}$$

The subscripts 'B' and 'A' imply the evaluation of q at the cell boundary from 'below' and 'above,' respectively. The structure of M is depicted in figure 2, using equations (5) and (6) and similar formulas for the state-variable interpolation in the y-direction. Any individual element on the non-zero diagonals represents a 4 x 4 matrix. Note that vertical line Gauss-Seidel relaxation sweeping through lines in the positive x-direction approximates M by assuming zero entries on the upper diagonals labeled U1 and U2.

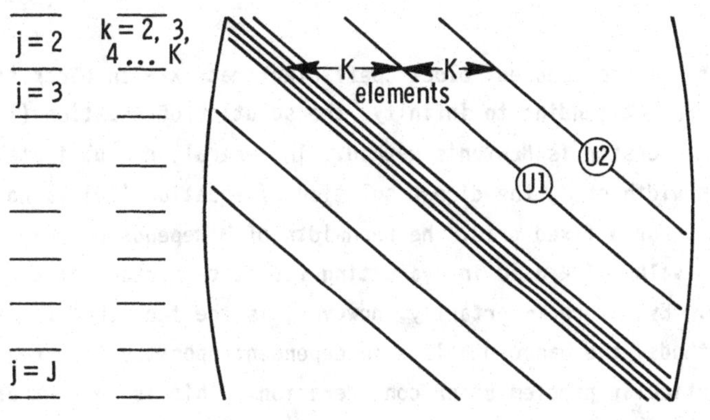

Figure 2.-
The structure of the coefficient matrix M.

With the ordering shown in figure 2, the elements on the upper diagonals are given by,

$$U1 = \frac{1}{\Delta x} \left[A^{+}_{j+1/2,k} \frac{\phi(1+\kappa_x)}{4} + A^{-}_{j+1/2,k}\left(1 - \frac{\phi\kappa_x}{2}\right) + A^{-}_{j-1/2,k} \frac{\phi(1-\kappa_x)}{4}\right] \tag{12}$$

$$U2 = -\frac{1}{\Delta x} \left[A^{-}_{j+1/2,k} \frac{\phi(1-\kappa_x)}{4}\right] \tag{13}$$

For a fully supersonic flow in the streamwise (x) direction A^- is zero everywhere, and thus U2 is zero. However, U1 is in general non-zero due to the presence of A^+. By choosing $\phi = 0$ (first-order scheme) or $\phi = 1$ and $\kappa_x = -1$ (fully upwind second-order scheme), then the elements on U1 become zero and vertical line Gauss-Seidel (VLGS) becomes a direct solver of the linear problem resulting in rapid convergence to the steady-state. Even for subsonic flows, when the upper diagonals are non-zero, the line relaxation algorithm results in an efficient solution method as has been shown by van Leer and Mulder[7] and Chakravarthy[6].

For supersonic problems, the classical implementation of VLGS (referred to as global iteration), that of solving one block tri-diagonal (first-order scheme) or penta-diagonal (second-order scheme) system on a line and then proceeding to the next line of the computational grid is not the most efficient approach. A superior implementation of line Gauss-Seidel is to start at the first interior line of the grid, next to the inflow (left) boundary, solve the linearized problem; i.e., obtain a new estimate for the state vector along the line, update the residual and continue the process until the machine zero steady-state solution on the first line is obtained. One can then proceed to the next line and continue as before. Thus when the last line is completed the steady-state solution to the problem is obtained. In this approach, as $\delta t \rightarrow \infty$, Newton's method is being implemented by lines so that the solution along each line will converge quadratically. This type of strategy is referred to as local, rather than global iteration. Note that the global strategy attempts to obtain the solution at a downstream field point before a fully converged solution has been obtained upstream. Based on the mathematical theory of characteristics, this is not possible since the steady solution along downstream lines depend on the steady state upstream. This manifests itself in the fact that local iteration is more efficient than global iteration for this class of problems.

A hybrid strategy combining local and global iterations can be developed for problems containing separate regions of supersonic and subsonic flow, such as inlet/diffuser problems with terminal shocks and also dual throat inlets. For this type of problem, where large subsonic regions can exist, one can use global iteration. The more efficient hybrid strategy can be developed if the steady-state flow regime interfaces can be located. As an example, consider the dual throat configuration sketched in figure (3a), a possible Mach number distribution for the inlet in figure (3b), and the coefficient matrix that would be obtained from a first-order upwind discretization of the problem in figure (3c), all of which are related to the iteration strategy that follows.

Assume, for the moment, that the steady-state location of the two interfaces separating the supersonic/subsonic states are known. Since the upper diagonal in the coefficient matrix has non-zero elements only in the subsonic region and also along the preceding supersonic/subsonic interface (but not vice versa), then the problem can be solved by lines in the supersonic region, converging each line before proceeding to the next, until one reaches a line that has an upstream propagating influence from a downstream grid point; i.e., until a supersonic/subsonic interface is reached. Then, the large sub-matrix depicted in figure (3c) can be solved iteratively until convergence, knowing the beginning and ending lines belonging to the sub-matrix. The last line included is the subsonic line just preceding the subsonic/supersonic cell interface at the second throat. It is sufficient to close the sub-matrix with this line becuase there is no influence on the solution from the downstream supersonic line. Solving the sub-matrix in this fashion is referred to as block iteration because it involves the iterative solution of a group or block of lines in a global fashion. Having obtained the steady-state solution in this region, the remainder of the field can be solved by lines (in this example, the last line).

The automatic implementation of the above strategy is very straightforward. All that is needed is a switch that determines if a line can be solved locally or must be included in a block iteration group. In figure (3c), the parameter ν serves this purpose and for the first-order scheme is defined by

$$\nu_j = \begin{cases} 1, \text{ if } M(q_R)_{j+1/2,k} > 1 \text{ for all } k \\ 0, \text{ otherwise} \end{cases} \tag{14}$$

If $\nu_j = 1$, the line can be solved locally, otherwise it must be

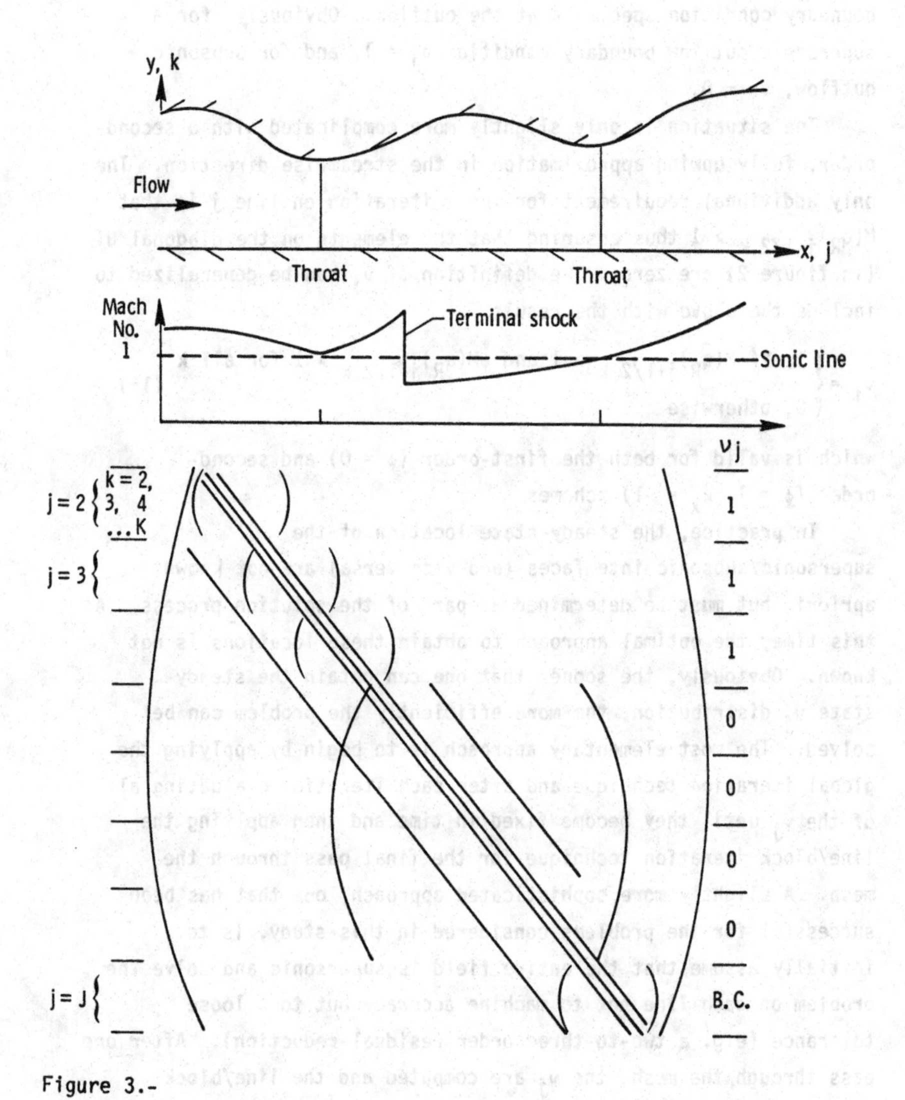

Figure 3.-

a) Sketch of dual throat inlet; b) Mach number distribution; and c) coefficient matrix structure.

included in a larger sub-matrix. Note, that the value of ν_j at the last line (the line next to the outflow boundary) has been labeled 'B.C.' indicating that the value of ν_j here depends on the type of boundary condition specified at the outflow. Obviously, for a supersonic outflow boundary condition $\nu_j = 1$, and for subsonic outflow, $\nu_j = 0$.

The situation is only slightly more complicated with a second-order, fully upwind approximation in the streamwise direction. The only additional requirement for local iteration on line j is that $M(q_R)_{j-1/2,k} > 1$ thus ensuring that the elements on the diagonal U1 (in figure 2) are zero. The definition of ν_j can be generalized to include the above with the result

$$\nu_j = \begin{cases} 1, \text{ if } M(q_R)_{j+1/2,k} > 1 \text{ and } \phi M(q_R)_{j-1/2,k} > 1 \text{ for all } k \\ 0, \text{ otherwise} \end{cases} \quad (15)$$

which is valid for both the first-order ($\phi = 0$) and second-order ($\phi = 1$, $\kappa_x = -1$) schemes.

In practice, the steady-state location of the supersonic/subsonic interfaces (and vice versa) are not known apriori, but must be determined as part of the solution process. At this time, the optimal approach to obtain these locations is not known. Obviously, the sooner that one can obtain the steady-state ν_j distribution, the more efficiently the problem can be solved. The most elementary approach is to begin by applying the global iteration technique and after each iteration evaluating all of the ν_j until they become fixed in time and then applying the line/block iteration technique for the final pass through the mesh. A slightly more sophisticated approach, one that has been successful for the problems considered in this study, is to initially assume that the entire field is supersonic and solve the problem on each line not to machine accuracy but to a loose tolerance (e.g. a two-to-three order residual reduction). After one pass through the mesh, the ν_j are computed and the line/block iteration strategy is applied as appropriate on the next pass, again with a loose tolerance. Subsequent passes are made with the line-block strategy until the ν_j become independent of time. A machine zero tolerance is then set and the steady-state solution is recovered on the next sweep through the grid. It is not known under what conditions (tolerance level, problem severity, etc.) this latter strategy will succeed or fail. Further investigation is necessary, because as shown in the results section, this latter strategy is more efficient than the first approach mentioned. Additional discussion concerning the iteration strategy can be found in the results section.

In summary, the upwind/relaxation algorithms that admit a marching strategy are described by the two-parameter family (ϕ, κ_y) with κ_x fixed at -1. ϕ takes on the discrete values 0 and 1, and κ_y is continuous on the closed interval [-1,1].

Initial Conditions

In order to start a problem, a simple set of initial conditions are usually specified; for example, uniform flow at free-stream conditions everywhere. However, in supersonic regions where marching is appropriate, a better choice resulting in a CPU savings can be made. A very simple choice that will reduce the computational effort is to use the solution from the previous line as an initial condition. While this is an improvement over that of the free stream, it is generally not a good enough guess to put one inside the domain of attraction of Newton's method without requiring additional iteration. This can be at least partially alleviated with only a minor coding effort by using an initial condition generated by solving the steady-state Euler equations approximately. As shown in the results section, significant savings are obtained by this approach.

Implementation of this procedure is as follows. Consider the differential form of the steady Euler equations in conservation law form;

$$\frac{\partial f}{\partial x} + \frac{\partial g}{\partial y} = 0 \tag{16}$$

and let

$$\Delta_j f = f_{j+1} - f_j \quad \text{etc.} \tag{17}$$

where the "k" subscript has been suppressed for simplicity.

Expanding f_{j+1} in a Taylor series about f_j and making use of equation (16) yields

$$\Delta_j f + \Delta x \frac{\partial \Delta_j g}{\partial y} = \Delta x \left(\frac{\partial f}{\partial x}\right)_j \tag{18}$$

Linearization of $\Delta_j f$ and $\Delta_j g$ with respect to q gives,

$$\left[\left(\frac{\partial f}{\partial q}\right)_j + \Delta x \frac{\partial}{\partial y}\left(\frac{\partial g}{\partial q}\right)_j\right]\Delta_j q = - \Delta x\left(\frac{\partial g}{\partial y}\right)_j \tag{19}$$

In the above, the Jacobian matrices, $\frac{\partial f}{\partial q}$ and $\frac{\partial g}{\partial q}$ are already formed (see equation 11a), consequently the implementation and solution of equation (19) is very straightforward. The spatial discretization is of course upwinded as previously described. One then sets

$$q_{j+1} = q_j + \Delta_j q \tag{20}$$

as the initial condition on the j+1 line. Note that this is similar to parabolizing the Navier-Stokes equations. However, the steady Euler equations with $M > 1$ are hyperbolic; i.e, the eigenvalues of $\partial f/\partial q$ are all greater than zero, hence this is not a parabolization procedure.

Operation Count, Time Stepping, and Boundary Conditions

Before proceeding to the test problems, a few additional remarks on overall implementation are in order.

Frequently, questions arise concerning the computational effort associated with the solution of a system of equations characterized by a block penta-diagonal coefficient matrix as opposed to a block tri-diagonal solution process. This is pertinent, since all of the discretizations with $\phi = 1$ require a penta-diagonal solver. The operation count, without pivoting, is approximately $5nM^3$ for a block tridiagonal solver and $11nM^3$ for a block penta-diagonal solver where n is the number of unknowns (mesh points) and M is the block size.[11] Thus, 2.2 times more effort is required for that part of the iteration associated with the penta-diagonal solver versus a tridiagonal solver. The actual increase in effort has been timed on a CY-855 computer and a 2.154 ratio was obtained. Solving the linear system, however, represents only 30% of the computational time per iteration. The remaining 70% represents the effort required to set up the linear problem and the boundary conditions. The total increase in CPU time per iteration is therefore only 35%. This is certainly worth the effort if quadratic convergence, as opposed to linear convergence, can be obtained.

The general time-stepping scheme is a variation of that described in reference 8. At any level of iteration, the time step is obtained from

$$\delta t = \frac{\delta t_o}{\| R^n \|_2} \tag{21}$$

where $\| R^n \|_2$ is the L_2-norm of the residual (the RHS of equation 2) normalized by the initial residual and δt_o is the initial time step. A common choice for δt_o would correspond to a maximum Courant number of the order of ten, allowing the initial solution to adjust to the enforcement of the steady-state boundary conditions without incurring severe oscillations. As the intermediate solution becomes attracted to the root, the residual drops rapidly resulting in very large time steps, thus mimicing a steady-state Newton method.

Finally, it has been tacitly assumed that all boundary conditions are treated implictly which is required if Newtonian convergence is to be obtained. Nonlinear, delta-form boundary conditions can be readily linearized by the techniques already presented and implictly built into the discrete algebraic problem, thus yielding a completely consistent formulation of Newton's method.

Results

In order to assess the relative performance of the various iteration strategies, the same test problem will be solved by different means and CPU times compared. The first problem considered is the well-known inviscid shock reflection problem[12] in which a shock wave of prescribed strength reflects off of a flat plate. The problem is simulated by prescribing free-stream conditions at the inflow boundary ($M_\infty = 2.9$), extrapolating conditions at the supersonic outflow boundary (first order), and along the upper boundary, all conditions are fixed by setting the incident shock angle to 29°. The boundary conditions on the plate are the simplest possible; i.e., $\rho_1 = \rho_2$, $u_1 = u_2$, $v_1 = 0$, $e_1 = e_2$. The grid contained 61 equally spaced points in the x-direction, $0 < x < 4.1$, and 21 equidistant points in the y-direction, $0 < y < 1$. Pressure contours for the first-order scheme ($\phi = 0$), (hereinafter referred to as method 1) the fully upwind second-order scheme ($\phi = 1$, $\kappa_x = -1$, $\kappa_y = -1$; method 2) and the hybrid scheme ($\phi = 1$, $\kappa_x = -1$, $\kappa_y = 1/3$; method 3) are compared in figure 4. Clearly, a dramatic resolution improvement is obtained with the $\phi = 1$ schemes.

The convergence rates of the three methods using the global iteration procedure are depicted in figure 5. This problem was then solved by lines using the following initial conditions on each line; free stream (type 1), the solution from the previous line (type 2), the solution of equation (19) with a first-order discretization in y (type 3), and finally, the solution of equation (19) with a y-discretization consistent with the solution algorithm (type 4).

Table 1

Comparison of CPU Time (in Seconds) for Shock Reflection Problem*

Method	Global	Line			
		Type 1	Type 2	Type 3	Type 4
1	20.6	7.9	7.8	3.8	3.8
2	30.8	**	10.6	6.5	**
3	40.8	**	11.8	11.9	12.4

* 1) Tolerance = 10^{-13} (L_2-norm of residual)
2) VPS 32 with Fortran 2.1.5 OPT = BE Compilation, Scalar Code

**Did not converge due to state vector interpolation near shock reflection - no limiters

It is apparent from the results in Table 1 that local iteration is significantly better than global iteration for supersonic flow. Using free-stream initial conditions on a line does lead to difficulty with the higher-order methods if discontinuities are present due to the nature of the interpolation. Using the solution from the previous line as the initial guess alleviates this problem but this can be further improved by using the type 3 initial conditions. Finally, for the higher-order methods, one might expect that using a consistent y-discretization in equation (19) to generate the initial condition (type 4) would be superior to the first-order discretization but this has not been found to be true. The third-order discretization resulted in a marginal CPU increase and the fully upwind second-order y-discretization resulted in failure. This failure to achieve a further CPU reduction is quite likely due to that fact that equation (19) is only first-order accurate in x. It is believed that additional efficiency gains could be made for the higher-order methods by modifying equation (19) such that it is $O(\Delta x^2)$, but this involves considerable extra effort and has not been done in this study.

The next problem considered is the supersonic and transonic flow in a dual throat inlet. The nozzle contour is very mild and was generated by specifying a parabolic Mach number distribution and then solving for the area distribution from the one-dimensional area-Mach number relations. The converged centerline Mach number distribution for both supersonic ($M_\infty = 1.1$) and transonic ($M_\infty = 1.07$) flow are presented in figure 6, for the first-order scheme. The global convergence histories for these two cases are also shown in figure 7a and b. For illustrative purposes, figure (7a) has two curves, one with all implicit boundary conditions and the other with an explicit boundary treatment on the nozzle surface. The difference in the convergence is dramatic, clearly indicating the need for fully implicit boundary condition treatment.

The transonic case was solved using three different approaches: type 1) global iteration only, type 2) global iteration for the first 160 iterations followed by one pass with the block iteration strategy, and type 3) one pass through the mesh assuming a fully supersonic flow followed by two passes with the block iteration strategy; the first two passes with a loose tolerance (10^{-3}) and the final pass with a tight (10^{-12}) tolerance. The results of these three approaches are summarized in Table 2 along with the supersonic results.

Table 2

Timing Comparison (CPU Seconds) for Dual Throat Problem*
(Tolerance = 10^{-12})

Iteration	Supersonic	Transonic
Type 1	68.9	489
Type 2	----	381
Type 3	----	151
by lines only	7.7	---

*CYBER-203 time with Fortran 2.1.5 cycle 607 compilation

The results again emphasize the fact that local iteration is signicantly more efficient than global iteration for supersonic flows. It is also apparent that major gains for transonic flow can be made using block iteration and as previously discussed, the sooner that the ν_j distribution can be obtained, the more efficiently the problem can be solved.

Another simple problem that indicates the efficiency of the block iteration strategy is the Mach reflection problem in an inlet/diffuser. The geometry and computational details are described in refs. 13 and 14. With an inflow Mach number of 1.95, a Mach reflection occurs, as shown in figure 8. The flow is almost entirely supersonic, with only a small subsonic region existing behind the Mach stem. The global convergence of the first-order scheme is given in Figure 9. The CPU time to obtain a 10^{-12} residual was 48.9 seconds and with the three pass block iteration strategy previously described, the CPU time was reduced to 17.8 seconds. This, however, was obtained using the solution from the previous line as the initial condition and so a further improvement could likely be realized.

The previous two problems with only first-order results were presented in order to indicate the relative efficency gains that can be obtained with the block iteration strategy in a transonic problem. The last problem to be considered makes use of the geometry from the previous problem but is far more difficult from a computational point of view. In this test case, the inflow Mach number is 2.5 and an exit-to-inlet static pressure ratio of 8.0 is specified which results in the formation of a strong terminal shock. Both first- and second-order results have been obtained. The non-dimensional pressure contours obtained from the second-order solution to this problem are shown in figure 10. The physics of the problem; the oblique shock, expansion, reflected shock and terminal shock, have been well represented by the simulation. A comparison of the first- and second-order accurate pressure distribution along the nozzle wall and centerline are given in figure 11. Clearly, the first-order scheme is signficantly more dissipative than its second-order counterpart.

The global convergence history of the first- and second-order schemes are compared in figure 12. The second-order scheme has ended up in a limit cycle even though the computed results are quite good. This is due to the fact that a very simple, but discontinuous, limiter was invoked for the higher-order method. For problems with strong shocks, all but the first-order method require some form of limiter in order to succeed. This is apparent if one considers the effect of interpolating data across a discontinuity. For this problem, the state variables at a cell interface are obtained by interpolation and then the Mach number is computed at the interface in order to determine the appropriate flux splitting. If this interpolation results in a negative value for $\gamma p/\rho$ then the second-order interpolation is rejected and the first-order scheme is used at this interface. In this simulation, the only place that the first-order scheme is invoked is in the immediate vicinity of the terminal shock. While this results in an acceptable solution, it does unfortunately also result in a limit cycle. It is known that the limit cycle behavior can be overcome by using a continuous limiter but this has not been done here. Finally, a comparison of CPU times is made with and without the block-line iteration strategy in Table 3. As before, this strategy

Table 3

Timing Comparison (CPU Seconds) for Terminal Shock Problem

Iteration Strategy	First Order ($TOL = 2x10^{-11}$)	Second Order ($TOL = 4x10^{-4}$)
Global	406.0	682.6
Block-Line	253.6	402.1

is implemented by passing through the mesh with a loose tolerance until the ν_j became fixed at which time a final tolerance is set for the last sweep.

Conclusions

The family of algorithms presented in this study result in a very efficient iterative technique for inviscid flow problems, particularly in the supersonic regime. The new iteration approach for problems containing separate regions of supersonic and subsonic flow also represents a significant savings in computational effort. The optimal approach of implementing this strategy is still unknown.

Acknowledgements

The authors would like to sincerely acknowledge several fruitful discussions on upwind differencing with J. E. Thomas, NASA Langley Research Center and Bram van Leer, Delft University of Tehcnology, the Netherlands.

References

1. Briley, W. R.; and McDonald, H. Solution of the Multidimensional Compressible Navier-Stokes Equations by a Generalized Implicit Method. Journal of Computational Physics, vol. 24, 1977, pp. 372-397.

2. Beam, R. M.; and Warming, R. F.: An Implicit Factored Scheme for the Compressible Navier-Stokes Equations. AIAA Journal, vol. 16, no. 4, 1978, pp. 393-402.

3. Jameson, A.; and Baker, T. J.: Multigrid Solution of the Euler Equations for Aircraft Configurations. AIAA Paper 84-0093, 1984.

4. MacCormack, R. W.: The Effect of Viscosity in Hypervelocity Impact Cratering. AIAA Paper 69-354, 1969.

5. Abarbanel, S. S.; Dwoyer, D. L.; and Gottlieb, D.: Improving the Convergence Rate of Parabolic ADI Methods. AIAA Paper No. 83-1897, 1983.

6. Chakravarthy, S. R.: Relaxation Methods for Unfactored Implicit Upwind Schemes. AIAA Paper No. 84-0165, 1984.

7. van Leer, B.; and Mulder, W. A.: Relaxation Methods for Hyperbolic Equations. T. H. D. Report 84-20 Delft University of Technology, The Nethelands, 1984.

8. Mulder, W. A.; and van Leer, B.: Implicit Upwind Methods for the Euler Equations. AIAA Paper 83-1930, 1983.

9. Buning, P. G.; and Steger, J. L.: Solution of the Two-Dimensional Euler Equations with Generalized Coordinate Transformation Using Flux Vector Splitting. AIAA Paper 82-0971, 1982.

10. van Leer, B.: Flux-Vector Splitting for the Euler Equations. Proceedings of the 8th International Conference on Numerical Methods in Fluid Dynamics, Germany, June 28-July 2, 1982.

11. Dahlquist, G.; and Bjorck, A.: Numerical Methods. Prentice-Hall, Inc.

12. Yee, H. C.; Warming, R. F.; and Harten, A.: Implicit Total Variation Diminishing (TVD) Schemes for Steady-State Calculations. AIAA Paper 83-1902, 1983.

13. Bogar, T. J.; Sajben, M.; and Kroutil, J. C.: Characteristic Frequency and Length Scales in Transonic Diffuser Flow Oscillations. AIAA Paper 81-1291, 1981.

14. Walters, R. W.: LU Methods for the Compressible Navier-Stokes Equations. Ph.D. Thesis, North Carolina State University, April 1984.

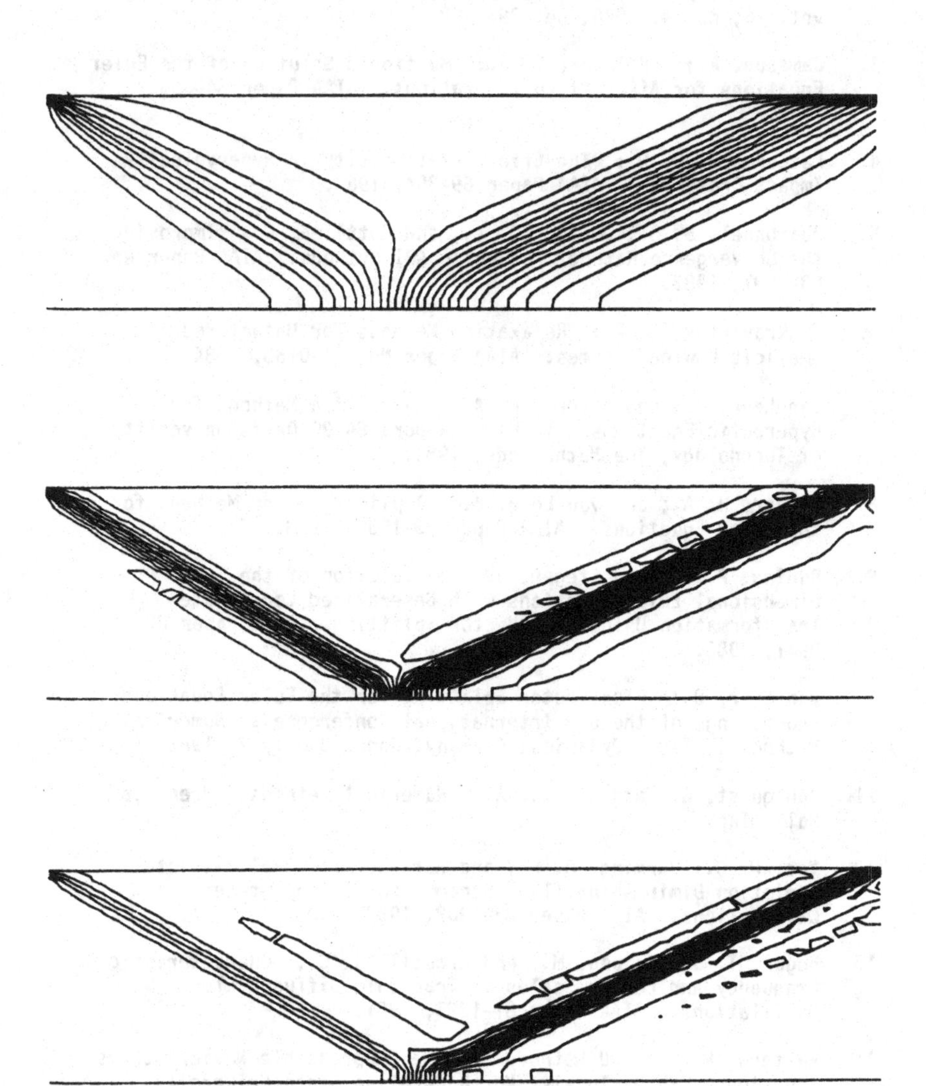

Figure 4. Pressure contours for the shock reflection problem: a) 1st order x & y; b) 2nd order x & y; c) 2nd order x, 3rd order y.

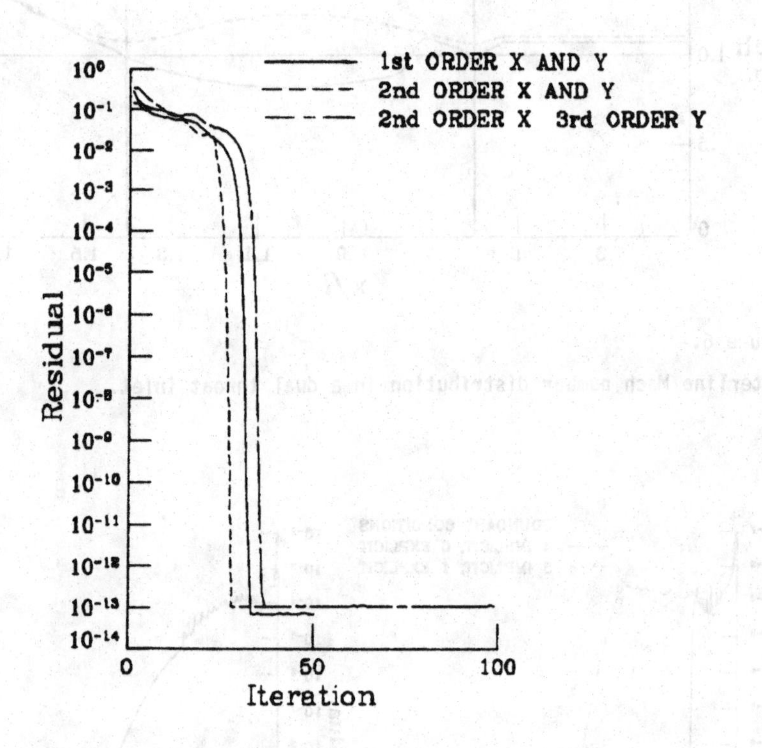

Figure 5.-

Convergence history for the shock reflection problem.

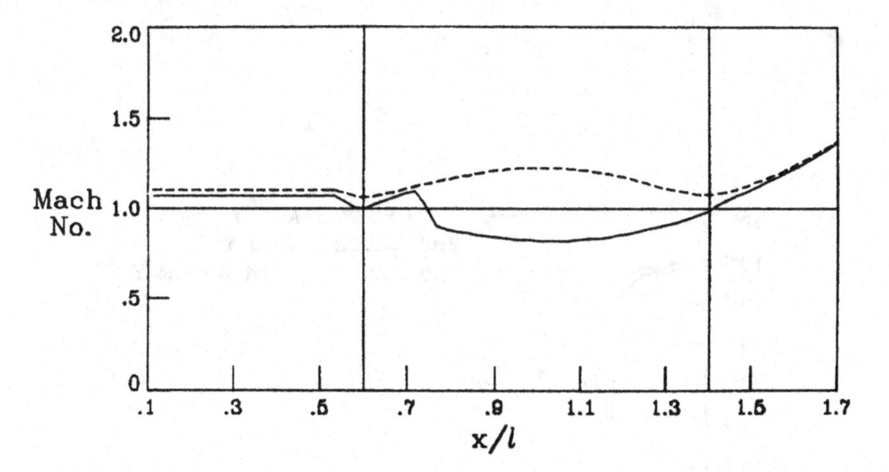

Figure 6.-

Centerline Mach number distribution in a dual throat inlet.

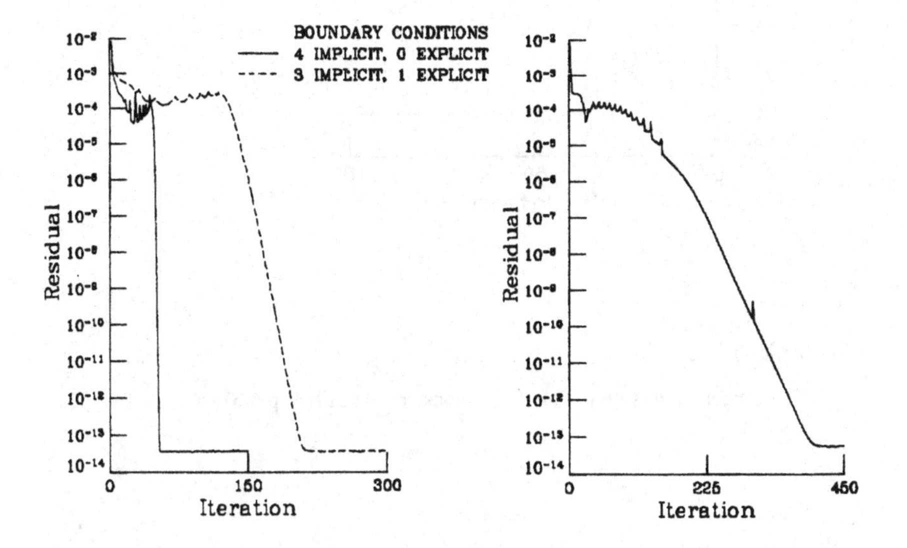

Figure 7.-

Convergence history for the dual throat problem. a) supersonic, b) transonic.

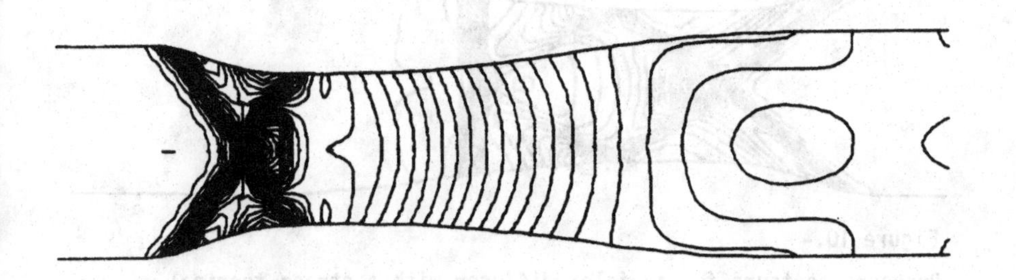

Figure 8.-

Pressure contours for the Mach reflection problem.

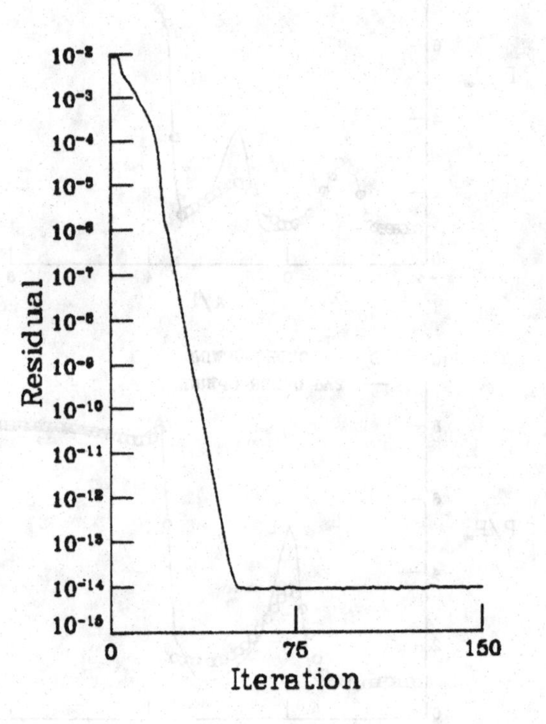

Figure 9.-

Convergence history for the Mach reflection problem--1st order scheme.

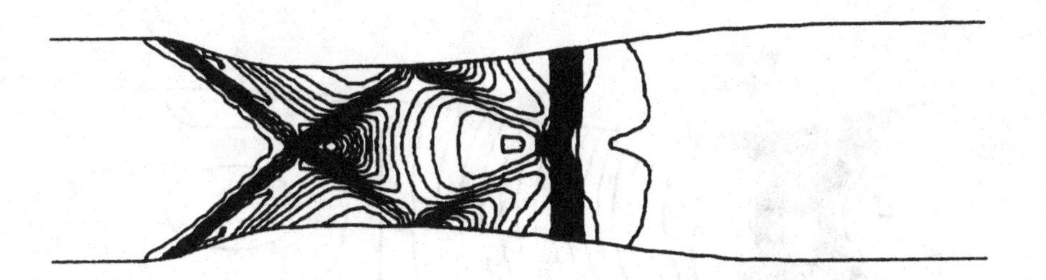

Figure 10.-

Pressure contours for an inlet/diffuser with a strong terminal shock.

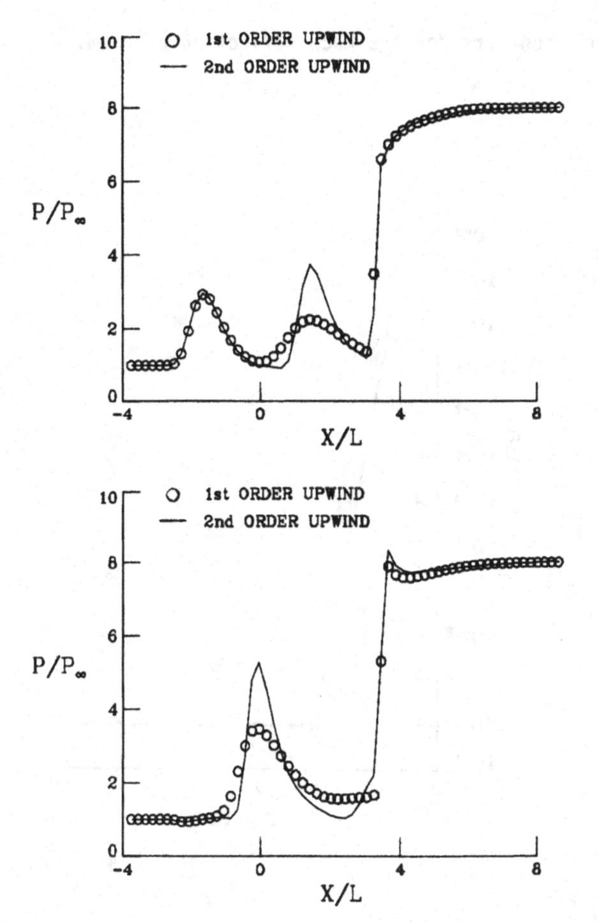

Figure 11.-

Wall (a) & centerline (b) pressure distribution for the terminal shock problem.

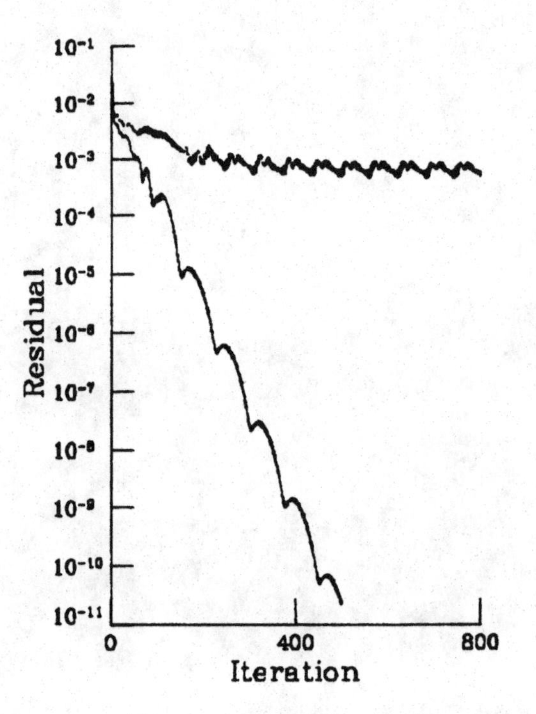

Figure 12.-

Comparison of first and second order convergence histories for the terminal shock problem.

Progress in Scientific Computing, Vol. 6
Proceedings of U.S.-Israel Workshop, 1984

NUMERICAL METHODS FOR THE NAVIER-STOKES EQUATIONS

Robert W. MacCormack
University of Washington
Seattle, WA 98195 USA

1. Introduction

In the early 1970's Murman and Cole in a landmark paper [1] set the groundwork for the development of computational fluid dynamics for years to come. Their paper demonstrated the use of type dependent finite difference approximations, chosen according to the characteristic speeds of the flow, and Gauss-Seidel line relaxation to solve the transonic small disturbance equation. Following their success, Jameson [2] introduced a "rotated" difference scheme and extended the Murman-Cole procedure to solve the full potential equation. These early contributions were responsible for the rapid growth and widespread application of computational fluid dynamics to aerodynamic design in the 1970's. Last year Chakravarthy [3] applied flux-split type-dependent difference approximations and Gauss-Seidel line relaxation to solve the Euler equations. And this year Napolitano and Walters [4] and this author [5] used these procedures to solve the Navier-Stokes equations. On the one hand it's amazing that the same key features of the Murman-Cole scheme can be applied to the more complete sets of governing equations, and on the other it's amazing that it took a decade and a half to realize it. This paper outlines the use of these features for solving the Navier-Stokes equations and presents some computed results demonstrating high numerical efficiency.

2. The Navier-Stokes Equations

The Navier-Stokes equations written in two-dimensional conservation form are

$$\frac{\partial U}{\partial t} + \frac{\partial F}{\partial x} + \frac{\partial G}{\partial y} = 0 \qquad (1)$$

where

$$U = \begin{pmatrix} \rho \\ \rho u \\ \rho v \\ e \end{pmatrix},$$

$$F = \begin{pmatrix} \rho u \\ \rho u^2 + \sigma_x \\ \rho uv + \tau_{xy} \\ (e + \sigma_x)u + \tau_{yx} v - k \frac{\partial T}{\partial x} \end{pmatrix}, \quad G = \begin{pmatrix} \rho v \\ \rho uv + \tau_{yx} \\ \rho v^2 + \sigma_y \\ (e + \sigma_y)v + \tau_{xy} u - k \frac{\partial T}{\partial y} \end{pmatrix}$$

and where

$$\sigma_x = p - \lambda \left(\frac{\partial u}{\partial x} + \frac{\partial v}{\partial y}\right) - 2\mu \frac{\partial u}{\partial x}, \quad \tau_{xy} = \tau_{yx} = -\mu \; \frac{\partial u}{\partial y} + \frac{\partial v}{\partial x},$$

$$\sigma_y = p - \lambda \left(\frac{\partial u}{\partial x} + \frac{\partial v}{\partial y}\right) - 2\mu \frac{\partial v}{\partial y}$$ in terms of density ρ, x and y velocity components u and v, viscosity coefficients λ and μ, total energy per unit volume e, coefficient of heat conductivity k, and temperature T. Finally, the pressure p is related to the specific internal energy ε and ρ by an equation of state, $p(\varepsilon, \rho)$, where $\varepsilon = e/\rho - (u^2 + v^2)/2$.

3. Numerical Method

The following two step, predictor-corrector, procedure can be used to solve Eq. (1)

$$p: \begin{cases} \Delta U^n_{i,j} = -\Delta t \left(\frac{D \cdot F}{\Delta x} + \frac{D \cdot G}{\Delta y}\right)^n_{i,j} \\ \left\{ I + \Delta t \left(\frac{D_+ \cdot A_-}{\Delta x} + \frac{D_- \cdot A_+}{\Delta x} - \frac{D_- \cdot M_x \frac{D_+ \cdot N}{\Delta x}}{\Delta x} \right) + \Delta t \left(\frac{D_+ \cdot B_-}{\Delta y} + \frac{D_- \cdot B_+}{\Delta y} - \frac{D_- \cdot M_y \frac{D_+ \cdot N}{\Delta y}}{\Delta y} \right) \right\} \delta U^{\overline{n+1}}_{i,j} = \Delta U^n_{i,j} \\ U^{\overline{n+1}}_{i,j} = U^n_{i,j} + \delta U^{\overline{n+1}}_{i,j} \end{cases}$$

and

$$
C: \begin{cases}
\Delta U_{i,j}^{\overline{n+1}} = -\Delta t \left(\dfrac{D \cdot F}{\Delta x} + \dfrac{D \cdot G}{\Delta y} \right)_{i,j}^{\overline{n+1}} \\
\left\{ I + \Delta t \left(\dfrac{D_+ \cdot A_-}{\Delta x} + \dfrac{D_- \cdot A_+}{\Delta x} - \dfrac{D_- \cdot M_x \dfrac{D_+ \cdot N}{\Delta x}}{\Delta x} \right) \right. \\
\left. + \Delta t \left(\dfrac{D_+ \cdot B_-}{\Delta y} + \dfrac{D_- \cdot B_+}{\Delta y} - \dfrac{D_- \cdot M_y \dfrac{D_+ \cdot N}{\Delta y}}{\Delta y} \right) \right\} \delta U_{i,j}^{n+1} = \Delta U_{i,j}^{\overline{n+1}} \\
U_{i,j}^{n+1} = \dfrac{1}{2} \left\{ U_{i,j}^{n} + U_{i,j}^{\overline{n+1}} + \partial U_{i,j}^{n+1} \right\}
\end{cases}
$$

where

(1) the superscripts n, $\overline{n+1}$, and n+1 refer to present, predicted and new solution values,

(2) the subscripts i and j refer to a special set of mesh point locations, $(i\Delta x, j\Delta y)$,

(3) ΔU is the temporal change in the solution during time interval Δt and is evaluated by a local explicit finite difference approximation to the right-hand-side of Eq. (1),

(4) the difference operators D_+ and D_- represent forward and backward difference approximations applied to all factors to the right of the operator with respect to the direction indicated in the denominator of the difference operator quotient, i.e.

$$\left\{ \frac{D_+ A_-}{\Delta x} + \ldots \right\} \delta U_{i,j} \quad \text{is equivalent to}$$

$$\frac{A_{-_{i+1,j}} \delta U_{i+1,j} - A_{-_{i,j}} \delta U_{i,j}}{\Delta x} \ldots$$

(5) the difference operator D is unspecified as yet and will be discussed subsequently.

(6) δU represents the solution change after solution of the implicit (middle) equation of each step which contains a larger (global) domain of dependence, and

(7) A_+, A_-, B_-, B_+, M_x, M_y, and N are Jacobian matrices and are defined as follows.

3.1 The Jacobian Matrices

The flux vector F can be written as the sum of its inviscid and viscous parts.

$$F = F'_{inviscid} + F''_{viscous}$$

$$F' = \begin{pmatrix} \rho u \\ \rho u^2 + p \\ \rho v u \\ (e + p)u \end{pmatrix}$$

Following the theory of Steger and Warming [6]

$$F' = A'U \quad \text{with} \quad A' = \partial F/\partial U \ .$$

The Jacobian matrix A' can be diagonalized by a similarity transformation S, [5]

$$A' = S^{-1} \Lambda S$$

where Λ is a diagonal matrix containing the characteristic speeds as elements. Separating the elements of Λ according to their sign we can form diagonal matrices Λ_+ and Λ_- and define

$$F'_+ = A_+U \quad \text{and} \quad F'_- = A_-U$$

where

$$A_+ = S^{-1}\Lambda_+S \quad \text{and} \quad A_- = S^{-1}\Lambda_-S \ .$$

The Jacobian matrices B_+ and B_- are similarly defined for the flux vector G.

We can write the viscous part of the flux vector F as

$$F'' = \frac{\partial M_x \frac{\partial V}{\partial x}}{\partial x} + \text{mixed derivative terms}$$

where

$$M_x = \begin{pmatrix} 0 & 0 & 0 & 0 \\ 0 & \lambda+2\mu & 0 & 0 \\ 0 & 0 & \mu & 0 \\ 0 & u(\lambda+2\mu) & v\mu & k \end{pmatrix}$$

and

$$V = \begin{pmatrix} \rho \\ u \\ v \\ T \end{pmatrix}$$

The derivatives of V and U are related by the Jacobian matrix N as follows

$$\frac{\partial V}{\partial x} = N \frac{\partial U}{\partial x} \quad \text{with} \quad N = \frac{\partial V}{\partial U}$$

The matrix M_y is similarly defined.

3.2 The Operators D

The operators D appearing in the first, explicit, equations of each step are as yet unspecified. In the author's 1969 paper [7] they were given as either two point forward or backward difference approximations in the predictor step. What ever choice was used in the predictor step, the correspondingly opposite choice was used in the corrector step. The result was a forward and a backward approximation for each derivative, resulting in second order accuracy upon completion of the two steps.

In addition to the above selection the following flux-split option can also be used

$$\frac{D \cdot F}{\Delta x} = \frac{D_+ \cdot F_-}{\Delta x} + \frac{D_- \cdot F_+}{\Delta x}$$

where a forward difference is used for the negatively traveling flux and a backward difference is used for the positively traveling flux. The operators D_+ and D_- can be either first or second order accurate. For example,

$$\frac{D_+ \cdot F_-}{\Delta x} = \begin{cases} \dfrac{F_{-_{i+1,j}} - F_{-_{i,j}}}{\Delta x}, & \text{first order} \\ \text{or} & \\ \dfrac{-F_{-_{i+2,j}} + 4F_{-_{i+1,j}} - 3F_{-_{i,j}}}{2\Delta x}, & \text{second order} \end{cases}$$

3.3 Gauss-Seidel Line Relaxation

The solution of the implicit middle equation of each step to obtain δU requires the inversion of a block matrix. If simple two-point one-sided difference approximations are used in this equation, a block pentadiagonal matrix must be inverted. Each block matrix element is itself a 4×4 matrix and the five diagonals are not all adjacent to each other. There is no straightforward efficient procedure for inversion. The standard numerical procedure is to approximately factor the block pentadiagonal matrix into two block tridiagonal matrices, each of which has a simple direct inversion. However, the factorization introduces a numerical error proportional to the product of the two one-dimensional CFL (Courant, Friedrichs and Lewy) numbers. Accuracy considerations then limit the size of the allowable time step and can lead to expensively long calculations.

To avoid the penalty imposed by approximate factorization, iterative indirect inversion procedures have been tried in the hope that the block matrix can be inverted accurately enough in only a few iterations. The choice of flux-split difference operators in the implicit equation was made to make the resulting matrix as diagonally dominant as possible to reduce the number of required iterations. The iterative procedure to be described (as in [3, 4, and 5]) is Gauss-Seidel line relaxation. The stencil of mesh points used is shown in Figure 1. A line at a time is solved for simultaneously using the latest available values ahead of and behind the line. The parameter k is the iteration index. For the test case to be discussed, the line was swept once counter to the flow direction and then was followed by a single sweep of the line in the flow direction.

Two point one-sided difference operators are only first order accurate. Spacial accuracy can be improved by using three point difference approximations. In the results to be shown, both first and second order approximations were used in the stream directions.

3.4 Newton Iteration

For flow converging to a steady state it is desirable to choose the time step size as large as possible. As the time step approaches infinity the method described above becomes a Newton iteration procedure for finding the steady state solution. Care should be taken to use the same difference approximations in the explicit and implicit sides of the equation in order to achieve a stable balanced scheme for large values of Δt.

4. Results

Figure 2 shows a transonic converging-diverging nozzle. Initially the flow was at rest with temperature and pressure set to their total values except at the nozzle exit plane where they were set to 1/3 their total values. The Reynolds number of the flow, based on stagnation conditions and nozzle throat height, was 9.3×10^5. The flow field was experimentally obtained by Mason, Putnam, and Re [8] and was computed by Kneile [9] using an implicit approximately factored method in approximately 1000 time steps and by this author [5] using both Gauss-Seidel line relaxation with $CFL_x = 12.8$ and $CFL_y = 1.7 \times 10^4$ in approximately 200 time steps and Newton iteration in approximately 24 steps. The computed results for surface pressure with both first and second order accurate difference approximations in the streamwise direction are compared with experiment in Figure 3. The second order calculation converged slower probably because the resulting matrix was less diagonally dominant. The rate of convergence of the velocity profile at the throat is shown for the second order calculation in Figure 4.

The mesh contained 24 points in the stream direction and 16 across the nozzle. It was suggested that the rapid convergence resulted because of the coarseness of the mesh and that if a mesh were used twice as fine in each direction convergence would be four times slower. The first order calculation was repeated on the suggested finer mesh and converged within 16 steps, the same as obtained on the original mesh.

The computer time used per step using Gauss-Seidel line relaxation with two sweeps is approximately twice that used by an approximately factored method. The vast reduction in the number of iterations required for convergence makes the Gauss-Seidel line relaxation method highly efficient.

5. Conclusion

The same techniques, type dependent difference approximations and Gauss-Seidel line relaxation, used to solve the transonic small disturbance equation appear to be also highly efficient for solving the Navier-Stokes equations as well.

6. References

[1] Murman, E. M. and Cole, J.D., "Calculation of Plane Steady Transonic Flows," AIAA Journal, Vol. 9 No. 1, January 1971, pp 114-121.

[2] Jameson, A., "Numerical Calculation of the Three Dimensional Transonic Flow Over a Yawed Wing." Proceedings, AIAA Computational Fluid Dynamics Conference, Palm Springs, CA, July 19-20, 1973.

[3] Chakravarthy, S. R., "Relaxation Methods for Unfactored Implicit Upwind Schemes," AIAA Paper No. 84-0165, 1984.

[4] Napolitano, M. and Walters, R. W., "An Incremental Block-Line Gauss-Seidel Method for the Navier-Stokes Equations" AIAA Paper No. 85-0033, 1985.

[5] MacCormack, R. W., "Current Status of Numerical Solutions of the Navier-Stokes Equations," AIAA Paper No. 85-0032, 1985.

[6] Steger, J. and Warming, R. F., "Flux Vector Splitting of the Inviscid Gasdynamics Equations with Application to Finite Difference Methods," NASA TM-78605, 1979.

[7] MacCormack, R. W., "The Effect of Viscosity in Hypervelocity Impact Cratering," AIAA Paper No. 69-354, 1969.

[8] Mason, M. L., Putnam, L. E. and Re, R. J., "The Effect of Throat Contouring on Two-Dimensional Converging-Diverging Nozzles at Sonic Conditions," NASA Technical Paper 1704, 1980.

[9] Kneile, K. R. and MacCormack, R. W., "Implicit Solution of the 3-D Compressible Navier-Stokes Equations for Internal Flows," Proceedings of the 9th International Conference on Numerical Methods in Fluid Dynamics, Saclay, France, 1984.

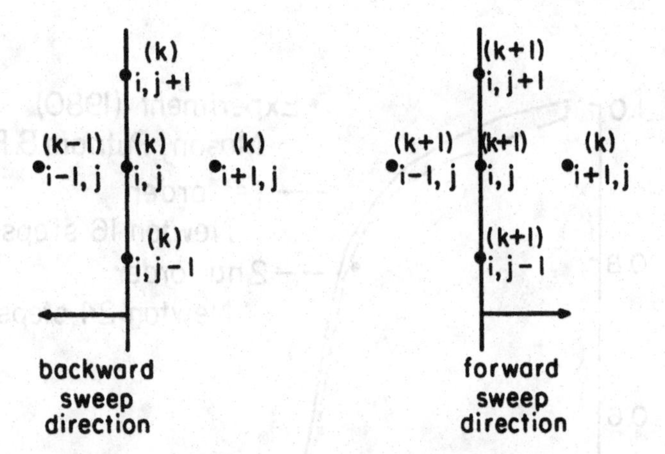

Figure 1 Sweep Directions for Gauss-Seidel Line Relaxation

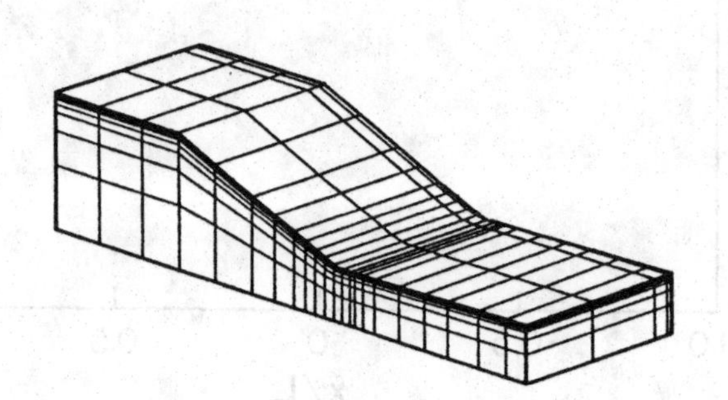

Figure 2 Transonic Converging-Diverging Nozzle

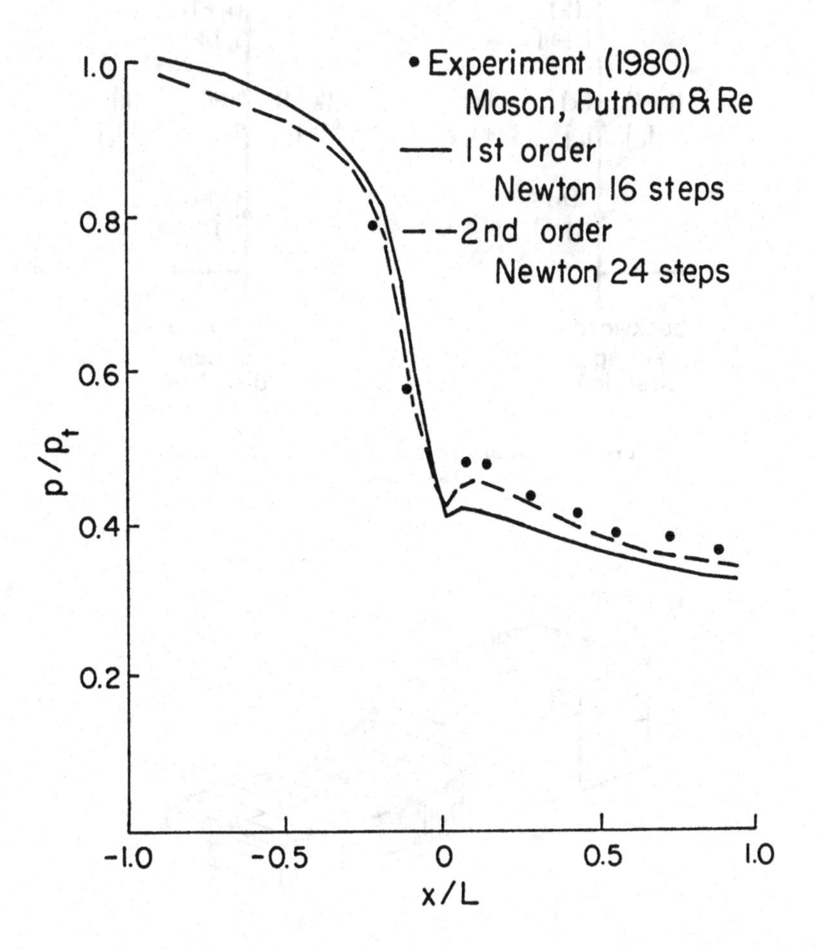

Figure 3 Surface Pressure

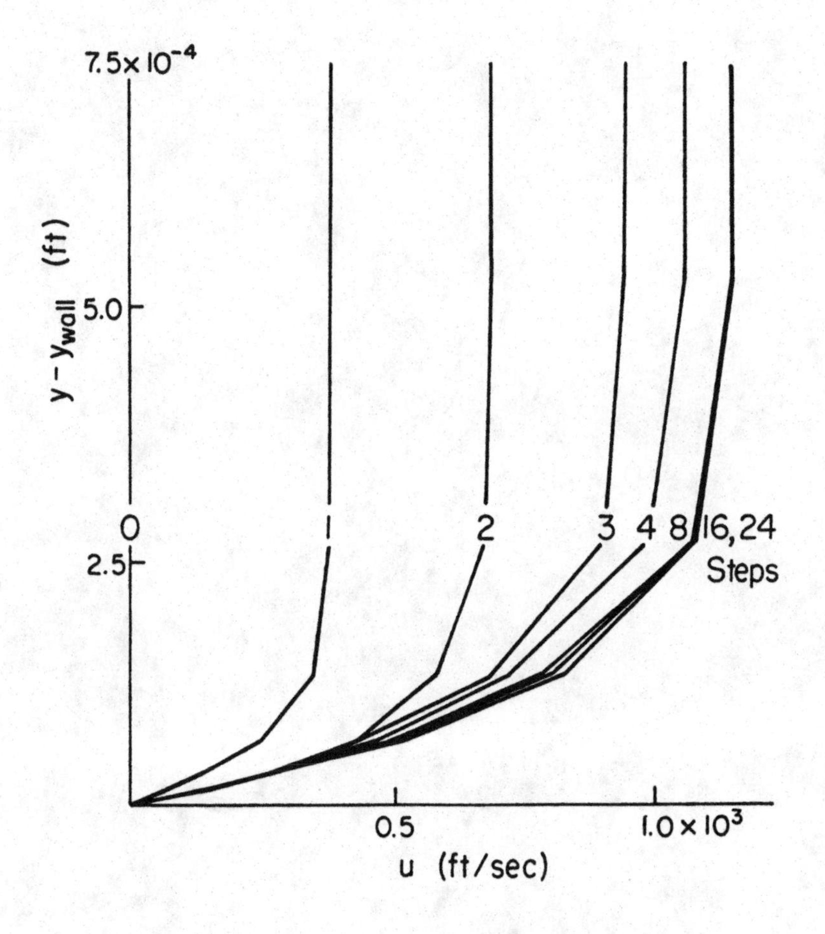

Figure 4 Convergence of Velocity Profiles at Nozzle Throat Using 2nd Order Newton Iteration

Progress in Scientific Computing, Vol. 6
Proceedings of U.S.-Israel Workshop, 1984

ALGORITHMS FOR THE EULER AND NAVIER-STOKES EQUATIONS FOR SUPERCOMPUTERS

Eli Turkel
Tel-Aviv University, Israel
Institute for Computer Applications to Science and Engineering

1. INTRODUCTION

With the introduction of the latest class of supercomputers, e.g. the Gray XMP and Cyber 205, it has begun to be feasible to solve the Euler and Navier-Stokes equations for three dimensional configurations. The major added difficulty in solving the Navier-Stokes equations is in the need to resolve the boundary layers. This is especially difficult for turbulent flow. Most codes rely on algebraic turbulence models but even these require an extremely fine mesh to resolve the sublayers. The use of one or two equation turbulence models require even finer meshes [15]. Hence, a Navier-Stokes code about a wing-body configuration requires a mesh that the new computers can just meet both in terms of speed and memory. Even with the new generation of supercomputers it is not feasible to routinely run three dimensional codes. It is therefore necessary to introduce new algorithms that will reduce the storage requirements and the running time compared with present schemes. Since several sophisticated schemes already exist it would be advantageous if the new algorithms could be incorporated within the presently existing codes.

In this paper we will only consider steady state problems. This will enable us to change the time dependent equations in any way that does not change the steady state. Thus, the approach that we use can be classified as a pseudo-unsteady approach to the steady state [19]. In addition we shall only consider conservation equations as this gives us greater flexibility in the problems that can solved.

2. PRECONDITIONING

We now consider two dimensional equations in conservation form

$$f_x + g_y = 0 \quad \text{in } D \tag{1}$$

with appropriate boundary conditions. We consider schemes that are pseudo time dependent. This approach allows the same code to treat true time dependent problems by removing the pseudo time elements. In addition the pseudo time changes can all be done locally. The present analysis is based on constant coefficient equations. However, both the Euler and Navier-Stokes equations are nonlinear equations. Hence, the preconditioners that will be developed will, in practice, vary at each mesh point. It will also be necessary to blend different regions together, which will not be discussed in this paper. As a result when we consider subsonic flow there is no need for the flow to be subsonic everywhere. Hence, even when discussing very slow flow we wish the equations to be in conservation form since there may be shocks in other regions of the domain. Similarly, when we consider supersonic flow one can not march in space as there may be regions of subsonic flow.

According to our philosophy of having the applications as general as possible the analysis will be done at the differential equation level. Hence, the results are scheme independent and apply to both explicit and implicit methods. Though we are interested in the steady state we shall use a time-like approach. Hence, we consider the system

$$w_t + f_x + g_y = 0 \tag{2}$$

where (x,y) represent general curvilinear coordinates. Since we are only interested in the steady state we replace (2) with the system

$$E^{-1}w + f_x + g_y = 0 \tag{3}$$

The minimum requirements on E are that E be nonsingular and that (3) be a well posed problem with boundary conditions that are consistent with those imposed on (1). It is straightforward to solve (3) with an explicit method. Using an implicit method only the diagonal block of the matrix to be inverted is changed compared with (2). We first consider the case that (2) is a hyperbolic system. Though the code solves (2) we will only consider the constant coefficient problem. Thus (2) is replaced by

$$w_t + Aw_x + Bw_y = 0 \qquad (4)$$

while the preconditioned system (3) is replaced by

$$E^{-1}w_t + Aw_x + Bw_y = 0 \qquad (5)$$

where A and B are the Jacobians of f and g with respect to w respectively. Also A,B and E are frozen at constant values. Let $w = Tv$ then (4) becomes

$$Tv_t + ATv_x + BTv_y = 0 \; .$$

Multiplying this equation by S we find that (4) is equivalent to

$$STv_t + A_o v_x + B_o v_y = 0 \qquad (6)$$

with $A_o = SAT$ and $B_o = SBT$. We will choose S and T to be nonsingular and such that A_o and B_o are "nice" matrices. We stress that there is no need for the transformation from (4) to (6) to be an equivalence transformation. Since (4) can be transformed into (6) it is sufficient to analyze (6). We now precondition the system (6) and consider

$$E_o^{-1}v_t + A_o v_x + B_o v_y = 0 \qquad (7)$$

with an appropriate E_o. Returning to the original w variables we find that (7) can be transformed into (5) and hence (3) with $E^{-1} = S^{-1}E_o^{-1}T^{-1}$ or $E = TE_oS$. Using an explicit scheme we wish to find the matrix E in (3) while for an implicit scheme we wish to construct E^{-1}. We thus wish

<u>Objective No. 1</u> Choose E_o so that

1. E_o is invertible,
2. (7) is well posed with appropriate boundary conditions,
3. (7) approaches the steady state as rapidly as possible.

We really wish to analyze (3) rather than (7) but we ignore nonlinear effects in this paper. The first property is straightforward. Implementation of the second and third properties will be discussed in the coming sections.

3. BOUNDARY CONDITIONS

Though the question of well-posedness is not the objective of this paper nevertheless we wish to point out several difficulties. By well-posed we mean that the solution exists, is unique and depends continuously on the data. In discussing appropriate boundary conditions we must distinguish between three systems. First is the transformed steady state equations

$$A_o v_x + B_o v_y = 0 \quad . \tag{8}$$

Next there is the transformed time dependent system (6) and finally there is the preconditioned system given by (7). We assume that the matrices A_o and B_o are symmetric and that ST and E_o are symmetric positive definite. Then both (6) and (7) form a symmetric hyperbolic system as considered by Friedrichs [7] and hence both are well posed for appropriate boundary data. If the boundary data are dissipative for (6) in L with weight ST then the same data will be dissipative and hence well posed for (7) in L with weight E_o. For more general boundary data it is not clear that data which makes (6) well posed will also make (7) well posed.

Furthermore, it is not known if data that makes (6) well posed will also make (8) well posed when a steady state is achieved. Thus, for example, one must rule out the possibility that the Helmholtz equation can be the steady state solution of a hyperbolic system. Even though the Helmholtz equation is well posed in the sense of Lopatinski this is not enough to yield uniqueness.

When the system (6) is strictly hyperbolic then one only needs analyze solutions to (6) of the form

$$v(x,y,t) = e^{\omega t} f(x,y) \; .$$

Since, we assume that a steady state is reached we must have that Re $\omega < 0$. Hence, all the eigenvalues of (8) are in the left half plane. This is enough to ensure well posedness in the sense of Lopatinski [17]. When the steady state is elliptic this guarantees regularity but not uniqueness. To show well posedness in the sense of Hadamard one must also get uniform bounds on how close to the imaginary axis the eigenvalues can be . In particular (8) may have a zero eigenvalue so that there are solutions to (8) that cannot be achieved by a time dependent process in addition to the solutions that are steady states of (6). Hence, we conclude that steady state solutions to (6) or (7) are solutions to (8) but we have no guarantee, even for constant coefficients, that these are the only solutions to (8) or that (8) is wellposed in the sense of Hadamard under the same boundary conditions.

We also wish to point out that if one begins with the steady state equations (8) then there are many possible boundary conditions that yield solutions but not all of them are physically relevant boundary conditions. One way of choosing the relevant boundary conditions is to demand that the solution be the limit of an appropriate time dependent problem. An alternative approach is to demand that the solution to (8) be the smooth limit of an appropriate viscous problem problem. As an example we consider the simple steady state

$$u_x = f \qquad 0 \leq x \leq 1 \quad . \tag{9}$$

the differential equation (9) is well posed if we impose

$$u(0) = \text{given} \tag{10a}$$

or if we impose

$$u(1) = \text{given} \quad . \tag{10b}$$

To decide which boundary condition is physically relevant we must determine, physically, whether (9) is the limit of

$$u_t + u_x = f \tag{11a}$$

or

$$-u_t + u_x = f \tag{11b}$$

as t goes to infinity. Equivalently, we can choose (11a) and decide whether (9) is the limit as the time goes to plus infinity or backwards to minus infinity. Since, a hyperbolic equation is reversible in time both possibilities are legitimate. For a nonlinear problem reversing time will reverse the entropy inequality.

An alternative method to choose between the boundary conditions

(10a) and (10b) is to claim that (9) is the smooth limit of a viscous system. Hence, (9) is the limit of either

$$u_x = \varepsilon u_{xx} + f \qquad \varepsilon > 0 \tag{12a}$$

or

$$u_x = -\varepsilon u_{xx} + f \qquad \varepsilon > 0 \ . \tag{12b}$$

Equation (12a) will have a boundary layer near $x = 1$. By eliminating the boundary condition at $x = 1$ for $\varepsilon = 0$ the boundary layer does not appear in the limiting solution. Equivalently we can eliminate the boundary layer for (12a) by specifying a Neumann type boundary condition rather than a Dirichlet condition. In the limit of small ε the Dirichlet condition at $x = 0$ remains while the boundary condition at $x = 1$ disappears. Of course, the roles of the two boundaries are interchanged when we choose (12b) instead of (12a).

In this case everything is obvious. A physically more relevant case is to consider flow through a nozzle. If the flow is subsonic then one should specify two conditions at inflow and one boundary condition at outflow. However, the steady state is unique if one specifies the total mass, the total enthalpy and the entropy. It makes no difference where these quantities are specified [26]. Thus, for example one could specify two of these quantities at outflow and only one at inflow. Nevertheless, the physically appropriate conditions are to specify two at inclow and one at outflow. This follows from the time dependent Euler equations or the steady Navier-Stokes equations.

We hence conclude that the matrix E_o in (7) must be chosen as positive definite whenever A_o and B_o are symmetric and ST is positive definite. This guarantees that we do not change the direction of the characteristics and so information flows in the same direction as before. Therefore, the number of boundary conditions is not changed.

4. ACCELERATION TO A STEADY STATE.

We wish to choose E_o in (7) so that we reach a steady state as fast as possible. When the equation is parabolic we can choose the free parameters so as to maximize the rate of decay to the steady state. This was first done by Garabedian [8] in analysing SOR. For a hyperbolic equation with constant coefficients energy is conserved except for boundary effects. Hence, the only way to introducing dissipation is through the artificial surfaces [1]. We shall therefore ignore dissipative mechanisms. Instead we consider explicit schemes and then we reach a steady state faster by choosing a larger time step within the stability limits. The methods to be developed are also effective for implicit methods using space factorization such as A.D.I. type methods. In order to compare different preconditionings we must normalize the time. Consider

$$u_t + u_x = f \quad . \tag{13}$$

Let $\tau = at$ then (13) becomes

$$au_\tau + u_x = f \quad . \tag{14}$$

When a is less than one then we reach a steady state faster in terms of absolute quantities. However, we do not achieve the steady state faster in terms of physical time scales. For example, using a typical explicit scheme one requires that $\Delta\tau/\Delta x \leq a$. Thus, the smaller a is the less time it takes to reach a steady state but at the same time the time steps are correspondingly smaller. The number of time iterations to reach a steady state is independent of a.

We therefore conclude that we can not compare the absolute time step allowed by different preconditioners. Instead we must scale all speeds by a given reference speed. Hence, we rephrase the third condition of Objective No. 1 as

Objective No. 2

Choose E_o so that we minimize the ratio of the fastest speed to the slowest speed of (7). Equivalently, choose E_o, positive definite, to minimize the condition number of

$$K[E_o(\omega_1 A_o + \omega_2 B_o)] \quad \text{with} \quad \omega_1^2 + \omega_2^2 = 1 \quad . \tag{15}$$

We now consider the question of minimizing (15) when (6) has different time scales. Kreiss [14] has developed a normal form for symmetric hyperbolic systems with three equations. The two dimensional Euler equations has four equations. However, since the entropy equation essentially decouples from the other three equations the two dimensional Euler equations are included in the theory. Tadmor [21] has extended the normal form to systems with more equations. Browning and Kreiss [3] have also analyzed nonlinear equations. In this study we wish to do the opposite of what Kreiss did. Instead of treating the initial conditions to filter the fast waves we wish to precondition the equations so that there is only one time scale. We shall choose E_o so as to equilibrate the time scales for the Euler equations. The normal form of Kreiss demonstrates that once we have accomplished this we have done the general two dimensional symmetric hyperbolic system with three equations.

It also follows from [14] that this approach will work only if the two time scales separate uniformly in the Fourier variables (ω_1,ω_2). The simplest case where the time scales are uniform in the Fourier variables is one dimensional flow, since there is only one Fourier variable,

$$u_t + A_o u_x = f \quad . \tag{16}$$

In this case (15) becomes: find E_o, positive definite, so that $K(E_o A_o^{-1})$ is minimum. The obvious choice is $E_o = |A^{-1}|$ where the absolute value of a matrix is found by going to diagonal form, taking absolute values and then transforming back. Thus, the optimal preconditioned form for (16) is

$$|A_o| u_t + A_o u_x = f \quad . \tag{17}$$

All the speeds of (17) are ± 1 and so the condition number is equal to 1. In two space dimensions this recipe doesn't work since $E_o = |\omega_1 A_o + \omega_2 B_o|$ implies that E_o is a pseudodifferential operator. Furthermore, since neither ω_1 nor ω_2 are small, in general, there are no obvious expansions. One possibility is to minimize this quantity in a root mean square sense over all ω_1 and ω_2.

Another possibility is to minimize the condition number in physical space rather than in Fourier space. If we replace the derivatives in space by central differences on a uniform periodic mesh then we wish to choose a $(4n) \times (4n)$ matrix so as so minimize the condition number of

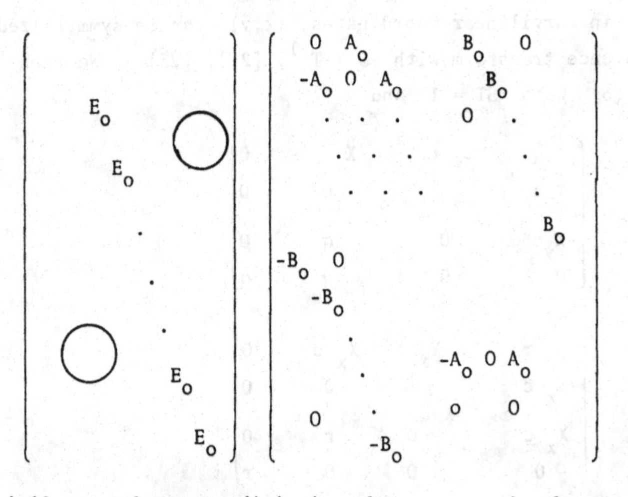

This is similar to the preconditioning that appears in the use of the conjugate gradient method [6]. However, now A_o and B_o are themselves matrices. Furthermore the matrix to be conditioned is not symmetric but antisymmetric. Hence, this approach is not very useful for general A_o and B_o. We therefore abandon the attempt to find a general solution to objective No. 2. Instead we shall consider specific cases for the Euler equations.

5. LOW SPEED FLOWS

When $u^2 + v^2 << c$ standard explicit schemes are inefficient. The time step is governed by 1/c while most important phenomena move at the convective speed. Implicit methods, especially A.D.I. type methods, also slow down due to the presence of different time scales. One possibility is to use a semi-implicit method but this is hard to implement in a conservative manner. If one is interested in time accuracy then one also needs to filter the high frequency content and then use an implicit method on the incompressible portion [10]. We shall instead precondition the Euler equations to remove the dependence on the sound speed, c. Viviand [25], and Briley, McDonald and Shamroth [2] have considered similar problems for the reduced isoenergetic equations. We shall also discuss this case in a later section. We now consider the full Euler equations so that we can easily extend the results to both the compressible and incompressible Navier-Stokes equations. The conservative Euler equations in curvilinear coordinates (x,y) can be symmetrized by an equivalence transform with $S = T^{-1}$, [22], [23]. We then recover (6) with $ST = I$ and

$$A_o = \begin{pmatrix} q & Y_y c & -X_y c & 0 \\ Y_y c & q & 0 & 0 \\ -X_y c & 0 & q & 0 \\ 0 & 0 & 0 & q \end{pmatrix}$$

$$B_o = \begin{pmatrix} r & -Y_x c & X_x c & 0 \\ -Y_x c & r & 0 & 0 \\ X_x c & 0 & r & 0 \\ 0 & 0 & 0 & r \end{pmatrix} \tag{18}$$

$$T_o = \begin{pmatrix} \rho/c & 0 & 0 & -1/c \\ u/c & \rho & 0 & -u/c \\ v/c & 0 & \rho & -v/c \\ h/c & u & v & (u^2 + v^2)/2c \end{pmatrix}$$

where (X,Y) are the cartesian coordinates and q and r are the contravariant components of velocity given by

$$q = Y_y u - X_y v \qquad\qquad r = X_x v - Y_x u \ . \tag{19}$$

We then choose the preconditioner E_o in (7) as

$$E_o = \begin{pmatrix} z^2/c^2 & 0 & 0 & 0 \\ 0 & 1 & 0 & 0 \\ 0 & 0 & 1 & 0 \\ 0 & 0 & 0 & 1 \end{pmatrix} \tag{20}$$

where $z^2 = \max(\varepsilon^2 , u^2 + v^2)$ is introduced so that E_o is non-singular at stagnation points. Typically ε is chosen as $.001\ c$ so that $z^2/c^2 > .001$. Transforming back to (3) we find that [24]

$$E^{-1} = I + dQ \qquad\qquad E = I + eQ$$
$$d = (\gamma - 1)(c^2/z^2 - 1)/c^2 \qquad\qquad e = \{c^2/(\gamma - 1) + z^2/2\}d \approx hd$$

where h is the enthalpy $h = c^2/(\gamma-1) + (u^2 + v^2)/2$,
$s^2 = (u^2 + v^2)/2$ and

$$Q = \begin{pmatrix} s^2 & -u & -v & 1 \\ us^2 & -u^2 & -uv & u \\ vs^2 & -uv & -v & v \\ hs^2 & -uh & -vh & h \end{pmatrix} \tag{21}$$

We note that the lower three rows of Q are obtained by multiplying the first row of Q by u,v, and h respectively. Hence, Q times a vector can be computed using six multiplies.

Let M be the Mach number defined by $M^2 = z^2/c^2$. Then the largest eigenvalue of $D = A\omega_1 + B\omega_2$ is given by

$$2\lambda = |w|(1+M^2) + \sqrt{w^2(1-M^2) + 4(a^2 + b^2)z^2} \tag{22}$$

where

$$w = q\omega_1 + r\omega_2 \;, \quad a = Y_y\omega_1 - Y_x\omega_2 \;, \quad b = X_x\omega_2 - X_y\omega_1 \;.$$

Hence, near a stagnation point $M = 0(\varepsilon)$ and $\lambda = 0(\varepsilon)$. It follows, that at low speeds $\Delta t/\Delta x = K/\max(\sqrt{u^2 + v^2},\varepsilon)$ and so Δt is independent of c. Briley et.al. [2] present results for the Navier-Stokes equations with turbulence using an implicit method. They show the advantages for a similar preconditioning for the isoevergetic equations.

6. Isoenergetic Equations

The steady state Euler equations have the property that the total specific enthalpy, $h = (E+p)/\rho$ is constant along streamlines. Hence, when the flow comes from a common reservoir the total enthalpy is constant throughout the entire field. Thus, various authors have replaced the energy equation in the inviscid equations by the algebraic condition that $h = h_o$. This system is no longer time consistent but gives the correct solution in the steady state.

In two space dimensions the isoenergetic equations form a 3 x 3 hyperbolic system. The theory of these equations was first discussed in [9]. Since there are several errors in that paper we shall derive the pertinent results. The nondimensional isentropic equations can be written in the form (4), [9], with

$$A = \begin{pmatrix} u & \rho & 0 \\ c^2/\gamma p & (1-2R)u & -2Rv \\ 0 & 0 & u \end{pmatrix} \quad B = \begin{pmatrix} v & 0 & \rho \\ 0 & v & 0 \\ c^2/\gamma p & -2Ru & (1-2R)v \end{pmatrix} \qquad (23)$$

where $R = (\gamma-1)/2\gamma$ and $c^2 = \gamma p/\rho$

Note, that the definition given for c differs slightly from that given in [9]. We now define

$$a_{\pm} = Ru \pm \sqrt{R^2u^2 + c^2/\gamma} \quad .$$

It is easily seen that a_+ is always positive while a_- is always negative. Using the technique described in [9] we let

$$T = \begin{pmatrix} \frac{c^2}{\gamma(\gamma-1)}\sqrt{\frac{-a_-}{a_+-a_-}} & \frac{c^2}{\gamma(\gamma-1)}\sqrt{\frac{a_+}{a_+-a_-}} & 0 \\ 0 & 0 & c^2/\gamma(\gamma-1) \\ \frac{-\rho c^2}{\gamma(\gamma-1)a_+}\sqrt{\frac{-a_-}{a_+-a_-}} & \frac{-\rho c^2}{\gamma(\gamma-1)a_-}\sqrt{\frac{a_+}{a_+-a_-}} & 2vR \end{pmatrix}$$

It can then be verified that

$$T^{-1}AT = \begin{pmatrix} u-a_+ & 0 & 0 \\ 0 & u-a_- & 0 \\ 0 & 0 & u \end{pmatrix}$$

and

$$T^{-1}BT = vI - \frac{2}{a_+-a_-}\begin{pmatrix} Rva_+ & Rvc/\delta\gamma & \sqrt{\frac{a_+(a_+-a_-)}{\gamma}}\,\frac{c}{2} \\ Rvc/\delta\gamma & -Rva_- & \sqrt{\frac{-a_-(a_+-a_-)}{\gamma}}\,\frac{c}{2} \\ \sqrt{\frac{a_+(a_+-a_-)}{\gamma}}\,\frac{c}{2} & \sqrt{\frac{-a_-(a_+-a_-)}{\gamma}}\,\frac{c}{2} & 0 \end{pmatrix} \qquad (24)$$

and so A and B can be simultaneously symmetrized. This property is also necessary if we wish to construct an entropy function [16]. Since, the isoenergetic equations are a symmetric hyperbolic system we can use energy methods to determine wellposed boundary conditions as well as the normal mode approach used in [9]. Furthermore, we define a state as supersonic if there exists numbers ω_1 and ω_2

such that $\omega_1 A + \omega_2 B$ is positive definite. It can then be shown that the isoenergetic equations are supersonic if and only if $u^2 + v^2 > c^2$.

Since the isoenergetic equations are symmetrizable we can use the theory developed in the previous sections. If we choose E_o as

$$E_o = \begin{pmatrix} z^2/c^2 & 0 & 0 \\ 0 & 1 & 0 \\ 0 & 0 & 1 \end{pmatrix} \quad z \text{ as in (20)} \tag{25}$$

then the condition number of $E^{-1}(\omega_1 A_o + \omega_2 B_2)$ is independent of c. This is similar to the preconditioning previously considered and also similar to that considered in [2].

7. INCOMPRESSIBLE FLOW

We next consider the steady state, inviscid, incompressible fluid dynamic equations. Klainerman and Majda [12] have proven that these equations are the asymptotic reduced equations of the Euler equations. Hence, one method of solving the incompressible equations is to consider the homentropic Euler equations or Navier Stokes e.g. [20] with a small Mach number and then use the preconditioning of section 5 to remove the stiffness of the equations. In this section we shall consider ways to directly integrate the incompressible equations. With both approaches the introduction of viscous terms does not introduce any fundamental difficulties especially with a high Reynolds number. Since we are interested in a pseudo time approach we consider the artificial density algorithm [5].

In conservation form the time dependent equations are

$$u_x + v_y = 0 \tag{26a}$$

$$u_t + (u^2 + p)_x + (uv)_y = 0 \tag{26b}$$

$$v_t + (uv)_x + (v^2 + p)_y = 0 \tag{26c}$$

Using the artificial density approach [18] we replace (26a)

$$p_t/c^2 + u_x + v_y = 0 \tag{26a'}$$

It is easy to verify that the resultant system is hyperbolic but not symmetrizable. Instead we replace (26) by

$$p_t/c^2 + u_x + v_y = 0$$
$$aup_t/c^2 + u_t + (u^2+p)_x + (uv)_y = 0 \qquad (27)$$
$$avp_t/c^2 + v_t + v_t + (uv)_x + (v^2+p)_y = 0$$

or equivalently

$$E^{-1}w_t + Aw_x + Bw_y = 0 \quad , \quad w = (p,u,v)$$

and so (27) can be considered as a preconditioning of the system (26). c is an artificial sound speed which need not be constant. We shall later discuss how to choose c.

When $a = 1$ in (27) then this system is equivalent to a symmetric hyperbolic system. In this case the eigenvalues of $E(A\omega_1 + B\omega_2)$ are

$$q \;;\; (q \pm \sqrt{q^2 + 4(\omega_1^2+\omega_2^2)\, c^2})/2 \;,\; q = u\omega_1 + v\omega_2 \qquad (28)$$

It thus follows that this system is always subsonic independent of the value we choose for c. This is to be expected as we do not wish an incompressible fluid to behave like a supersonic flow with shocks even in the nonphysical time dependent phase. When a differs from 1 the system is no longer symmetrizable though still hyperbolic. In this case the eigenvalues of (27) are

$$q \;;\; ((2-a)q \pm \sqrt{(2-a)^2 q + 4(\omega_1^2+\omega_2^2)c^2})/2 \qquad (29)$$

P. Roe (private communication) has noted that for $a = 2$ the eigenvalues have the simple form

$$q \;;\; \pm c \qquad (30)$$

Hence, in this case the speed of the sound waves are independent of the convective speed and hence the sound waves spread isotropically even in the presence of a flow. We shall later see that this allows a more optimal selection for the artificial speed of sound c. We therefore rewrite (27) with $a = 2$ in nonconservative form

$$p_t/c^2 + u_x + v_y = 0$$
$$up_t/c^2 + u_t + uu_x + vu_y + p_x = 0 \qquad (31)$$
$$vp_t/c^2 + v_t + uv_x + vv_y + p_y = 0$$

or equivalently

$$
\begin{aligned}
&p_t + c^2(u_x + v_y) = 0 \\
&u_t + vu_y - uv_y + p_x = 0 \qquad (32) \\
&v_t + uv_x - vu_x + p_y = 0
\end{aligned}
$$

The eigenvalues of this new artificial density equation are given by (30). The improvement in the sound speed is achieved at the expense of the loss of symmetry. It is not clear that this loss of symmetry is of any importance since all the coefficients that appear are well behaved and the system is strictly hyperbolic. The original pseudo density equations in nonconservative form are

$$
\begin{aligned}
&p_t/c^2 + u_x + v_y = 0 \\
&u_t + uu_x + vu_y + p_x = 0 \qquad (33) \\
&v_t + uv_x + vv_y + p_y = 0
\end{aligned}
$$

This is equivalent to (27) with $a = 1$, and so is symmetric.

There remains the question of how to choose the artificial sound speed c. As we have stressed, for inviscid flow we wish to reduce the ratio of the largest eigenvalue to the smallest eigenvalue. For the system (31) or (32) or, (27) with $a = 2$, the eigenvalues are given by (30). Hence, we would like to choose $c = q = \omega_1 u+\omega_2 v$. This choice would give us a condition number of one. However, we can not allow c to depend on the Fourier variables ω_1 and ω_2. Hence, an alternate choice is to set $c^2 = u^2 + v^2$.

For the original equations (33) or else, (27) with $a = 1$, we wish to minimize both

$(1 + \sqrt{1+4c^2/q^2})/(1 - \sqrt{1+4c^2/q^2})$ and $1 + \sqrt{1+4c^2/q^2}$. If we choose c small we enlarge the first ratio while if we choose c small we increase the second ratio. It is easy to calculate that the minimum of the maximum of both ratios is reached when $c^2 = 3q^2/4$. In that case the condition number is three. Hence, if we could choose this value for c the original pseudo density system (33) would be three times slower than the new version given by (31) or (32). As before this choice for c is not legitimate since it depends on the Fourier variables (ω_1,ω_2). As before an alternative is to choose $c^2 = 3(u^2 + v^2)/4$. In this analysis we have only considered the effect of the inviscid time step on c. In [4] the effect of the viscous terms is considered.

When the incompressible Navier-Stokes equations are considered the pseudo density approach can be easily modified to include these terms. When the Reynolds number is sufficiently large, for a given mesh, the time step is only governed by the inviscid part and the previous analysis is valid. For lower cell Reynolds number one can treat the viscous terms implicitly. Since the coefficients are constant for the viscous portion a backward Euler method even in several space dimensions is feasible.

ACKNOWLEDGEMENT

Research was supported by the National Aeronautics and Space Administration under NASA Contract Nos. NAS1-16394 and NAS1-17130 while the author was in residence at the Institute for Computer Applications in Science and Engineering, NASA Langley Research Center, Hampton Va 23665.

References

[1] A. Bayliss and E. Turkel, Far Field Boundary Conditions for Compressible Flow, J. Comput. Phys., 48[1982], pp. 182-199

[2] W.R. Briley, H. McDonald and S.J. Shamroth, A Low Mach Number Euler Formulation and Application to Time-Iterative LBI Schemes, AIAA J., 21[1983], pp. 1467-1469.

[3] G. Browning and H.O. Kreiss, Problems with Different Time Scales for Nonlinear Partial Differential Equations, SIAM J. Appl. Math., 42[1982], pp. 704-718.

[4] J.L.C. Chang and D. Kwak, On the Method of Pseudo Compressibility for Numerically Solving Incompressible Flows, AIAA paper 84-0252[1984].

[5] A.J. Chorin, A Numerical Method for Solving Incompressible Viscous Flow Problems, J. Comput. Phys., 2[1967], pp. 12-26.

[6] P. Concus, G.H. Golub and D.P. O'Leary, A Generalized Conjugate Gradient Method for the Numerical Solution of Elliptic Partial Differential Equations, Proc. Symp. Sparse Matrix Comput., J.R. Bunch, D.J. Rose Editor Academic Press, 1975.

[7] K.O. Friedrichs, Symmetric Positive Linear Differential Equations, Comm. Pure Appl. Math., 11[1958], pp. 333-418.

[8] P.R. Garabedian, Estimation of the Relaxation Factor for Small Mesh Size, Math. Tables Aids Comput., 10[1956], pp. 183-185.

[9] D. Gottlieb and B. Gustafsson, On the Navier-Stokes Equations with Constant Total Temperature, Stud. Appl. Math., 11[1976], pp. 167-185.

[10] J. Guerra, Numerical Solution of the Euler Equations for Small Mach Numbers, Uppsale Univ. Dept. Comput. Sci., Report No. 96 [1984].

[11] A. Jameson, W. Schmidt, E. Turkel, Numerical Solutions of the Euler Equations by Finite Volume Methods Using Runge-Kutta Time-Stepping Schemes, AIAA paper 81-1259 [1981].

[12] S. Klainerman, A. Majda, Compressible and Incompressible Fluids, Comm. Pure Appl. Math., 35[1982], pp. 629-651.

[13] H.O. Kreiss, Initial Boundary Valve Problems for Hyperbolic Systems, Comm. Pure Appl. Math., 23[1970], pp. 277-298.

[14] H.O. Kreiss, Problems with Different Time Scales for Partial Differential Equations, Comm. Pure Appl. Math., 33[1980], pp. 399-439.

[15] H. Lomax and U.B. Mehta, Some Physical and Numerical Aspects of Computing the Effects of Viscosity on Fluid Flow, Viscous Flow Comput. Meth., Recent Adv. Num. Meth. Fluids, W.G. Habashi Editor [1984].

[16] M.S. Mock, Systems of Conservation Laws of Mixed Type, J.Diff. Eq., 37[1980], pp. 70-88.

[17] S. Osher, Hyperbolic Equations in Regions with Characteristic Boundaries or with Corners, Num. Sol. PDE III Synspade 1975, B. Hubbard editor, pp. 413-441, Academic Press.

[18] R. Peyret and T.D. Taylor, Computational Methods for Fluid Flow, Springer-Verlag, N.Y., 1983.

[19] R. Peyret and H. Viviand, Pseudo-Unsteady Methods for Inviscid or Viscous Flow Computation, Recent Adv in in Aerospace Science.

[20] R.C. Swanson and E. Turkel, A Multistage Time-Stepping Scheme for the Navier-Stokes Equations, AIAA paper 85-0035, 1985.

[21] E. Tadmor, Hyperbolic Systems with Different Time Scales, Comm. Pure Appl. Math., 35[1982], pp. 839-866.

[22] E. Turkel, Symmetric Hyperbolic Difference Schemes and Matrix Problems, Lin. Alg. Applic., 16 [1977], pp. 109-129.

[23] E. Turkel, Acceleration to a Steady State for the Euler Equations, Proc. Workshop on Euler eq., [1983].

[24] E. Turkel, Fast Solutions to the Steady State Compressible and Incompressible Fluid Dynamic Equations, Ninth Int. Conf. Numerical Meth. Fluid Dynamics, 1984 Springer-Verlag Lecture Notes in Physics.

[25] H. Viviand, Pseudo-Unsteady Systems for Steady Inviscid Flow Flow Calculation, Proc. Workshop on Euler eq. [1983].

[26] S.F. Wornom, M.M. Hafez, A Rule for Selecting Analytical Boundary Conditions for the Conservative Quasi-one-Dimensional Nozzle Flow Equations, AIAA paper.

Progress in Scientific Computing, Vol. 6
Proceedings of U.S.-Israel Workshop, 1984

VISCOUS FLOW SIMULATION BY FINITE ELEMENT METHODS AND RELATED NUMERICAL TECHNIQUES

R. Glowinski*

Introduction

We would like to discuss in this paper some numerical methods for solving the Navier-Stokes equations modeling viscous flows. Since our experience with *incompressible fluids* is more important than with compressible ones, we shall discuss first the incompressible case and then show how some of the conclusions can be applied to the *compressible* case.

Actually some of the numerical methods discussed in the following sections can be applied not only to *finite element* approximations, but also to *finite difference* or *spectral* approximations.

Our motivation for considering finite element approximations is their capability to handle the *complicated geometries* encountered in most practical applications; it was that property which made finite element methods so successful in structural analysis, and in our opinion this point of view still holds in *Computational Fluid Dynamics (CFD)*, despite the difficulties due to the more complicated nature of the equations.

Comments will be done about the feasibility of the methods to be discussed in view of their implementation on vector machines and/or supercomputers.

Some numerical results will illustrate the methods described in this paper.

*University of Paris VI, 4 place Jussieu, 75230 PARIS CEDEX 05 and INRIA.

1. <u>Formulation of the Navier-Stokes Equations for Incompressible Viscous Fluids</u>.

Let us consider a Newtonian incompressible viscous fluid. If Ω is the flow domain ($\Omega \subset \mathbb{R}^N$, N=2,3 in practice) and Γ is its boundary, then this flow is governed by the following *Navier-Stokes equations*

(1.1) $$\frac{\partial \underset{\sim}{u}}{\partial t} - \nu\Delta \underset{\sim}{u} + (\underset{\sim}{u}.\underset{\sim}{\nabla})\underset{\sim}{u} + \underset{\sim}{\nabla} p = \underset{\sim}{f} \text{ in } \Omega ,$$

(1.2) $$\underset{\sim}{\nabla}.\underset{\sim}{u} = 0 \quad \text{in } \Omega \text{ (incompressibility condition).}$$

In (1.1), (1.2), we have

(a) $\underset{\sim}{\nabla} = \{\frac{\partial}{\partial x_i}\}_{i=1}^{N}, \Delta = \underset{\sim}{\nabla}^2 = \sum_{i=1}^{N} \frac{\partial^2}{\partial x_i^2}$,

(b) $\underset{\sim}{u} = \{u_i\}_{i=1}^{N}$ is the *flow velocity*,

(c) p is the *pressure*,

(d) ν is a *viscosity parameter*,

(e) $\underset{\sim}{f}$ is a *density of external forces*.

In (1.1), $(\underset{\sim}{u}.\underset{\sim}{\nabla})\underset{\sim}{u}$ is a symbolic notation for the nonlinear vector term

$$\{\sum_{j=1}^{N} u_j \frac{\partial u_i}{\partial x_j}\}_{i=1}^{N} .$$

Boundary conditions have to be added ; for example in the case of the airfoil A of Figure 1.1, we have (since the fluid is *viscous*) the following *adherence condition*

(1.3) $$\underset{\sim}{u} = \underset{\sim}{0} \quad \text{on } \Gamma_A = \partial A \quad .$$

Typical conditions at infinity are

(1.4) $$\underset{\sim}{u} = \underset{\sim}{u}_\infty$$

where $\underset{\sim}{u}_\infty$ is *constant* vector (with regards to the space variables at least).

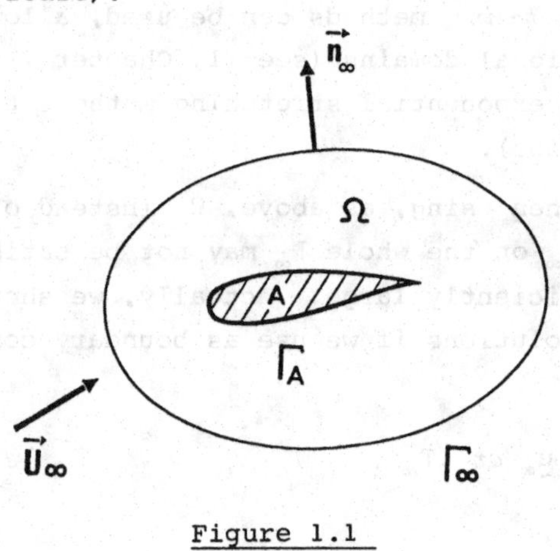

Figure 1.1

If Ω is a *bounded* region of $\mathbb{R}^N$, very usual boundary conditions are

(1.5) $\quad \underset{\sim}{u} = \underset{\sim}{g} \quad on\ \Gamma$

where (from the *incompressibility* of the fluid) the given function $\underset{\sim}{g}$ has to satisfy

(1.6) $\quad \int_\Gamma \underset{\sim}{g}.\underset{\sim}{n}\, d\Gamma = 0,$

where $\underset{\sim}{n}$ is the outward *unit* vector normal to Γ.

Finally for the time dependent problem (1.1), (1.2) an *initial condition* such as

(1.7) $\quad \underset{\sim}{u}(x,0) = \underset{\sim}{u}_o(x) \quad a.e.\ on\ \Omega$

with $\underset{\sim}{u}_o$ given, is usually prescribed.

In practice, for the problem corresponding to Fig. 1.1, we should replace Ω by a large *bounded* domain Ω_c (the *computational* domain) and on the external boundary Γ_∞ of Ω_c we should prescribe $\underset{\sim}{u} = \underset{\sim}{u}_\infty$, or some more complicated boundary conditions (see Remark 1.2, below).

<u>Remark 1.1</u>: For *two-dimensional* problems on unbounded domains Ω *exponential stretching* methods can be used, allowing *very large* computational domains (see [1, Chapter 7] for an application of exponential stretching methods to inviscid flow calculations).

<u>Remark 1.2</u>: When using, as above, Ω_c instead of Ω, prescribing $\underset{\sim}{u} = \underset{\sim}{u}_\infty$ on the whole Γ_∞ may not be satisfactory if Ω_c is not sufficiently large. Actually, we should improve the computed solutions if we use as boundary conditions

(1.8) $\quad \underset{\sim}{u} = \underset{\sim}{u}_\infty \text{ on } \Gamma_\infty^-$,

and either

(1.9)$_1$ $\quad \dfrac{\partial \underset{\sim}{u}}{\partial n} - \underset{\sim}{n}_\infty p = \underset{\sim}{0} \text{ on } \Gamma_\infty^+$,

or

(1.9)$_2$ $\quad \dfrac{\partial \underset{\sim}{w}}{\partial t} + c\,\dfrac{\partial \underset{\sim}{w}}{\partial n} = \underset{\sim}{0} \text{ on } \Gamma_\infty^+$,

where

(i) $\quad \Gamma_\infty^+ = \{x \mid x \in \Gamma_\infty \,,\, \underset{\sim}{u}_\infty . \underset{\sim}{n}_\infty(x) \geq 0\}$,

$\quad \Gamma_\infty^- = \{x \mid x \in \Gamma_\infty \,,\, \underset{\sim}{u}_\infty . \underset{\sim}{n}_\infty(x) < 0\}$,

(ii) $\quad \underset{\sim}{n}_\infty$ is the outward unit vector normal at Γ_∞ (see Fig. 1.1),

(iii) $\quad c$ is a *constant* ; a natural choice seems to be $c = |\underset{\sim}{u}_\infty|$,

(iv) $\quad \underset{\sim}{w} = \underset{\sim}{\nabla} \times \underset{\sim}{u}$ is the vorticity of the flow.

The main reason for using either (1.8), (1.9)$_1$ or (1.8), (1.9)$_2$, instead of $\underset{\sim}{u} = \underset{\sim}{u}_\infty$ on Γ_∞, is that the former boundary conditions are *less reflecting* (i.e are *more aborbing*) than the latter.

The Navier-Stokes equations for incompressible viscous fluids have motivated a large number of papers, books, reports and symposia. We shall limit our references to:

(i) [2] - [7] for the *theoretical* aspects,

(ii) the related papers in these proceedings and [1],[5], [7]-[11] for the *numerical* aspects.

In both cases, the interested reader may consult the references contained in the above books.

The difficulties with the Navier-Stokes equations (even for flows at low Reynolds number, in bounded regions Ω) are

(i) The nonlinear term $(\underset{\sim}{u}.\underset{\sim}{\nabla})\underset{\sim}{u}$ in (1.1),

(ii) The *incompressibility* condition (1.2),

(iii) The fact that the solutions of the Navier-Stokes equations are *vector-valued* functions of x,t, whose components are coupled by the nonlinear term $(\underset{\sim}{u}\cdot\underset{\sim}{\nabla})\underset{\sim}{u}$ and by the incompressibility condition $\nabla\cdot u = 0$.

Using convenient *operator splitting methods* for the *time discretization* of the Navier-Stokes equations, we shall be able to decouple the difficulties due to the nonlinearity and to the incompressibility, respectively.

For simplicity, we suppose from now on that Ω is *bounded* and that we have (1.5) as boundary conditions (with $\underset{\sim}{g}$ satisfying (1.6) and possibly depending upon t).

2. Time discretization by operator splitting methods.

2.1. Generalities. Description of the basic schemes.

Among the properties that we require for the time discretization methods, we shall insist on the following:

(i) A good accuracy when simulating *unsteady* flows,

(ii) Good convergence properties as $t \to +\infty$, in order to capture *steady* solutions, if such solutions exist and are stable. Moreover, we wish to have the steady state solutions, obtained through the above time dependent process, independent of Δt. Such a requirement excludes various fractional step methods, for example.

Let us start with some generalities. Consider a *real Hilbert space* H. We consider then in H the following initial value problem

$$(2.1)\qquad \begin{cases} \dfrac{du}{dt} + A(u) = f, \\ u(0) = u_o , \end{cases}$$

where A is an operator from H to H ; f is a *source term* and u_o the *initial value*. Let A_1 and A_2 be two operators such that

$$(2.2)\qquad A = A_1 + A_2 .$$

With Δt (> 0) a *time discretization* step, let us define several schemes for solving (2.1), taking advantage of the decomposition (2.2).

<u>A first scheme (of Peaceman-Rachford type)</u>. This scheme is defined as follows:

$$(2.3)\qquad u^o = u_o ,$$

then for $n \geq 0$, *with* u^n *known compute* $u^{n+1/2}$ *and then* u^{n+1} *by*

$$(2.4)\qquad \frac{u^{n+1/2}-u^n}{\Delta t/2} + A_1(u^{n+1/2}) + A_2(u^n) = f^{n+1/2},$$

$$(2.5)\qquad \frac{u^{n+1}-u^{n+1/2}}{\Delta t/2} + A_1(u^{n+1/2}) + A_2(u^{n+1}) = f^{n+1}.$$

In (2.4), (2.5) $u^{n+\alpha}$ denotes an approximation of $u((n+\alpha)\Delta t)$, and $f^{n+\alpha} = f((n+\alpha)\Delta t)$.

<u>A second scheme</u>. Let θ belong to the open interval (0,1/2). The idea behind the scheme is to split the time interval $[n\Delta t,(n+1)\Delta t]$ in three subintervals, instead of two, and integrate in time using a discretization scheme *implicit*

for A_1 (resp. *explicit* for A_2) on $[n\Delta t,(n+\theta)\Delta t]$, then switch the role of A_1 and A_2 on $[(n+\theta)\Delta t,(n+1-\theta)\Delta t]$, and on $[(n+1-\theta)\Delta t,(n+1)\Delta t]$ do like on $[n\Delta t,(n+\theta)\Delta t]$. Using these principles, we obtain the following scheme, some forms of which have been advocated in [12] - [15] (for $\theta = 1/4$, particularly) :

$$(2.6) \qquad u^o = u_o ,$$

then, for $n \geq 0$, *we obtain* $u^{n+\theta}$, $u^{n+1-\theta}$ *and* u^{n+1}, *from* u^n, *as follows*

$$(2.7) \qquad \frac{u^{n+\theta} - u^n}{\theta\Delta t} + A_1(u^{n+\theta}) + A_2(u^n) = f^{n+\theta} ,$$

$$(2.8) \qquad \frac{u^{n+1-\theta} - u^{n+\theta}}{(1-2\theta)\Delta t} + A_1(u^{n+\theta}) + A_2(u^{n+1-\theta}) = f^{n+1-\theta} ,$$

$$(2.9) \qquad \frac{u^{n+1} - u^{n+1-\theta}}{\theta\Delta t} + A_1(u^{n+1}) + A_2(u^{n+1-\theta}) = f^{n+1} .$$

The convergence of (2.3)-(2.4) has been proved in [16] (see also [17]) under quite general monotonicity assumptions on A_1 and A_2. We don't know, yet, similar results for (2.6)-(2.9) (but very likely, the methods in [16] , [17] still apply).

2.2. Convergence and stability properties of the basic schemes

Following the approach used in [18] , we consider for simplicity the case where $H = \mathbb{R}^N$, $f = 0$, $u_o \in \mathbb{R}^N$, A is a $N \times N$ matrix, *symmetric* and *positive definite*, and where

$$(2.10) \qquad A_1 = \alpha A,\ A_2 = \beta A,\ \textit{with}\ \alpha+\beta = 1,\ 0< \alpha, \beta < 1 .$$

In that case, the solution of (2.1) is clearly given by

$$u(t) = e^{-tA}\, u_o .$$

Analysis of scheme (2.3)-(2.5) (see also[18]) .
We have from (2.4), (2.5), (2.10)

$$(2.11)\qquad u^{n+1}=(I+\beta \frac{\Delta t}{2}A)^{-1}(I-\alpha \frac{\Delta t}{2}A)(I+\alpha \frac{\Delta t}{2}A)^{-1}(I-\beta \frac{\Delta t}{2}A)u^n.$$

Using a vector basis of $\mathbb{R}^N$, consisting of eigenvectors of A, we have from (2.11), and with obvious notation,

$$(2.12)\qquad u_i^{n+1}=\frac{(1-\alpha \frac{\Delta t}{2}\lambda_i)(1-\beta \frac{\Delta t}{2}\lambda_i)}{(1+\alpha \frac{\Delta t}{2}\lambda_i)(1+\beta \frac{\Delta t}{2}\lambda_i)}\, u_i^n ,$$

where $\lambda_i (> 0, \forall i = 1,\ldots,N)$ is the ith eigenvalue of A ; we suppose that $\lambda_1 \le \lambda_2 \le \ldots \le \lambda_N$. Consider now the rational function R_1 defined by

$$(2.13)\qquad R_1(x) = \frac{(1-\frac{\alpha}{2}x)(1-\frac{\beta}{2}x)}{(1+\frac{\alpha}{2}x)(1+\frac{\beta}{2}x)} .$$

We observe that $|R_1(x)| <1, \forall x > 0$, implying, in that simple case, the *unconditional stability* of scheme (2.3)-(2.5). Since

$$(2.14)\qquad \lim_{x\to +\infty} R_1(x) = 1 ,$$

we observe that for stiff problems, i.e. problems such that $\lambda_N / \lambda_1 >> 1$, scheme (2.3)-(2.5) is not very good to damp simultaneously the components of u^n associated with the large and small eigenvalues of A. From this observation we can expect that the above scheme is not well suited to "capture" the steady state solutions of stiff problems (like those obtained from the discretization of partial differential equations). This has been confirmed by numerical experiments. Let discuss now the accuracy of scheme (2.3)-(2.5). Since we have, in the neighborhood of $x = 0$,

$$(2.15)\qquad e^{-x} = 1-x + \frac{x^2}{2} + x^3\, 0(1),$$

and from (2.13),

$$(2.16)\qquad R_1(x) = 1-x + \frac{x^2}{2} + x^3\, 0(1),$$

we have that scheme (2.3)-(2.5) is *second order accurate* in the simple case that we have considered. We observe, from (2.11), that if one takes $\alpha = \beta = 1/2$, then the two linear systems which have to be solved at each full step, are in fact associated to the same matrix, namely $I + \Delta t\ A/4$.

<u>Analysis of scheme (2.6)-(2.9).</u> We have (with $\theta' = 1-2\theta$)

$$(2.17)\qquad u^{n+1} = (I+\alpha\theta\Delta tA)^{-1}(I-\beta\theta\Delta tA)(I+\beta\theta'\Delta tA)^{-1}(I-\alpha\theta'\Delta tA)\times$$
$$\times(I+\alpha\theta\Delta tA)^{-1}(I-\beta\theta\Delta tA)u^n,$$

which implies

$$(2.18)\qquad u_i^{n+1} = \frac{(1-\beta\theta\Delta t\lambda_i)^2(1-\alpha\theta'\Delta t\lambda_i)}{(1+\alpha\theta\Delta t\lambda_i)^2(1+\beta\theta'\Delta t\lambda_i)}u_i^n .$$

Consider now the rational function R_2 defined by

$$(2.19)\qquad R_2(x) = \frac{(1-\beta\theta x)^2(1-\alpha\theta' x)}{(1+\alpha\theta x)^2(1+\beta\theta' x)} .$$

Since

$$(2.20)\qquad \lim_{x\to+\infty} |R_2(x)| = \beta/\alpha ,$$

we should prescribe

$$(2.21)\qquad \alpha \geq \beta$$

to have, from (2.17), (2.18), the stability of scheme (2.6)-(2.9) for the *large eigenvalues* of A. We discuss now the accuracy of scheme (2.6)-(2.9). We can show that in the neighborhood of $x = 0$

$$(2.22)\quad R_2(x) = 1-x+\frac{x^2}{2}\{1+(\beta^2-\alpha^2)(2\theta^2-4\theta+1)\} + x^3\ O(1).$$

It follows from (2.22) that scheme (2.6)-(2.9) is *second order accurate*, if either

$$(2.23)\quad \alpha = \beta \ (= 1/2 \textit{ from } (2.10)),$$

or

$$(2.24)\qquad \theta = 1-\sqrt{2}/2 = .29289\ldots .$$

Scheme (2.6)-(2.9) is only *first order accurate* if neither (2.23) nor (2.24) holds. If one takes $\alpha = \beta = 1/2$, it follows from (2.18) that scheme (2.6)-(2.9) is *unconditionaly stable*, $\forall\theta \in]0,1/2[$ However, since (from (2.20)) we have in that case

$$\lim_{x \to +\infty} |R_2(x)| = 1,$$

the remark done for scheme (2.3)-(2.5), about the integration of stiff systems, still holds. In practice, we shall choose α and β in order to have the same matrix for all the partial steps of the integration procedure. Therefore α, β, θ have to satisfy

$$\alpha\theta = \beta(1-2\theta) \quad , \tag{2.25}$$

which implies

$$\alpha = (1-2\theta)/(1-\theta), \ \beta = \theta/(1-\theta). \tag{2.26}$$

Combining (2.21) and (2.26) we obtain

$$0 < \theta \leq 1/3. \tag{2.27}$$

For $\theta = 1/3$, (2.26) implies $\alpha = \beta = 1/2$; the resulting scheme is just a variant of scheme (2.3)-(2.5).

If $0 < \theta < 1/3$, and if α and β are given by (2.26) we have then

$$\lim_{x \to +\infty} |R_2(x)| = \beta/\alpha = \theta/(1-2\theta) < 1 . \tag{2.28}$$

Actually, we can prove that $\theta \in]\theta^*, 1/3]$ (1) and α, β given by (2.26), imply the *unconditional stability* of scheme (2.6) (2.9). Moreover, if $\theta \in]\theta^*, 1/3[$, property (2.28) makes that scheme (2.6)-(2.9) has good asymptotic properties as $n \to +\infty$, and, for example, is well suited to compute steady

---

(1) With $\theta^* = .087385580$ ... Another interesting value is $\theta = 1/6$ for which we have $\max_{x \geq 1/\alpha\theta'} |R_2(x)| = \beta/\alpha = .25$.

state solutions. If $\theta = 1-\sqrt{2}/2$ (resp. $\theta = .25$), we have $\alpha = 2-\sqrt{2}$, $\beta=\sqrt{2}-1$, $\beta/\alpha = 1/\sqrt{2}$ (resp. $\alpha = 2/3$, $\beta = 1/3$, $\beta/\alpha = 1/2$).

2.3. Application to the solution of the time dependent Navier-Stokes equations.

2.3.1. A first operator splitting method

This method which is directly derived from the *Peaceman-Rachford scheme* (2.3)-(2.5) is described as follows :

$$\text{(2.29)} \qquad \underset{\sim}{u}^{o} = \underset{\sim}{u}_{o} \quad ,$$

then for $n \geq 0$, *compute* $\{\underset{\sim}{u}^{n+1/2}, p^{n+1/2}\}$ *and* $\underset{\sim}{u}^{n+1}$, *from* $\underset{\sim}{u}^{n}$, *by solving*

$$\text{(2.30)} \qquad \begin{cases} \dfrac{\underset{\sim}{u}^{n+1/2} - \underset{\sim}{u}^{n}}{\Delta t/2} - \dfrac{\nu}{2}\Delta \underset{\sim}{u}^{n+1/2} + \underset{\sim}{\nabla} p^{n+1/2} = \\ \underset{\sim}{f}^{n+1/2} + \dfrac{\nu}{2}\Delta \underset{\sim}{u}^{n} - (\underset{\sim}{u}^{n}.\underset{\sim}{\nabla})\underset{\sim}{u}^{n} \quad in\ \Omega, \\ \underset{\sim}{\nabla}.\underset{\sim}{u}^{n+1/2} = 0 \quad in\ \Omega, \\ \underset{\sim}{u}^{n+1/2} = \underset{\sim}{g}^{n+1/2} \quad on\ \Gamma \end{cases}$$

and

$$\text{(2.31)} \qquad \begin{cases} \dfrac{\underset{\sim}{u}^{n+1} - \underset{\sim}{u}^{n+1/2}}{\Delta t/2} - \dfrac{\nu}{2}\Delta \underset{\sim}{u}^{n+1} + (\underset{\sim}{u}^{n+1}.\underset{\sim}{\nabla})\underset{\sim}{u}^{n+1} = \\ \underset{\sim}{f}^{n+1} + \dfrac{\nu}{2}\Delta \underset{\sim}{u}^{n+1/2} - \underset{\sim}{\nabla} p^{n+1/2} \quad in\ \Omega, \\ \underset{\sim}{u}^{n+1} = \underset{\sim}{g}^{n+1} \quad on\ \Gamma \end{cases}$$

respectively.

2.3.2. A second operator splitting method

This method is derived from scheme (2.6)-(2.9) and is described as follows :

$$\text{(2.32)} \qquad \underset{\sim}{u}^{o} = \underset{\sim}{u}_{o},$$

then for $n \geq 0$ *and starting from* $\underset{\sim}{u}^{n}$ *we solve*

$$(2.33)\quad \begin{cases} \dfrac{\tilde{u}^{n+\theta}-\tilde{u}^{n}}{\theta\Delta t} - \alpha\nu\,\Delta\tilde{u}^{n+\theta} + \nabla\tilde{p}^{n+\theta} = \tilde{f}^{n+\theta} + \beta\nu\Delta\tilde{u}^{n} - (\tilde{u}^{n}.\nabla)\tilde{u}^{n}\ \text{in } \Omega, \\ \tilde{\nabla}.\tilde{u}^{n+\theta} = 0\ \text{in } \Omega, \\ \tilde{u}^{n+\theta} = \tilde{g}^{n+\theta}\ \text{on } \Gamma, \end{cases}$$

$$(2.34)\quad \begin{cases} \dfrac{\tilde{u}^{n+1-\theta}-\tilde{u}^{n+\theta}}{(1-2\theta)\Delta t} - \beta\,\nu\,\Delta\tilde{u}^{n+1-\theta} + (u^{n+1-\theta}.\ )u^{n+1-\theta} = \\ \tilde{f}^{n+1-\theta} + \alpha\,\nu\Delta\tilde{u}^{n+\theta} - \nabla\tilde{p}^{n+0}\ \text{in } \Omega, \\ \tilde{u}^{n+1-\theta} = \tilde{g}^{n+1-\theta}\ \text{on } \Gamma \end{cases}$$

$$(2.35)\quad \begin{cases} \dfrac{\tilde{u}^{n+1}-\tilde{u}^{n+1-\theta}}{\theta\,\Delta t} - \alpha\nu\Delta\tilde{u}^{n+1} + \nabla\tilde{p}^{n+1} = \\ \tilde{f}^{n+1} + \beta\nu\Delta\tilde{u}^{n+1-\theta} - (\tilde{u}^{n+1-\theta}.\tilde{\nabla})\tilde{u}^{n+1-\theta}\ \text{in } \Omega, \\ \tilde{\nabla}.\tilde{u}^{n+1} = 0\ \text{in } \Omega, \\ \tilde{u}^{n+1} = \tilde{g}^{n+1}\ \text{on } \Gamma. \end{cases}$$

2.3.3. Some comments and remarks concerning schemes (2.29)-(2.31) and (2.32)-(2.35).

Using the two above operator splitting methods, we have been able to decouple *nonlinearity* and *incompressibility* in the Navier-Stokes equations (1.1), (1.2). We shall discuss, briefly, in the following sections the specific treatment of the subproblems encountered at each step of (2.29)-(2.31) and (2.32)-(2.35). We shall consider only the case where the subproblems are still continuous in space (since the formalism of the continuous problems is much simpler). For the fully discrete case, see [1], [19] where *finite element approximations* of (1.1), (1.2) are discussed.

We observe that $\{\tilde{u}^{n+1/2}, p^{n+1/2}\}$ and $\{\tilde{u}^{n+\theta}, p^{n+\theta}\}$, $\{\tilde{u}^{n+1}, p^{n+1}\}$ are obtained from the solution of *linear problems* very close to the *steady Stokes problem*. Despite its greater complexity, scheme (2.32)-(2.35) is almost as economical to use as scheme (2.29)-(2.31). This is mainly due to the fact that the "quasi" steady Stokes problems (2.30) and (2.33), (2.35) (actually convenient *finite dimensional approximations* of them) can be solved by very efficient solvers so that most of the computer time used to solve a full **step is in fact used** to solve the nonlinear subproblem.

The good choice for α and β is given by (2.26) if one uses scheme (2.32)-(2.35). With such a choice, many computer subprograms can be used for both the linear and nonlinear subproblems, resulting therefore in quite substantial core memory savings.

3. Solution methods for the nonlinear subproblems.

3.1. Classical and variational formulations. Synopsis

At each full step of the splitting methods (2.29)-(2.31) and (2.32)-(2.35) we have to solve a nonlinear elliptic system of the following type

$$(3.1) \qquad \begin{cases} \alpha \underset{\sim}{u} - \nu \Delta \underset{\sim}{u} + (\underset{\sim}{u}.\underset{\sim}{\nabla})\underset{\sim}{u} = \underset{\sim}{f} \textit{ in } \Omega \\ \\ \underset{\sim}{u} = \underset{\sim}{g} \textit{ on } \Gamma, \end{cases}$$

where α and ν are two *positive* parameters and where $\underset{\sim}{f}$ and $\underset{\sim}{g}$ are two given functions, defined on Ω and Γ, respectively. We do not discuss here the existence and uniqueness of solutions for problem (3.1). We introduce now the following functional spaces of *Sobolev's* type (see, e.g. [20]-[23] for information on Sobolev spaces) :

$$(3.2) \qquad H^1(\Omega) = \{\phi | \phi \in L^2(\Omega), \frac{\partial \phi}{\partial x_i} \in L^2(\Omega), \forall i=1,\dots,N\},$$

$$(3.3) \qquad H_0^1(\Omega) = \{\phi | \phi \in H^1(\Omega), \phi = 0 \textit{ on } \Gamma\},$$

$$(3.4) \qquad V_0 = (H_0^1(\Omega))^N, \; V_g = \{\underset{\sim}{v} | \underset{\sim}{v} \in (H^1(\Omega))^N, \underset{\sim}{v} = \underset{\sim}{g} \textit{ on } \Gamma\}.$$

If $\underset{\sim}{g}$ is sufficiently smooth, then V_g is *nonempty*.

We shall use the following notation

$$dx = dx_1 \dots dx_N,$$

and if $\underset{\sim}{u} = \{u_i\}_{i=1}^N$, $\underset{\sim}{v} = \{v_i\}_{i=1}^N$, then

$$\underset{\sim}{u}.\underset{\sim}{v} = \sum_{i=1}^{N} u_i v_i, \quad \underset{\sim}{\nabla}\,\underset{\sim}{u}.\underset{\sim}{\nabla}\,\underset{\sim}{v} = \sum_{i=1}^{N}\sum_{j=1}^{N} \frac{\partial u_i}{\partial x_j}\frac{\partial v_i}{\partial x_j}.$$

Using *Green's formula* we can prove that for sufficiently smooth functions $\underset{\sim}{u}$ and $\underset{\sim}{v}$ belonging to $(H^1(\Omega))^N$ and V_0, respectively, we have

$$(3.5) \qquad -\int_\Omega \Delta\underset{\sim}{u}.\underset{\sim}{v}\,dx = \int_\Omega \underset{\sim}{\nabla}\underset{\sim}{u}.\underset{\sim}{\nabla}\underset{\sim}{v}\,dx .$$

It can also be proved that $\underset{\sim}{u}$ is a solution of the *nonlinear variational problem*

$$(3.6) \qquad \begin{cases} \underset{\sim}{u} \in V_g , \\ \alpha \displaystyle\int_\Omega \underset{\sim}{u}.\underset{\sim}{v}\,dx + \nu \int_\Omega \underset{\sim}{\nabla}\underset{\sim}{u}.\underset{\sim}{\nabla}\underset{\sim}{v}\,dx + \int_\Omega ((\underset{\sim}{u}.\underset{\sim}{\nabla})\underset{\sim}{u}).\underset{\sim}{v}\,dx = \\ \displaystyle\int_\Omega \underset{\sim}{f}.\underset{\sim}{v}\,dx, \ \forall \underset{\sim}{v} \in V_0 , \end{cases}$$

and *conversely*. We observe that (3.1), (3.6) *is not equivalent* to a problem of the *calculus of variations* since there is no functional of $\underset{\sim}{v}$ with $(\underset{\sim}{v}.\underset{\sim}{\nabla})\underset{\sim}{v}$ as differential. Using, however, a convenient *least-squares formulation* we shall be able to solve (3.1), (3.6) by iterative methods originating from *nonlinear programming*, such as *conjugate gradient*, for example.

3.2. Least squares formulation of (3.1), (3.6)

Let $\underset{\sim}{v} \in V_g$; from $\underset{\sim}{v}$ we define $\underset{\sim}{y}$ $(= \underset{\sim}{y}(\underset{\sim}{v})) \in V_0$ as the solution of

$$(3.7) \qquad \begin{cases} \alpha\underset{\sim}{y} - \nu\Delta\underset{\sim}{y} = \alpha\underset{\sim}{v} - \nu\Delta\underset{\sim}{v} + (\underset{\sim}{v}.\underset{\sim}{\nabla})\underset{\sim}{v} - \underset{\sim}{f} \ \textit{in} \ \Omega \\ \underset{\sim}{y} = \underset{\sim}{0} \ \textit{on} \ \Gamma \end{cases}$$

We observe that $\underset{\sim}{y}$ is obtained from $\underset{\sim}{v}$ via the solution of N uncoupled linear Poisson problems (one for each component of $\underset{\sim}{y}$). Using (3.5) it can be shown that problem (3.7) is actually *equivalent* to the *linear variational problem*

$$(3.8)\quad \begin{cases} \textit{Find } \underset{\sim}{y} \in V_0, \textit{ such that } \forall \underset{\sim}{z} \in V_0, \textit{ we have} \\ \alpha \int_\Omega \underset{\sim}{y}.\underset{\sim}{z}\, dx + \nu \int_\Omega \nabla \underset{\sim}{y}.\nabla \underset{\sim}{z}\, dx = \alpha \int_\Omega \underset{\sim}{v}.\underset{\sim}{z} dx + \nu \int_\Omega \nabla \underset{\sim}{v}.\nabla \underset{\sim}{z}\, dx \\ + \int_\Omega ((\underset{\sim}{v}.\nabla)\underset{\sim}{v}).\underset{\sim}{z}\, dx - \int_\Omega \underset{\sim}{f}.\underset{\sim}{z}\, dx, \end{cases}$$

which has a unique solution. Suppose now that v is a solution of the nonlinear problem (3.1), (3.6). The corresponding $\underset{\sim}{y}$ (obtained from the solution of (3.7), (3.8)) is clearly $\underset{\sim}{y} = \underset{\sim}{0}$. From this observation, it is quite natural to introduce the following *(nonlinear) least squares formulation* of (3.1), (3.6):

$$(3.9)\quad \begin{cases} \textit{Find } \underset{\sim}{u} \in V_g \textit{ such that} \\ J(\underset{\sim}{u}) \le J(\underset{\sim}{v}),\ \forall \underset{\sim}{v} \in V_g, \end{cases}$$

where $J : (H^1(\Omega))^N \to \mathbb{R}$ is the function of $\underset{\sim}{v}$ defined by

$$(3.10)\qquad J(\underset{\sim}{v}) = \frac{1}{2}\int_\Omega \{\alpha|\underset{\sim}{y}|^2 + \nu|\nabla \underset{\sim}{y}|^2\}dx\ ,$$

where $\underset{\sim}{y}$ is defined from $\underset{\sim}{v}$ by (3.7), (3.8). We observe that if $\underset{\sim}{u}$ is solution of (3.1), (3.6), then it is also a solution of (3.9) such that $J(\underset{\sim}{u}) = 0$. Conversely, any solution of (3.9) such that $J(\underset{\sim}{u}) = 0$ is also a solution of (3.1), (3.6).

3.3. Iterative solution of the least squares problem (3.9)

The iterative solution of the least square problem (3.9) by a *preconditioned conjugate gradient* method is described in [24] (see also [1] and [25]). Without going into the details of the conjugate gradient algorithm discussed there we can say that each iteration requires the solution of *three linear elliptic systems* associated to operator $\alpha I-\nu\Delta$. Three is an *optimal* number in this case, since the solution of a linear problem by a least squares-preconditioned conjugate gradient algorithm would require the solution of two linear systems associated to the preconditioning operator (see [24] for more details.) If a *Cholesky factorization* is used for solving the linear systems approximating the above linear elliptic problems, we may improve by a factor of five to ten the CPU time on **CRAY** type machine by simply carefully coding the *inner products* of two vectors, the product of a vector by a matrix, the solution of the triangular systems associated to the Cholesky factorization, etc. (see [25] for more details).

Actually we can improve the convergence of *conjugate gradient* methods by using those methods discussed in e.g. [26],[27], and which combine their main features with those of *quasi-Newton* methods (see [28] for a general discussion of quasi-Newton methods). The basic idea behind these new methods is simple.

Consider the *minimization* problem

$$(M) \quad \begin{cases} \text{Find } \underset{\sim}{x} \in \mathbb{R}^N \text{ such that} \\ F(\underset{\sim}{x}) \le F(\underset{\sim}{y}),\ \forall \underset{\sim}{y} \in \mathbb{R}^N, \end{cases}$$

where $F : \mathbb{R}^N \to \mathbb{R}$; $\underset{\sim}{x}$ is clearly a solution of $F'(\underset{\sim}{x}) = \underset{\sim}{0}$ where F' denotes the differential of F. The algorithm constructs a *minimizing sequence* $\{\underset{\sim}{x}^n\}_{n \ge 0}$ as follows :

Starting from $\underset{\sim}{x}^n$ and from the identity matrix I we use few iterations (say p) of a *quasi-Newton method (BFGS,* for example) to build

(i) Further iterates $\underset{\sim}{x}^{n+j/2p}$ of the minimizing sequence,

(ii) A sparse, *local* approximation of the Hessian matrix $F''(\underset{\sim}{x})$, say $\underset{\sim}{S}_n$.

Starting from $\underset{\sim}{x}^{n+1/2}$ we use then a conjugate gradient method preconditioned by $\underset{\sim}{S}_n$ to decrease F, until $\underset{\sim}{S}_n$ is no more active as a preconditioner ; we go back then to the quasi-Newton iterations, and so on.

The algorithms that we have sketched above are in our opinion related to the so-called *algebraic multigrid* methods, discussed in the paper of A. Brandt, S. McCormick, and J. Ruge [29].

3.4 Subcycling treatment of the nonlinear subproblems

Instead of using an iterative method like those mentioned in Sec. 3.3 we can proceed with several steps if an *explicit method for initial value problems* (this supposes that a space approximation has been used in order to reduce the original problem to a finite dimensional one). This approach has been advocated by several authors; among them we shall mention Gresho [30] (see also the references therein). This idea applied to the methods discussed here leads to the approach now discussed. Consider the splitting scheme (2.29)-(2.31) (the same conclusion would hold for scheme (2.32)-(2.35).) We can associate to (2.31) the following initial value problem on the time interval $(0, \Delta t/2)$:

$$(3.11) \qquad \underset{\sim}{w}(0) = \underset{\sim}{u}^{n+1/2},$$

$$(3.12) \qquad \begin{cases} \dfrac{\partial \underset{\sim}{w}}{\partial \tau} - \dfrac{\nu}{2} \Delta \underset{\sim}{w} + (\underset{\sim}{w}.\underset{\sim}{\nabla})\underset{\sim}{w} = \underset{\sim}{f} + \dfrac{\nu}{2} \Delta \underset{\sim}{u}^{n+1/2} - \underset{\sim}{\nabla} p^{n+1/2}, \\ \underset{\sim}{w}(\tau) = \underset{\sim}{g}((n+1/2)\Delta t + \tau) \text{ on } \Gamma, \end{cases}$$

and once $\underset{\sim}{w}$ is known, we set

$$(3.13) \qquad \underset{\sim}{u}^{n+1} = \underset{\sim}{w}(\Delta t/2).$$

In practice we should compute $\underset{\sim}{w}$ using *explicit Runge-Kutta* methods, with good stability properties, like those discussed in, e.g., [31] and [32] (see also [30]). The numerical results obtained using this approach seem to be very promising. Actually we have obtained very good results using it for the *incompressible and compressible Navier-Stokes equations* and also for the *compressible Euler equations*.

4. Solution of the "quasi" Stokes linear subproblems

At each full step of the splitting methods discussed in Sec. 2.3 we have to solve one or two linear problems of the following type :

$$(4.1)\quad \begin{cases} \alpha \underset{\sim}{u} - \nu\Delta\underset{\sim}{u} + \underset{\sim}{\nabla}p = \underset{\sim}{f} \quad in\ \Omega, \\ \underset{\sim}{\nabla}.\underset{\sim}{u} = 0\ in\ \Omega, \\ \underset{\sim}{u} = \underset{\sim}{g}\ on\ \Gamma\ \left(with \int_\Gamma \underset{\sim}{g}.\underset{\sim}{n}\, d\Gamma = 0\right), \end{cases}$$

where α and ν are two *positive* parameters, and where $\underset{\sim}{f}$ and $\underset{\sim}{g}$ are two given functions, defined on Ω and Γ, respectively. We recall that if $\underset{\sim}{f}$ and $\underset{\sim}{g}$ are sufficiently smooth, then (4.1) has a unique solution in $V_g \times (L^2(\Omega)/\mathbb{R})$ (with V_g still defined by (3.4) ; $p \in L^2(\Omega)/\mathbb{R}$ means that p is defined only to within an arbitrary constant). We shall find in [1], [19], [24] several iterative methods for solving (4.1), quite easy to implement using finite element methods. In this paper we shall consider a method which appears quite efficient for solving time dependent Navier-Stokes problems at *high Reynold's numbers* (i.e. for *small* values of ν). The starting remark is that if ν is small and if we consider a *wall* we have, due to the *friction*,

$$(4.2)\quad \underset{\sim}{u} = \underset{\sim}{0}$$

which, combined to (1.1), implies (at least formally)

$$(4.3)\quad \nu\Delta\underset{\sim}{u} = \underset{\sim}{\nabla}p$$

which implies in turn

$$(4.4)\quad \nu(\Delta\underset{\sim}{u}).\underset{\sim}{n} = \frac{\partial p}{\partial n}\ .$$

If $\nu(\Delta\underset{\sim}{u}).\underset{\sim}{n}$ converges to 0 (in some sense) as $\nu \to 0$, then we can expect $\frac{\partial p}{\partial n}$ to be "small" at a wall at high Reynolds numbers. Actually this conjecture is supported by numerical experiments (see for example the results shown in Sec. 6). From that property we can use the following algorithm (of *conjugate gradient* type) to solve (4.1) :

$$(4.4)\quad p^o \in L^2(\Omega),\ given,$$

$$(4.5)\quad \begin{cases} \alpha\underset{\sim}{u}^o - \nu\Delta\underset{\sim}{u}^o = \underset{\sim}{f} - \underset{\sim}{\nabla}p^o, \\ \underset{\sim}{u}^o = \underset{\sim}{g}\ on\ \Gamma, \end{cases}$$

$$(4.6)\qquad \begin{cases} \text{Find } z^o \in H^1(\Omega) \text{ such that} \\ \int_\Omega \nabla z^o . \nabla q \, dx = \int_\Omega \nabla . \underset{\sim}{u}^o q \, dx, \ \forall q \in H^1(\Omega), \end{cases}$$

$$(4.7)\qquad w^o = z^o .$$

Then for $n \geq 0$, p^n, $\underset{\sim}{u}^n$, $\underset{\sim}{z}^n$, w^n *being known, compute* p^{n+1}, $\underset{\sim}{u}^{n+1}$, z^{n+1}, w^{n+1} *as follows :*

Solve

$$(4.8)\qquad \begin{cases} \alpha \underset{\sim}{\chi}^n - \nu \Delta \underset{\sim}{\chi}^n = -\underset{\sim}{\nabla} w^n, \\ \underset{\sim}{\chi}^n = \underset{\sim}{0} \ \text{ on } \Gamma, \end{cases}$$

$$(4.9)\qquad \rho_n = \frac{\int_\Omega |\underset{\sim}{\nabla} z^n|^2 dx}{\int_\Omega \underset{\sim}{\nabla} . \underset{\sim}{\chi}^n w^n dx},$$

$$(4.10)\qquad p^{n+1} = p^n - \rho_n w^n,$$

$$(4.11)\qquad \underset{\sim}{u}^{n+1} = \underset{\sim}{u}^n - \rho_n \underset{\sim}{\chi}^n ,$$

$$(4.12)\qquad \begin{cases} \int_\Omega \underset{\sim}{\nabla} z^{n+1} . \underset{\sim}{\nabla} q dx = \int_\Omega \underset{\sim}{\nabla} z^n . \underset{\sim}{\nabla} q dx - \rho_n \int_\Omega \underset{\sim}{\nabla} . \underset{\sim}{\chi}^n q dx, \forall q \in H^1(\Omega), \\ z^{n+1} \in H^1(\Omega), \end{cases}$$

$$(4.13)\qquad \gamma_n = \frac{\int_\Omega |\underset{\sim}{\nabla} z^{n+1}|^2 dx}{\int_\Omega |\underset{\sim}{\nabla} z^n|^2 dx},$$

$$(4.14)\qquad w^{n+1} = z^{n+1} + \gamma_n w^n .$$

Do $n = n+1$ *and go to* (4.8).

The Poisson problems encountered at each iteration can be solved by various methods, such as *multigrid algorithms*, once a convenient approximation has been used to reduce (4.1) to a *finite dimensional* problem. We would like to insist however on the fact that the coding of multigrid methods for non regular grids is complicated, particularly in view of taking advantage of the very special architecture of vector and/or parallel computers.

Pure gradient variants of algorithm (4.4)-(4.14) are discussed in [33] and [34] and have been advocated by several authors. To our knowledge it is the first time that the conjugate method described above appears in the literature. Calculations done at INRIA by F. Hecht, for unsteady flows around a car, seem to show that algorithm (4.4)-(4.14) (in fact *finite element* variants of it) is quite efficient.

5. Finite Element Approximations of the Unsteady Navier-Stokes Equations

We shall describe in this section a specific class of *finite element approximations* for the time dependent Navier-Stokes equations. Actually, these methods which lead to *continuous approximations* for both pressure and velocity are fairly simple, and some of them have been known for years. They have been advocated for example by Hood and Taylor (see [35]). Other finite element approximations of the incompressible Navier-Stokes equations can be found in [1], [5], [7], [8], [9], [36], [37] (see also the references therein).

5.1. Basic hypotheses. Fundamental discrete spaces

We suppose that Ω is a *bounded polygonal domain* of $\mathbb{R}^2$. With $\mathcal{T}_h$ a standard finite element triangulation of Ω, and h the maximal length of the edges of the triangle of $\mathcal{T}_h$, we introduce the following discrete spaces (with P_k = space of the polynomials in two variables of degree $\le k$) :

$$H_h^1 = \{q_h | q_h \in C^0(\bar{\Omega}), q_h|_T \in P_1, \forall T \in \mathcal{T}_h\} , \tag{5.1}$$

$$V_h = \{\underset{\sim}{v}_h | \underset{\sim}{v}_h \in C^0(\bar{\Omega}) \times C^0(\bar{\Omega}), \underset{\sim}{v}_h|_T\ P_2 \times P_2, \forall T \in \mathcal{T}_h\}, \tag{5.2}$$

$$V_{oh} = \{\underset{\sim}{v}_h | \underset{\sim}{v}_h \in V_h, \underset{\sim}{v}_h = \underset{\sim}{0} \text{ on } \Gamma\} = V_h \cap V_o . \tag{5.3}$$

Two useful variants of V_h (and V_{oh}) are obtained as follows :

either

$$V_h = \{\underset{\sim}{v}_h \mid \underset{\sim}{v}_h \; C^0(\bar{\Omega})\times C^0(\bar{\Omega}), \underset{\sim}{v}_h|_T \in P_1\times P_1, \; \forall T \in \tilde{\mathcal{T}}_h\}, \tag{5.4}$$

or (this space has been introduced in [38])

$$V_h = \{\underset{\sim}{v}_h \mid \underset{\sim}{v}_h \in C^0(\bar{\Omega})\times C^0(\bar{\Omega}), \underset{\sim}{v}_h|_T \in P^*_{1T}\times P^*_{1T}, \forall T \in \mathcal{T}_h\}. \tag{5.5}$$

In (5.4), $\tilde{\mathcal{T}}_h$ is that triangulation of Ω obtained from $\mathcal{T}_h$ by joining the midpoints of the edges of $T \in \mathcal{T}_h$, as shown on Fig. 5.1. We have the same global number of unknowns if we use V_h defined by either (5.2) or (5.4), however the matrices encountered in the second case are more compact and sparse.

In (5.5), P^*_{1T} is the subspace of P_3 defined as follows:

$$P^*_{1T} = \{q \mid q = q_1 + \lambda\phi_T \; \textit{with} \; q_1 \in P_1, \; \lambda \in \mathbb{R}, \; \textit{and} \; \phi_T \in P_3, \; \phi_T = 0 \; \textit{on} \; \partial T, \; \phi_T(G_T) = 1\}, \tag{5.6}$$

where, in (5.6) G_T is the centroid of T (see Fig. 5.2).

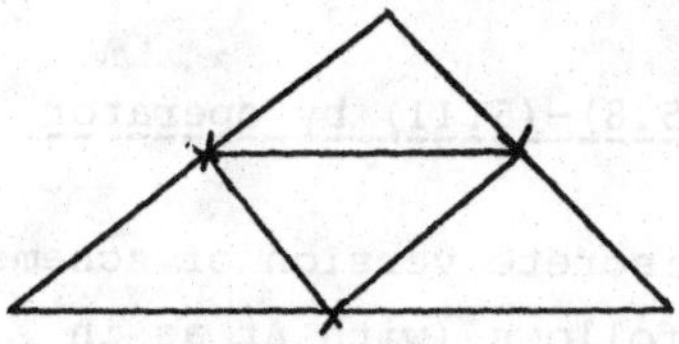

Figure 5.1.

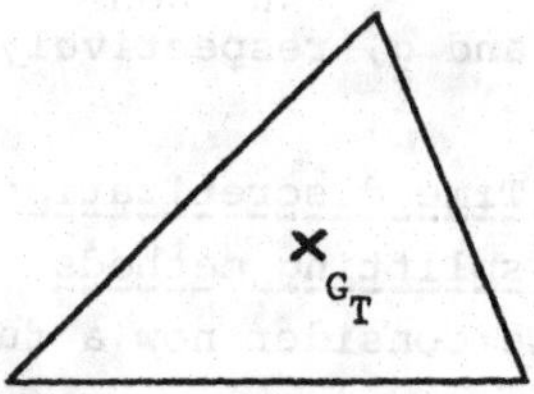

G_T

Figure 5.2.

A function like ϕ_T is usually called a *bubble-function*.

5.2. Approximation of the boundary conditions

If the boundary conditions are defined by

$$\underset{\sim}{u} = \underset{\sim}{g} \; \textit{on} \; \Gamma \qquad \textit{with} \int_\Gamma \underset{\sim}{g}.\underset{\sim}{n} \, d\Gamma = 0 \; , \tag{5.7}$$

it is of fundamental importance to approximate $\underset{\sim}{g}$ by $\underset{\sim}{g}_h$ such that $\int_\Gamma g_h \cdot n \, d\Gamma = 0$. (see [1], Appendix 3).

5.3. Space approximation of the time dependent Navier-Stokes equations

Using the spaces H_h^1, V_h and V_{oh}, we approximate the time dependent Navier-Stokes equations as follows :

Find $\{u_h(t), p_h(t)\} \in V_h \times H_h^1$, $\forall t \geq 0$, *such that*

$$(5.8) \quad \begin{cases} \int_\Omega \frac{\partial u_h}{\partial t} \cdot v_h dx + \nu \int_\Omega \nabla u_h \cdot \nabla v_h dx + \int_\Omega (u_h \cdot \nabla) u_h \cdot v_h \, dx \\ + \int_\Omega \nabla p_h \cdot v_h dx = \int_\Omega f_h \cdot v_h dx, \ \forall v_h \in V_{oh}, \end{cases}$$

$$(5.9) \quad \int_\Omega \nabla \cdot u_h q_h dx = 0, \ \forall q_h \ \ H_h^1 ,$$

$$(5.10) \quad u_h = g_h \ on \ \Gamma,$$

$$(5.11) \quad u_h(x,0) = u_{oh}(x) \ (with \ u_{oh} \in V_h) \ ;$$

in (5.8)-(5.11), f_h, u_{oh} and g_h are convenient approximations of f, u_o and g, respectively.

5.4. Time discretization of (5.8)-(5.11) by operator splitting methods

We consider now a fully discrete version of scheme (2.29)-(2.31). It is defined as follows (with Δt as in Sec. 2.3) :

$$(5.12) \quad u_h^o = u_{oh},$$

then for $n \geq 0$, *compute (from* u_h^n*)* $\{u_h^{n+1/2}, p_h^{n+1/2}\} \in V_h \times H_h^1$, *and then* $u_h^{n+1} \in V_h$, *by solving*

$$(5.13)_1 \quad \begin{cases} \int_\Omega \frac{u_h^{n+1/2}-u_h^n}{\Delta t/2}.v_h dx + \frac{\nu}{2}\int_\Omega \nabla u_h^{n+1/2}.\nabla v_h dx \\ + \int_\Omega \nabla p_h^{n+1/2}.v_h\, dx = \\ = \int_\Omega f_h^{n+1/2}.v_h dx - \frac{\nu}{2}\int_\Omega \nabla u_h^n.\nabla v_h dx - \\ - \int_\Omega (u_h^n.\nabla)u_h^n.v_h dx, \ \forall v_h \in V_{oh}, \end{cases}$$

$$(5.13)_2 \quad \int_\Omega \nabla.u_h^{n+1/2} q_h dx = 0, \ \forall q_h \in H_h^1,$$

$$(5.13)_3 \quad u_h^{n+1/2} \in V_h,\ p_h^{n+1/2} \in H_h^1,\ u_h^{n+1/2} = g_h^{n+1/2} \ \mathit{on}\ \Gamma$$

and then

$$(5.14)_1 \quad \begin{cases} \int_\Omega \frac{u_h^{n+1}-u_h^{n+1/2}}{\Delta t/2}.v_h dx + \frac{\nu}{2}\int_\Omega \nabla u_h^{n+1}.\nabla v_h dx + \\ \int_\Omega (u_h^{n+1}.\nabla)u_h^{n+1}.v_h\, dx = \\ \int_\Omega f_h^{n+1}.v_h dx - \frac{\nu}{2}\int_\Omega \nabla u_h^{n+1/2}.\nabla v_h dx - \\ \int_\Omega \nabla p_h^{n+1/2}.v_h dx, \ \forall v_h \in V_{oh}, \end{cases}$$

$$(5.14)_2 \quad u_h^{n+1} \in V_h,\ u_h^{n+1} = g_h^{n+1} \ \mathit{on}\ \Gamma,$$

respectively.

The same technique apply to the space discretization of scheme (2.32)-(2.35) (see [1] - where $\theta = .25$ - for more details).

The solution of the various subproblems encountered at each step of (5.12)-(5.14) can be done by discrete variants of the methods discussed in Secs. 3 and 4. See [1] (and also [39]) for more details about this part of the solution process.

6. Numerical experiments (simulation of incompressible viscous flows)

We illustrate the numerical methods of the above sections by the presentation of the results of numerical experiments, where these methods have been applied to simulate some incompressible viscous flows of practical interest. All the calculations which follow have been done using the finite element method associated to the discrete spaces defined by (5.1), (5.3), (5.4) (very good results have been obtained recently, using V_h defined by (5.5)).

6.1. A first class of test problems (flow in a channel with a backward facing step)

We consider the solution of the *Navier-Stokes Equations* for the flow of *incompressible viscous fluids*, entering the *channel with a step* of Fig. 6.1 by the *left* side extremity. We have selected this problem since it is a quite classical and significant test problem for Navier-Stokes solvers (see e.g. [36], [37]). The finite element triangulations used for these calculations are shown on Fig. 6.1. The coarse (resp. fine) one is used for approximating the *pressure* (resp. the *velocity*).

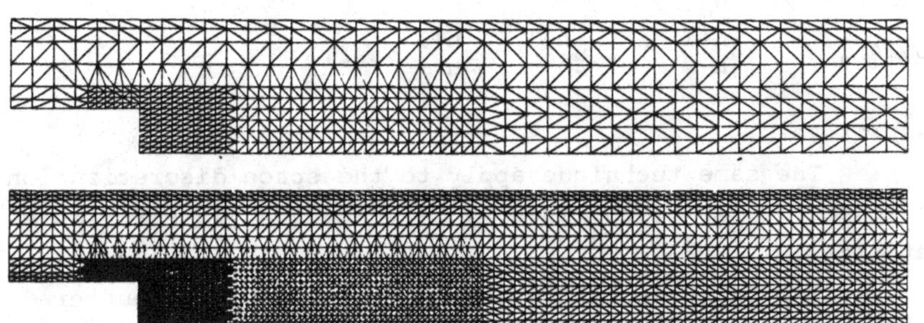

	$\mathcal{C}_h$	$\tilde{\mathcal{C}}_h$
Nodes	619	2346
Triangles	1109	4436
Cholesky's coefficients	21654	154971

Figure 6.1.

We have also indicated on Fig. 6.1, the number of nodes, triangles, and nonzero Cholesky coefficients of the matrix approximating $-\Delta$ (resp. $\alpha I-\nu\Delta$) on $\mathcal{T}_h$ (resp. $\tilde{\mathcal{T}}_h$). We have applied the methods described in the above sections to compute the *steady state* solutions of (1.1), (1.2), for the following boundary conditions

$$(6.1)\qquad \begin{cases} \underset{\sim}{u} \textit{ satisfies Poisseuille velocity distributions} \\ \textit{at the entrance and exit of the channel and is} \\ \underset{\sim}{0} \quad \textit{elsewhere on } \Gamma \end{cases}$$

We took Re = 100 and 191. The numerical results agree well with those in [36], [37] and show a clear superiority of the schemes derived from (2.32)-(2.35) over those derived from (2.29)-(2.31). Using the former we need less steps to obtain the steady states. For scheme (2.32)-(2.35), we have tested $\theta = .25$ and $\theta = 1-\sqrt{2}/2$ (for the same Δt). The convergence to the steady state is faster with the second value of θ (for this class of problems at least). We have shown on Fig. 6.2 (resp. 6.3) the *stream* lines (the *isobar* lines) of the steady state solutions corresponding to Re = 100 and Re = 191. We observe that the size of the *recirculation region* increases with Re, and also (from Fig. 6.3) that $\frac{\partial p}{\partial n}$ is "small" at the wall of the channel, justifying "a posteriori" the assumption made in Sec. 4 and therefore the possible application of discrete variants of algorithm (4.4)-(4.14), to solve the quasi-Stokes problem (4.1).

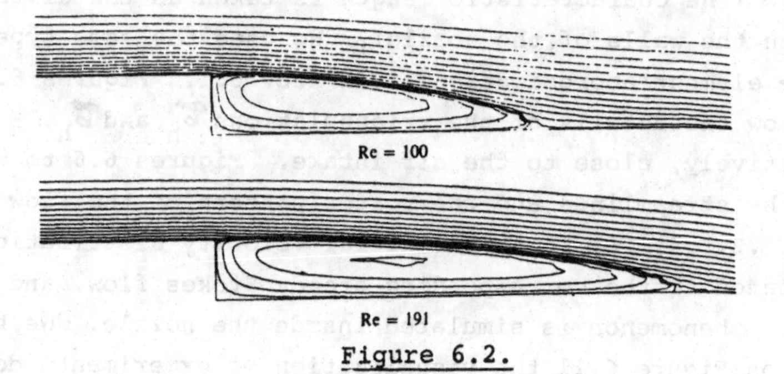

Figure 6.2.
Stream lines

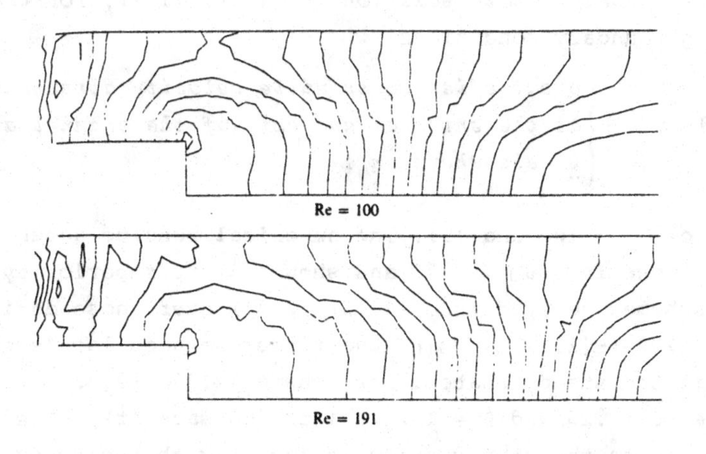

Figure 6.3. (Isobar lines)

6.2 A second class of test problems (flow in a nozzle at high incidence

The second test problem that we consider is much more complicated than the first one. It concerns the simulation of an incompressible viscous flow inside and around a (two-dimensional) nozzle at *high incidence* (40 degrees) and at $Re = 750$ (the characteristic length is taken as the distance between the walls of the nozzle). We used the same type of finite element approximation as in Sec. 6.1. Figures 6.4, 6.5 show the details of the triangulations $\mathcal{T}_h$ and $\mathcal{T}_h$, respectively, close to the air intake. Figures 6.6 to 6.10 show the streamlines and the vortex pattern of the flow at $t = .0, .2, .4, .6, .8$. The initial velocity distribution is associated to the corresponding steady Stokes flow, and a suction phenomenon is simulated inside the nozzle. We have shown on Figure 6.11 the visualization of experiments done at ONERA for the same Re and angle of attack. We observe the good agreement between the computed and experimental results.

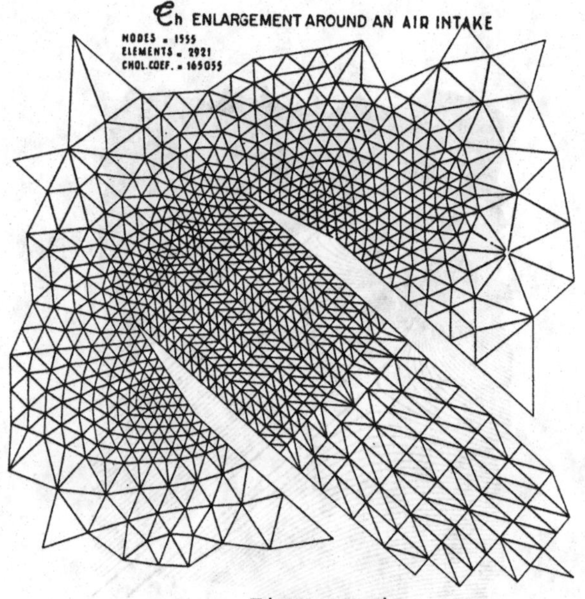

<u>Figure 6.4.</u>
(Pressure grid)

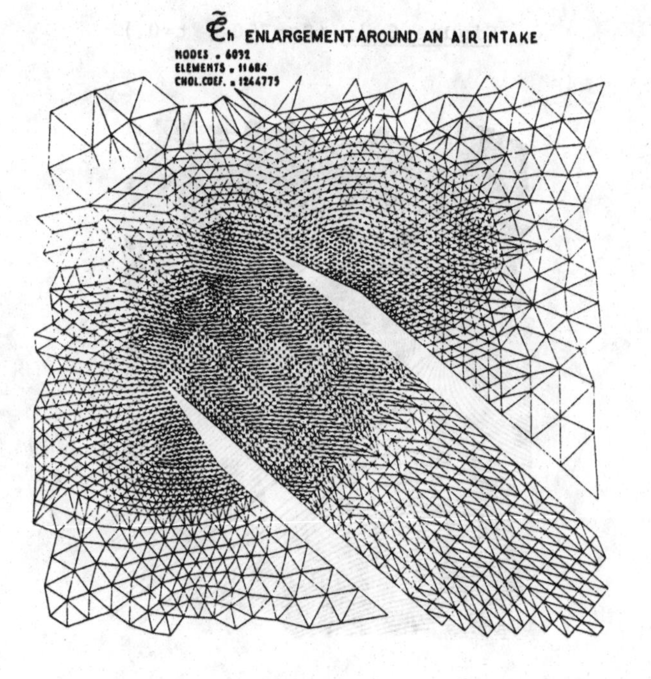

<u>Figure 6.5.</u>
(Velocity grid)

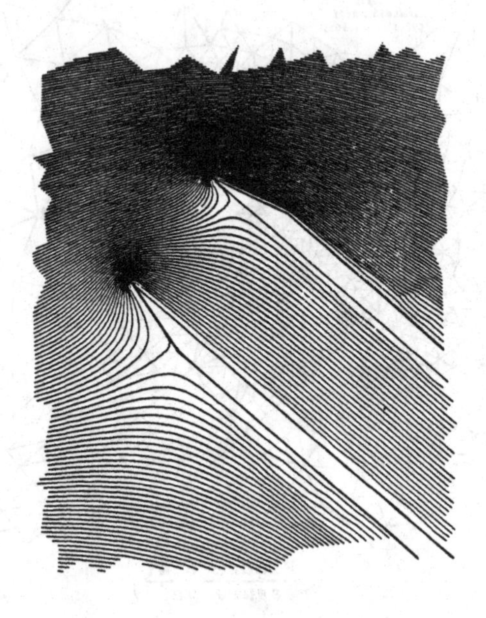

Figure 6.7. (Re=750 ; t=0.)

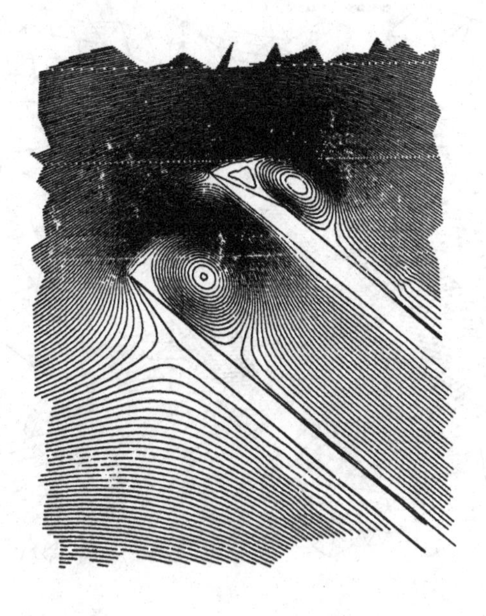

Figure 6.8. (Re=750 ; t=.2)

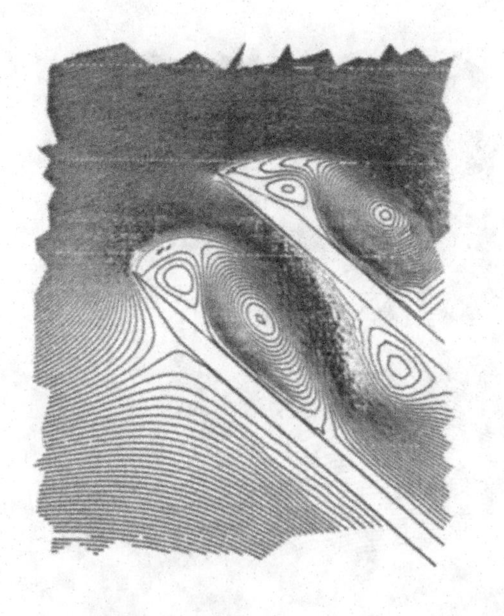

<u>Figure 6.9.</u> (Re=750 ; t=.4)

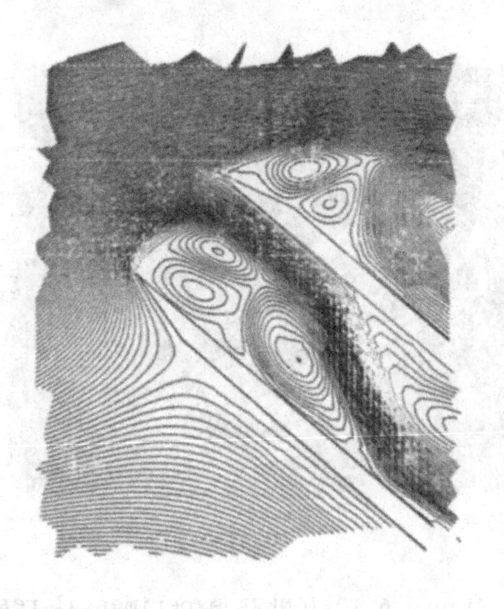

<u>Figure 6.10.</u> (Re=750 ; t=.6)

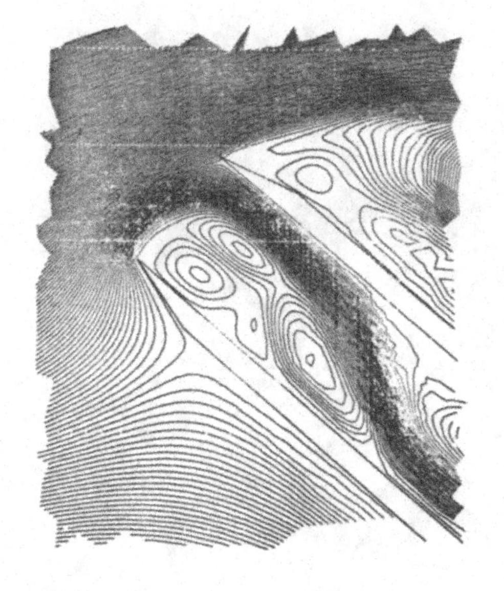

<u>Figure 6.11.</u> (Re=750 ; t=.8)

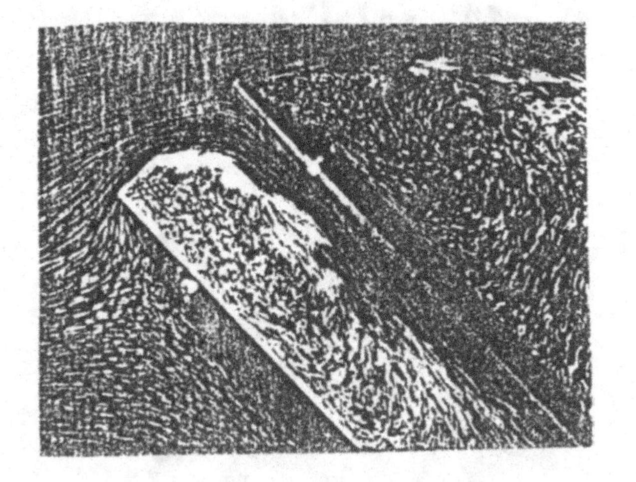

<u>Figure 6.12.</u>(ONERA experimental results : courtesy of DRET and H. Werle).

7. Formulation of the Navier-Stokes equations for compressible viscous fluids. Some comments about their numerical solution

Let consider the flow of a *compressible* viscous fluid. As in Sec. 1, Ω is the flow domain, Γ is its boundary and $x = \{x_i\}_{i=1}^{N}$ is the generic point of $\mathbb{R}^N$. If we suppose that the fluid satisfies the *perfect gas law*

$$p = (\gamma-1)\rho T, \tag{7.1}$$

the flow is governed by the following set of (Navier-Stokes) equations, written in *non-conservative form* :

Continuity equation :

$$\frac{\partial \rho}{\partial t} + \underset{\sim}{\nabla} \cdot \rho \underset{\sim}{u} = 0, \tag{7.2}$$

Momentum equation :

$$\rho \frac{\partial \underset{\sim}{u}}{\partial t} + \rho(\underset{\sim}{u}.\underset{\sim}{\nabla})\underset{\sim}{u} + (\gamma-1)\underset{\sim}{\nabla}\rho T = \frac{1}{Re}[\Delta \underset{\sim}{u} + \frac{1}{3}\underset{\sim}{\nabla}(\underset{\sim}{\nabla}.\underset{\sim}{u})] , \tag{7.3}$$

Energy equation (in two-dimension) :

$$\left\{\begin{aligned} &\rho \frac{\partial T}{\partial t} + \rho\underset{\sim}{u}.\underset{\sim}{\nabla}T + (\gamma-1)\rho T\underset{\sim}{\nabla}.\underset{\sim}{u} = \frac{1}{Re}\{\frac{\gamma}{Pr}\Delta T + \frac{4}{3}[(\frac{\partial u_1}{\partial x_1})^2 + \\ &+ (\frac{\partial u_2}{\partial x_2})^2 - \frac{\partial u_1}{\partial x_1}\frac{\partial u_2}{\partial x_2}] + (\frac{\partial u_1}{\partial x_2} + \frac{\partial u_2}{\partial x_1})^2\}. \end{aligned}\right. \tag{7.4}$$

In (7.1)-(7.4), ρ is the *density*, p is the *pressure*, and T is the *temperature* of the fluid. The constants Re, Pr and γ are the *Reynolds number*, the *Prandtl-number* and the *ratio of specific heats*, respectively ($\gamma = 1.4$ in *air*).

A classical simplification is provided by the *constant total enthalpy* assumption

$$\gamma T + \frac{|\underset{\sim}{u}|^2}{2} = H_o , \tag{7.5}$$

where H_o is defined from the boundary conditions.

For a discussion of the boundary conditions to be associated with the compressible Navier-Stokes equations see [24] for more details. We refer also to [24] for references concerning the numerical solution by various methods such as *finite elements, finite differences and spectral methods.* Recent solution methods by *finite differences* are discussed in these proceedings (see the papers by MacCormack [40], Steger-Buning [41], Turkel [42]. Let us mention also that *theoretical results* are proved in Masmura-Nishida [43], assuming that the data are small.

8. Transformation of the equations. Further comments

In order to apply the methods used for *incompressible* fluids (or for the *shallow-water equations* ; see [44], [45]) it is convenient to introduce a new function σ

$$\sigma = \ln \rho \,. \tag{8.1}$$

With this new variable, the governing equations become

$$\frac{\partial \sigma}{\partial t} + \underset{\sim}{\nabla} . \underset{\sim}{u} + \underset{\sim}{u} . \underset{\sim}{\nabla} \sigma = 0, \tag{8.2}$$

$$\frac{\partial \underset{\sim}{u}}{\partial t} + (\underset{\sim}{u} . \underset{\sim}{\nabla}) \underset{\sim}{u} + (\gamma - 1)(T \underset{\sim}{\nabla} \sigma + \underset{\sim}{\nabla} T) = \frac{e^{-\sigma}}{Re} (\Delta \underset{\sim}{u} + \frac{1}{3} \underset{\sim}{\nabla} (\underset{\sim}{\nabla} . \underset{\sim}{u})), \tag{8.3}$$

$$\frac{\partial T}{\partial t} + \underset{\sim}{u} . \underset{\sim}{\nabla} T + (\gamma - 1) T \underset{\sim}{\nabla} . \underset{\sim}{u} = \frac{e^{-\sigma}}{Re} (\frac{\gamma}{Pr} \Delta T + F(\underset{\sim}{\nabla} \underset{\sim}{u})), \tag{8.4}$$

with

$$F(\underset{\sim}{\nabla} \underset{\sim}{u}) = \frac{4}{3} \{ (\frac{\partial u_1}{\partial x_1})^2 + (\frac{\partial u_2}{\partial x_2})^2 - \frac{\partial u_1}{\partial x_1} \frac{\partial u_2}{\partial x_2} \} + (\frac{\partial u_1}{\partial x_2} + \frac{\partial u_2}{\partial x_1})^2 \,. \tag{8.5}$$

We refer to [24] for the *time* and *space discretization* of (8.2)-(8.4) with associated boundary conditions (see also [46] - [48] for further details). Numerical results are also presented in [24], [46]-[48], including (good) comparisons with laboratory experiments.

As a brief comment, we would like to say that the methods obtained using the new set of equations (8.2)-(8.4) seem to be more robust than those obtained from (7.2)-(7.4).

9. Navier-Stokes equations and supercomputers

If we wished to solve three-dimensional problems modeled by the Navier-Stokes equations at Reynolds numbers required by industrial applications, we definitely need computers much more efficient than the ones operational at the moment. In our opinion, we have not yet reached to time where wind tunnels will be replaced by computers. However, it is very likely that by the end of the century we shall be able to solve some of the complicated problems that we cannot even dream to solve *at the moment* (it is clear that human ingenuity will find applied problems too complicated for the supercomputers of year 2000).

Back to the present situation, it is clear that the finite element methods, if applied to problems with complicated geometries, lead to discrete problems lacking the structures which, when combined with the special architectures of supercomputers, result in important factors of performances compared to sequential machines. However, some important gain can be obtained by a careful coding of basic operations such as scalar products, products of matrices, solution of tridiagonal or triangular systems. Another possibility is to use domain decompositions in which on each subdomain a regular mesh can be used with methods such as multigrid, spectral, etc. The remaining problem is of course the coupling between the local solutions. Such domain decomposition methods are discussed in [41] and also in [49], [50]. Their implementation on a computer system consisting of a powerful sequential host machine and two array processors is discussed in [49]. The reader interested in the impact of supercomputers on scientific computing will find in Lichnewsky-Lions [51] a most interesting discussion combining the numerical and computer science aspects of the problem.

Finally, we should comment on the software aspect when it comes to the coding of sophisticated methods for solving complicated problems (particularly in three dimension). In that direction we can expect support from artificial intelligence methods, and we firmly believe that tools such as *expert systems* may have a great impact on the scientific computing of the future.

References

[1] R. Glowinski, *Numerical Methods for Nonlinear Variational Problems*, Springer-Verlag, New York, 1984

[2] J.L. Lions, *Quelques Methodes de Resolution des Problemes aux Limites Non Lineaires*, Dunod, Paris, 1969.

[3] O.A. Ladysenskaya, *The Mathematical Theory of Viscous Incompressible Flows*, Gordon and Breach, New York, 1969.

[4] L. Tartar, *Topics in Nonlinear Analysis*, Publications Mathematiques d'Orsay, Universite Paris-Sud, Departement de Mathematiques, Orsay, 1978.

[5] R. Temam, *Navier-Stokes Equations*, North Holland, Amsterdam, 1977.

[6] R. Temam, *Navier-Stokes Equations and Nonlinear Functional Analysis*, CBMS 41, SIAM, Philadelphia, PA, 1983.

[7] R. Rautmann (ed.), *Approximation Methods for Navier-Stokes Problems*, Lecture Notes in Mathematics, 771, Springer-Verlag, Berlin, 1980.

[8] V. Girault, P.A. Raviart, *Finite Element Approximation of Navier-Stokes Equations*, Lecture notes in Mathematics, 749, Springer-Verlag, Berlin, 1979.

[9] F. Thomasset, *Implementation of Finite Element Methods for Navier-Stokes Equations*, Springer-Verlag, New York, 1981.

[10] R. Peyret, T.D. Taylor, *Computational Methods for Fluid Flow*, Springer-Verlag, New York, 1982.

[11] P.J. Roache, *Computational Fluid Dynamics*, Hermosa, Albuquerque, N.M., 1972.

[12] G. Strang, On the construction and comparison of difference schemes, *SIAM J.Num.Anal.*, 5 (1968), pp 506-517.

[13] J.T. Beale, A. Magda, Rates of convergence for viscous splitting of the Navier-Stokes equations, *Math.Comp.*, 37, (1981), pp 243-260.

[14] R. Leveque, *Time-Split Methods for Partial Differential Equations*, Ph.D. Thesis, Computer Science Dept., Stanford University, Stanford, CA, 1982.

[15] R. Leveque, J. Oliger, *Numerical methods based on additive splitting for hyperbolic partial differential equations*, Manuscript NA-81-16, Numerical Analysis Project, Computer Science Dept., Stanford University, Stanford, CA, 1981.

[16] P.L. Lions, B. Mercier, Splitting algorithms for the sum of two nonlinear operators, *SIAM J.Num.Anal.*, 16, (1979), pp 964-979.

[17] E. Godlewski, *Methodes a pas multiples et de directions alternees pour la discretisation d'equations d'evolution*, These de 3e cycle, Universite P. et M. Curie, Paris, 1980.

[18] G.I. Marchuk, *Methods of Numerical Mathematics*, Springer-Verlag, New York, 1975.

[19] R. Glowinski, B. Mantel, J. Periaux, Numerical solution of the time dependent Navier-Stokes equations for incompressible viscous fluids by finite element and alternating direction methods, in *Numerical Methods in Aeronautical Fluid Dynamics*, P.L. Roe, Ed., Academic Press, London, 1982, pp. 309-336.

[20] J.L. Lions, *Problemes aux limites dans les equations aux derivees partielles*, Presses de l'Universite de Montreal, Montreal, P.Q., Canada, 1962.

[21] J. Necas, *Les Methodes Directes en Theorie des Equations Elliptiques*, Masson, Paris, 1967.

[22] R.A. Adams, *Sobolev Spaces*, Academic Press, New York, 1975.

[23] J.T. Oden, J.N. Reddy, *An introduction to the mathematical theory of finite elements*, Wiley, New York, 1976.

[24] M.O. Bristeau, R. Glowinski, B. Mantel, J. Periaux, P. Perrier, Numerical methods for the simulation of viscous flows modeled by the Navier-Stokes equations, to appear in *Finite Elements in Fluids*, Vol. 6, Wiley, Chichester.

[25] R. Glowinski, H.B. Keller, L. Reinhart, Continuation-conjugate gradient methods for the least square solution of nonlinear boundary value problems, to appear in *Siam J.Scient.Stat.Comp.*

[26] D.F. Shanno, Conjugate gradient method with inexact line search, *Math. of Oper. Research*, 13, (1978), pp. 155-175.

[27] A. Buckley, A. Lenir, QN-like variable storage conjugate gradients, *Math. Programming*, 27, (1983), pp. 155-175.

[28] J.E. Dennis, R.B. Schnabel, *Quasi-Newton Methods for Nonlinear Problems*, Prentice Hall, Englewood Cliffs, NJ, 1983.

[29] A. Brandt, S. McCormick and J. Ruge, Algebraic multigrid (AMG) for sparse matrix equations. *Sparsity and Its Applications* (D.J. Evans, ed.), Cambridge University Press, 1984.

[30] P.M. Gresho, S.T. Chan, R.L. Lee, C.D. Upson, A modified finite element method for solving the time dependent incompressible Navier-Stokes equations. Part 1: Theory, *Int.J.Num.Meth.Fluids*, 4, (1984), pp. 557-598.

[31] N. Satofuka, Modified differential quadrature method for numerical solution of multi-dimensional flow problems, *Proc. of the Int. Symposium on Applied Mathematics and Information Science*, Kyoto, March 1982, pp. 5.7-5.14.

[32] A. Jameson, Numerical solution of the Euler equations for compressible inviscid fluids, to appear in *Proc. of INRIA Workshop on Numerical Methods for the Euler equation for compressible fluids, December 1983.* SIAM publication.

[33] J. Goussebaile, J.P. Gregoire, A. Hauguel, Iterative Stokes solvers and splitting techniques for industrial flows, in *Proc. of 5th Int. Symposium on Finite Elements and Flow Problems, January 23-26, 1984* (G.F. Carey, J.T. Oden eds.) University of Texas, Austin, 1984, pp. 213-217.

[34] J.P. Benque, J.P. Gregoire, A. Hauguel, M. Maxant, Application des methodes de decomposition aux calculs numeriques en hydraulique industrielle, in *Computing Methods in Applied Sciences and Engineering, VI*, (R. Glowinski, J.L. Lions eds.), North Holland, Amsterdam, 1984, pp. 471-483.

[35] P. Hood, C. Taylor, A numerical solution of the Navier-Stokes equations using the finite element technique, *Computer and Fluids*, 1, (1973), pp. 73-100.

[36] A.G. Hutton, *A general finite element method for vorticity and stream function applied to a laminar separated flow*, Central Elasticity Generating Board Report, Research Dept., Berkeley Nuclear Laboratories, U.K., 1975.

[37] K. Morgan, J. Periaux, F. Thomasset (eds.), *Numerical Analysis of Laminar Flow Over a Step, GAAM Workshop, Bievres, France, January 1983*, Vieweg-Verlag, Braunschmeig-Wiesbaden, 1984.

[38] D.N. Arnold, F. Brezzi, M. Fortin, *A stable finite element for the Stokes equations*, Instituto di Analisi Numerica del C.N.R., Pavia, Report 362, 1983 (to appear in *Calcolo*).

[39] M. Fortin, F. Thomasset, Application of Augmented Lagrangian methods to the Stokes and Navier-Stokes equations, *Augmented Lagrangian Methods*, M. Fortin, R. Glowinski eds., North Holland, Amsterdam, 1983, pp. 47-95.

[40] R.W. MacCormack, Current status of numerical solutions of the Navier-Stokes equations (these proceedings).

[41] J.L. Steger, P.G. Buning, Developments in the simulation of compressible inviscid and viscous flow on supercomputers (these proceedings).

[42] E. Turkel, Algorithms for the Euler and Navier-Stokes equations for supercomputers (these proceedings).

[43] A. Matsumura, T. Nishida, Initial boundary value problems for the equations of motion in general fluids, *Computing Methods in Applied Sciences and Engineering, V*, R. Glowinski, J.L. Lions eds., North Holland, Amsterdam, 1982, pp. 339-406.

[44] J.P. Benque, G. Labadie, B. Latteux, Une method d'elements finish pour les equations de Saint-Venant, *Proc. of 2nd Int. Conf. on Numerical Methods in Laminar and Turbulent Flows*, Venice, Italy, 1981.

[45] A. Hauguel, F. Hecht, L. Reinhart, Finite Element Solution of Shallow Water Equations by a Quasi-Direct Decomposition Procedure (to appear in *Numerical Methods in Fluids)*.

[46] M.O. Bristeau, R. Glowinski, B. Dimoyat, J. Periaux, P. Perrier, Finite element methods for the compressible Navier-Stokes equations, AIAA paper 83-1890, *AIAA CFD Conference, Danvers, MA., July 13-15, 1983.*

[47] M.O. Bristeau, R. Glowinski, B. Mantel, J. Periaux, P. Perrier, Numerical methods for the time dependent compressible Navier-Stokes equations, to appear in *Computing Methods in Applied Science and Engineering, VI*, R. Glowinski, J.L. Lions eds., North Holland, Amsterdam.

[48] M.O. Bristeau, R. Glowinski, B. Mantel, J. Periaux, P. Perrier, Finite element solution of the Navier-Stokes equations for compressible viscous fluids, *Proc. of 5th Int. Conf. on Finite Elements and Flow Problems*,

G.R. Carey, J.T. Oden eds., University of Texas, Austin, 1984, pp. 449-462.

[49] Q.V. Dinh, R. Glowinski, B. Mantel, J. Periaux, P. Perrier, Subdomain solution of nonlinear problems in fluid dynamics on parallel processors, *Computing Methods in Applied Sciences and Engineering, V*, R. Glowinski, J.L. Lions eds., North Holland, Amsterdam, 1982, pp. 123-164.

[50] O.B. Widlund, Iterative methods for elliptic problems on regions partitioned into substructures and the biharmonic Dirichlet problem, *Computing Methods in Applied Sciences and Engineering, VI*, R. Glowinski, J.L. Lions eds., North Holland, Amsterdam, 1984, pp. 33-45.

[51] A. Lichnewsky, J.L. Lions, Super-ordinateurs: evolutions et tendances, *La Vie des Sciences, Comptes-Rendus*, serie generale, 1, (1984), 4, pp. 263-284.

Progress in Scientific Computing, Vol. 6
Proceedings of U.S.-Israel Workshop, 1984

MARCHING ITERATIVE METHODS FOR THE PARABOLIZED AND THIN LAYER NAVIER-STOKES EQUATIONS

Moshe Israeli
Computer Science Department
Technion - Israel Institute of Technology
Haifa, Israel

Downstream marching iterative schemes for the solution of the Parabolized or Thin Layer (PNS or TL) Navier-Stokes equations are described. Modifications of the primitive equation global relaxation sweep procedure result in efficient second order marching schemes. These schemes take full account of the reduced order of the approximate equations as they behave like the SLOR for a single elliptic equation. The improved smoothing properties permit the introduction of Multi-Grid acceleration. The proposed algorithm is essentially Reynolds number independent and therefore can be applied to the solution of the subsonic Euler equations. The convergence rates are similar to those obtained by the Multi-Grid solution of a single elliptic equation; the storage is also comparable as only the pressure has to be stored on all levels. Extensions to three-dimensional and compressible subsonic flows are discussed. Numerical results are presented.

1. INTRODUCTION

Considerable evidence accumulated recently about the applicability of the Parabolized Navier-Stokes equations for high Reynolds number flows with a principal flow direction, see Rubin [1]. The PNS equations are obtained by neglecting the streamwise viscous terms in the Navier-Stokes (NS) equations. When the viscous terms in the circumferential direction are also neglected, one gets the Thin Layer approximation.

The steady PNS equations still have an elliptic nature and therefore the initial value problem in the marching direction is not well posed [2]. A well posed initial-boundary value problem can be formulated by specifying (for example) upstream and side conditions for the velocities and one downstream condition for the pressure. This coupled system of partial differential equations behaves like a single elliptic equation for the pressure. Therefore the PNS equations must be solved globally and cannot be solved by a single sweep marching. The reduced order of the PNS equation can be exploited by constructing an iterative marching method for updating the pressure field only. Such a multiple sweep iteration method has the advantage that the velocity fields are generated during the marching process and only the pressure field has to be stored from sweep to sweep. A considerable saving in storage results. However, simple minded marching does not result in good convergence properties and sometimes diverges. For the two-dimensional incompressible case, Israeli and Lin [3] devised a stable marching scheme that behaves like the Successive Line Over Relaxation (SLOR) method for a single elliptic equation. The good smoothing properties of the above-mentioned scheme can be used in a Multi-Grid (MG) framework in order to accelerate the convergence of the solution of the PNS (or TL) equations. The marching scheme is implemented using a new stable algorithm which is second order also in the marching direction. The same method can be used without modification for the subsonic Euler equations as the effect of the Reynolds number on the convergence rate is insignificant. In two dimensions the PNS and TL equations are identical and therefore the same analysis applies to both.

It turns out that the extension to three-dimensions is conceptually simple but the resulting algorithm, a successive plane over relaxation, is complicated by the requirement of the simultaneous solution of the equations in planes perpendicular to the marching direction. This problem can be alleviated by splitting of the equation of continuity from the momentum equations.

The extension of the method to compressible flows is conceptually non-trivial. The original iterative method is based on the concept that the convergence relies on the implicit relaxation of a single quantity, the pressure, which approximately satisfies a single elliptic equation. In the compressible case a viable approach is to eliminate the pressure and to derive an equation for $\tilde{p}$, the logarithm of the density. It can be shown that $\tilde{p}$ satisfies approximately equation (1.1), $(1-M^2)\tilde{p}_{ss} + \tilde{p}_{nn} = 0$, where M is the Mach number and s and n are coordinates along and perpendicular to the flow direction. Although this equation is never derived or used in the algorithm it reveals the fact that for $M < 1$ the upstream influence is transmitted through the quantity $\tilde{p}$ and therefore only this quantity should be stored or updated. The flow of information should be downstream for the velocity and temperature and upstream for the density, and the difference scheme must be built accordingly. For supersonic flows the flow of information should be only downstream and the marching method is non iterative. For supersonic flows with imbedded subsonic regions the iterative method should be used, combined with an appropriate switching at shock waves and sonic lines.

It should be pointed out here, that the present approach is very different conceptually from that of Reddy and Rubin [4]. Although they used our idea (Israeli [6]) of back shifting the pressure, one full mesh distance, with respect to the velocity for incompressible flows, their generalization to compressible flows is a Mach number dependent shift which vanishes for $M \geqslant 1$. This smooth transition from subsonic to supersonic flows is bootless since the change of type of equation (1.1) is sudden, at $M = 1$. Indeed only our full shift is used in their papers and properly results in a conservative scheme across a shock.

Another question raised by the above-mentioned paper is that of the distinction between the pressure which uses downstream data and the density which uses upstream data. This obscures the issue of the direction of flow of information and proper location of boundary conditions. This approach should result in inconsistency of boundary data and may eventually lead to ill posedness and divergence.

In the next sections we will summarize our previous theoretical results, present some new numerical results and the extensions to 3-D and compressible flows.

2. FORMULATION FOR THE INCOMPRESSIBLE CASE

For simplicity we will consider initially the case of the steady, incompressible and two-dimensional PNS (or TL) equations in cartesian coordinates [x;y]:

$$U_x + V_y = 0 \tag{2.1}$$

$$(U^2)_x + (UV)_y = -P_x + U_{yy}/Re \tag{2.2}$$

$$(UV)_x + (V^2)_y = -P_y + V_{yy}/Re \tag{2.3}$$

where x is the mainstream direction, Re is the Reynolds number. U and V are the nondimensional velocity components in the x and y direction, respectively. P is the nondimensional pressure.

The two-dimensional NS equations are elliptic of order four - Brandt and Dinar [5]. The PNS are elliptic only of order two like the Poisson equation (the mathematical nature of several two-dimensional and three-dimensional approximations to the Navier-Stokes equations was analyzed in [7]). This ellipticity is due to the pressure gradient terms via the continuity equation. A well posed problem can be formulated by defining the boundary conditions as described in Fig. 2. The following Dirichlet conditions may be specified:

* upstream boundary (AB): $U = U_{in} \; ; \; V = V_{in}$ (2.4)
* at a solid wall (AD): $U = U_{wall} \; ; \; V = V_{wall}$ (2.5)
* at the outer boundary (BC): $U = U_{out} \; ; \; V = V_{out}$ (2.6)
* at the downstream boundary (CD): $U = P_{down}$. (2.7)

Other boundary conditions can be used but the same number of conditions on each boundary must be kept.

In order to separate linear and non-linear effects, some of the convergence tests were performed with the following linear version of equations (2.1)-(2.3):

$$U_x + V_y = 0 \tag{2.8}$$

$$(aU)_x + (bU)_y = -P_x + U_{yy}/Re \tag{2.9}$$

$$(aV)_x + (bV)_y = -P_y + V_{yy}/Re \tag{2.10}$$

where a and b are known functions of x and y.

3. DISCRETIZATION AND MARCHING

Numerical solutions of Eqs. (2.1)-(2.3) are obtained by spreading a grid over the computational domain. Let us assume that the grid points are distributed evenly along the x and y coordinates with the spacing Δx and Δy respectively. When differencing these equations it should be remembered that their nature should be reflected [1,8] in the finite difference approximation. In order to be consistent with the boundary layer (parabolic) nature of the flow, the axial gradients of the velocities should be computed using only upstream values, while the elliptic nature is preserved by forward differencing the axial pressure gradient [1,8,9]. Consequently, it was assumed that a stable marching scheme must be of the first order in the marching direction. It turns out that this effect can be achieved by a judicious choice of the placement of the variables to be solved at each station. The choice can be explained most easily by taking $V = 0$ and $\frac{1}{Re} = 0$ in Eq. (2.2) for U, yielding

$$U^2_x = - p_x$$

A first order difference scheme then becomes

$$U^2_{m,j} - U^2_{m-1,j} = p_{m,j} - p_{m+1,j}$$

the unknowns are $U_{m,j}$ and $p_{m,j}$. The scheme first suggested by Israeli [9,10] is:

$$U^2_{m+1,j} - U^2_{m,j} = p_{m,j} - p_{m+1,j}$$

with the unknowns $U_{m+1,j}$ and $p_{m,j}$. The scheme is centered about $m + \frac{1}{2}$ and is second order. This approach was subsequently used by Rubin and Reddy [8] and Reddy and Rubin [4].

In addition, one may stagger the velocity V with respect to the other variables as shown in Fig. 3, where the centering points of the different difference equations are also plotted. The differential equations are approximated by central second-order approximations. Whenever needed averaging was used as is usually done for staggered grids.

Numerical experiments with a first order computer code show that the solution after one marching sweep is not close to the final solution of the PNS equations, when the initial pressure field is constructed using the boundary layer assumption $p_y = 0$. Since the p_x term is forward differenced, some global iterations over the whole solution's domain should be performed in order to converge the explicit contribution to this pressure term. The simplest global iterative technique to solve the equations is by multiple marching sweeps with the primitive equations where only the pressure field is kept from iteration to iteration [1]. Numerical experiments also show that for certain nets this procedure diverges. The divergence occurs also for the linearized version of Eqs. (2.8)-(2.10). Figure 1 presents the residuum of the pressure field as a function of the global iteration's sweep number for a 21×11 field. A jump is encountered every 10 iterations (probably related to the arrival of the boundary pressure pulse traveling at the numerical scheme speed) leading to ultimate divergence. However convergence was reported with different mesh and boundary conditions and also when combining the above procedure with a multigrid technique [4]. It was thought that the replacement of one of the momentum equations by the Poisson equation for the pressure will improve the convergence rate but the solution did not satisfy the replaced momentum equation. A successful implementation of the marching technique is derived in the next section. A short and reduced version of the analysis was presented first in [3].

4. A MULTI-GRID ALGORITHM

The Multi-Grid technique is a numerical strategy for substantially improving the convergence rate of an iterative procedure. In order to facilitate comparison with theory the accomodative C-cycle MG algorithm was chosen.*

Each MG process consists of three basic parts: relaxation, restriction and interpolation [5].

The Relaxation Scheme

The overall convergence rate of any MG process is greatly influenced by the smoothing properties of the relaxation scheme. It can be shown analytically and experimentally that the usual multiple sweep marching [1] does not have good convergence and smoothing properties because short wave errors are not efficiently smoothed. Israeli and Lin [3] showed that certain modifications in the streamwise momentum equation, which vanish upon convergence, give rise to an iterative scheme which is equivalent, in the linear case, to the SLOR method for one Poisson equation. In the general nonlinear case the modified iterative process is essentially equivalent to the relaxation of a single nonlinear Poisson-like equation for the pressure. The velocities can be viewed as auxiliary variables needed during the marching since they have no "memory" by themselves.

Furthermore, we have automatically gained the good smoothing properties of the line relaxation scheme of a single Poisson equation. The problems associated with the loss of ellipticity of the difference

---

* Some of the elements of the present approach were used independently by Rubin and Reddy [8]. Detailed comparisons cannot be made because convergence rates and storage estimates were not presented there.

approximation for the Navier-Stokes equations at high Reynolds number [5] are thus avoided and no upstream-weighting or artificial viscosity is required. There results a considerable saving in storage, as well as a simpler relaxation scheme (compare to the distributive relaxation [5]) where the convergence rate is essentially independent of the Reynolds number. We note that the same marching algorithm can thus be used for the (subsonic) Euler equation with the same favorable convergence rate. (For supersonic flows the marching method is non-iterative.)

A part of the analysis of [3] is repeated here to motivate the later extensions to three-dimensional and compressible flows. We start with the PNS equations (2.1)-(2.3) and linearize them about a constant state. We also introduce

$$\bar{L}(f) = \frac{\partial}{\partial x}(\bar{U}f) + \frac{\partial}{\partial y}(\bar{V}f) - \frac{1}{Re}\frac{\partial^2}{\partial y^2} f \tag{4.1}$$

where $\bar{U}$ and $\bar{V}$ are constant reference velocities. The next step is to discretize the equations only in the x direction to obtain:

$$-\frac{U_{m-1}-U_m}{\Delta x} + (V_y)_m = 0 \tag{4.2}$$

$$D(U_m) = -\frac{P_{m+1}-P_m}{\Delta x} \tag{4.3}$$

$$D(V_m) = -(P_y)_m \tag{4.4}$$

where U_m, V_m and P_m are functions of y. Here $D(f_m)$ is the semi-discretized form of $\bar{L}(f)$ at the marching station m. The semi-discretized system should be discretized also in the y direction before solution is attempted but since the specific form of this discretization is not important for the following argument, we postpone this step for the sake of transparency.

The marching iterative procedure assumes that $U_1^k(y)$, $V_1^k(y)$ are known as well as P_m^{k-1} for $m = 2,3,4,\ldots M$, where k is the current iteration index. Therefore, the marching scheme for $m \geqslant 2$ is:

$$-\frac{U^k_{m-1}-U^k_m}{\Delta x} + (V_y)_m = 0 \tag{4.5}$$

$$D(U^k_m) = -\frac{P^{k-1}_{m+1}-P^k_m}{\Delta x} \tag{4.6}$$

$$D(V^k_m) = -(P^k_y)_m \tag{4.7}$$

We now apply D to Eq. (4.5) and differentiate Eq. (4.7) with respect to y. Elimination of the V terms between Eqs. (4.5) and (4.7) gives:

$$D(U^k_{m-1} - U^k_m) = -(P^k_{yy})_m \Delta x \tag{4.8}$$

Now by substitution of Eq. (4.6) into Eq. (4.8) we get:

$$P^{k-1}_{m+1} - P^k_m - P^{k-1}_m + P^k_{m-1} + \Delta x^2 (P^k_{yy})_m = 0 \tag{4.9}$$

It follows that the marching scheme for the primitive system (4.2)-(4.4) can be viewed as a line iterative scheme for the semi-discretized Laplace equation, indeed upon convergence Eq. (4.9) will become:

$$\frac{P_{m+1} - 2P_m + P_{m-1}}{\Delta x^2} + (P_{yy})_m = 0 \tag{4.10}$$

In order to find out the rate of convergence of Eq. (4.9) to the final state (4.10) we Fourier transform Eq. (4.9) in y assuming appropriate boundary conditions in that direction:

$$P^k_m = \bar{Z}_m e^{I\,ny} \quad , \quad P^{k-1}_m = Z_m e^{I\,ny} \tag{4.11}$$

where $I^2 = -1$ and n is the Fourier wave number. After substituting these definitions into Eq. (4.9) we get:

$$Z_{m+1} - \bar{Z}_m - Z_m + \bar{Z}_{m-1} - \Delta x^2 n^2 \bar{Z} = 0 \tag{4.12}$$

Transforming in the x direction we define again

$$Z = Ae^{I\,\theta\,x} \tag{4.13}$$

where θ is the wave number in the x direction. By substitution of this definition into Eq. (4.12) we get:

$$\left|\frac{\bar{A}}{A}\right| = \left|\frac{1-e^{-I\theta}}{1+\Delta x^2 n^2 - e^{-I\theta}}\right| \tag{4.14}$$

This means that all the long waves (with small $\Delta x^2 n^2$) in the cross flow direction are only weakly damped irrespective of their structure in the marching direction. In particular the $n=0$ modes which exist for derivative boundary conditions in the cross flow direction are not affected at all by the relaxation, i.e.

$$\left|\frac{\bar{A}}{A}\right| = 1$$

On the other hand, the well known SLR scheme for the Poisson equation gives (after the same Fourier transformations):

$$Z_{m+1} - (2+\Delta x^2 n^2)\bar{Z}_m + \bar{Z}_{m-1} = 0 \tag{4.15}$$

and

$$\left|\frac{\bar{A}}{A}\right| = \frac{1}{q-e^{-I\theta}} \; ; \; q = 2 + \Delta x^2 n^2 \geq 2 \tag{4.16}$$

and also

$$\left|\frac{\bar{A}}{A}\right|^2 = \frac{1}{q^2+1-2q\cos\theta} \tag{4.17}$$

This quantity is less than 1 for all acceptable q's and $\cos\theta < 1$. Most waves are strongly damped and only the longest waves in both directions are weakly damped by the iteration. This behaviour was used to accelerate the convergence as is done by the SLOR technique, Chebychev acceleration or Multi-Grid method.

The question is how to generate an equivalent relaxation scheme for the primitive system in the marching form. This means that we may add terms which can be evaluated during the marching process but should vanish upon convergence.

A rational approach to the construction of the relaxation scheme is to retrace backwards the steps of the derivation of the discrete Laplace equation from the discrete primitive equations. We start from the SLOR equation (4.15):

$$Z_{m+1} - 2\bar{Z}_m + \bar{Z}_{m-1} = \Delta x^2 n^2 \bar{Z}_m$$

which we inverse Fourier transform with respect to y to get:

$$P^{k-1}_{m+1} - 2P^k_m + P^k_{m-1} = -\Delta x^2 (P^k_{yy})_m$$

Now, we substitute Eq. (4.8) for the right hand side of the last equation to get:

$$P_{m+1}^{k-1} - 2P_m^k + P_{m-1}^k = \Delta x D(U_{m-1}^k - U_m^k)$$

which can be written as:

$$(P_{m\,1}^{k-1} - P_m^k) + (P_m^{k-1} - P_m^k) - (P_m^{k-1} - P_{m-1}^k) = \Delta x D(U_{m-1} - U_m)$$

adding the equations from $m = 2$ and using the linear form of D we get:

$$-(P_{m+1}^{k-1} - P_m^k) - \sum_{i=2}^{m} (P_i^{k-1} - P_i^k) + (P_2^{k-1} - P_1^k) = \Delta x D(U_m - U_1)$$

but from Eq. (4.3):

$$P_2^{k-1} - P_1^k = -\Delta x D(U_1^k) \tag{4.18}$$

therefore we get for $m \geq 2$

$$D(\bar{U}_m) = -\frac{P_{m+1}^{k-1} - P_m^k}{\Delta x} - \frac{1}{\Delta x} \sum_{i=2}^{m} (P_i^{k-1} - P_i^k) \tag{4.19}$$

Eq. (4.19) contains all the modifications required in order to convert the iteration scheme of (4.5)-(4.7) into a scheme equivalent to the SLOR scheme for one Laplace equation with $\Omega = 1$. We see that in this approach only the x momentum equation is modified. The new added term can be generated easily during the marching process and is inexpensive in storage (one extra line vector) and computation (one subtraction per grid point). In what follows we will rederive Eq. (4.19) in a more general way and introduce the over-relaxation parameter $\Omega > 1$.

In practice we will use difference approximations and boundary conditions also in the y direction and the resulting scheme may not be amenable to the discrete analogue of the Fourier transform. It is therefore worthwhile to generalize the previous approach by using the matrix finite difference formulation.

Let the vectors U_m, V_m, P_m contain the N values of the corresponding variables on the m-th line (x = constant) of the marching sweep (including the specified boundary values). The u-momentum equation (1.5) can be written in the form:

$$P_{m+1} - P_m = -\Delta x D(U_m) \tag{4.20}$$

On the other hand elimination of V_m between the continuity and the v-momentum equations will result in:

$$FP_m = R_m - R_{m-1} \tag{4.21}$$

where $F = \Delta x^2 I \frac{\theta^2}{\Delta y^2}$. Substructing successively u-momentum equations (4.20) and using Eq. (4.21) gives:

$$P_{m+1} - (2I + F)P_m + P_{m-1} = 0, \quad m = 3,4,\ldots \tag{4.22}$$

which is Laplace's equation. The first equation of (4.20) can be used as a derivative condition at the left (inlet) boundary, namely:

$$P_3 - P_2 = R_2 \tag{4.23}$$

We now apply the SLOR scheme to the last two equations (ignoring temporarily the downstream boundary condition) to get the downstream marching form:

$$-P_2^* + P_3^{(k-1)} = R_2 \tag{4.24}$$

$$P_{m-1}^{(k)} - (2I + F)P_m^* + P_{m+1}^{(k-1)} = 0, \quad m = 3,4,\ldots \tag{4.25}$$

where $P_m^{(k)} = \Omega P_m^* + (1-\Omega)P_m^{(k-1)}$; Ω is the overrelaxation factor, and the superscript denotes the iteration sweep number. In order to recover the primitive variable formulation, we relate the velocity field in Eq. (4.21) to the starred pressure field, i.e.

$$FP_m^* = R_m - R_{m-1} \tag{4.26}$$

Substitution in Eq. (4.25) gives:

$$P_{m-1}^{(k)} - 2P_m^* + P_{m+1}^{(k-1)} = R_m - R_{m-1}, \quad m = 3,4,\ldots \tag{4.27}$$

Successive summations if Eqs. (4.24) and (4.25) gives:

$$P_{m+1}^{(k-1)} - P_m^* = R_m + S_m, \quad m = 2,3,4,\ldots \tag{4.28}$$

which is the primitive variable marching form of the u-momentum equation. The source term S_m in Eq. (4.28) satisfies:

$$S_m = S_{m-1} + (P_m^* - P_m^{(k-1)}) + (P_{m-1}^* - P_{m-1}^{(k)}), \quad m = 3,4,\ldots \tag{4.29}$$

with $S_2 = 0$. It can be seen that S_m vanishes upon convergence The computational form of Eq. (12) for $m = 3,4,\ldots$ is:

$$-2P_m^* = R_m + \bar{S}_m \tag{4.30}$$

$$\bar{S}_m = \bar{S}_{m-1} - P_{m+1}^{(k-1)} + 2P_{m-1}^* - P_{m-1}^{(k)}; \quad \bar{S}_2 = -P_2^* - P_3^* \tag{4.31}$$

Thus, the theory of overrelaxation can be applied exactly to the constant coefficient case of system (2.8)-(2.10). For the non-linear case this theory can serve as a guide to the choice of Ω. Alternately, one can choose $\Omega = 1$ and apply the Multi-Grid procedure.

Restriction and Storage Requirements

Let the finite difference approximation of equations (2.1)-(2.3) on the finest grid M be represented as in [5]:

$$L_j^M \tilde{W}^M(\tilde{x}) = F_j^M(\tilde{x}) \tag{4.32}$$

where $\tilde{x} = [x,y]$, $\tilde{W}^M = [U^M, V^M, P^M]^T$ is the exact solution of the difference equations, and j is the number of the differential equation, $j = 1,2,3$.

The problem is transferred from the current level k to a coarser level $k-1$, see Fig. 4, by correcting the right hand side of (4.31)

$$F_j^{k-1}(\tilde{x}) = L_j^{k-1}(\tilde{I}_{j,k}^{k-1}\tilde{w}^k(\tilde{x})) + I_{j,k}^{k-1}[F_j^k(\tilde{x}) - L_j^k\tilde{w}^k(\tilde{x})] \tag{4.33}$$

in the Full Approximation Storage (FAS) mode. $\tilde{w}^k(x)$ is an approximation to $\tilde{W}^k(\tilde{x})$ in the finer level. $I_{j,k}^{k-1}$ and $\tilde{I}_{j,k}^{k-1}$ are proper restriction operators for equation j.

The term in square bracket in equation (6) is the residual of the j-th equation. For the present marching scheme there is no residual in the continuity and in the y-momentum equations since they are solved exactly in each step. The residual of the x-momentum equation results only from the streamwise pressure gradient term and its computation needs only one subtraction. $\tilde{I}_{j,k}^{k-1}$ was chosen to be linear interpolation, which yields for the continuity equation: $L_1^{k-1}(\tilde{I}_{1,k}^{k-1}\tilde{w}^k(\tilde{x})) = 0$. $I_{j,k}^{k-1}$, $j = 1,2$ is computed by averaging in both the x and y directions. $I_{3,k}^{k-1}$ is a simple injection.

In summary, equation (4.33) takes the following terms:

$$F_1^{k-1}(\tilde{x}) = 0$$

$$F_2^{k-1}(\tilde{x}) = L_2^{k-1}(\tilde{I}_{2,k}^{k-1}\tilde{w}^k(\tilde{x})) + I_{2,k}^{k-1}(F_2^k - L_2^k\tilde{w}^k(\tilde{x}))$$

$$F_3^{k-1}(\tilde{x}) = L_3^{k-1}(\tilde{I}_{3,k}^{k-1}\tilde{w}^k(\tilde{x}))$$

Two consequences should be emphasized:

(a) Only two corrections $(F_2^{k-1}(\tilde{x}), F_3^{k-1}(\tilde{x}))$ have to be computed and stored.

(b) All the dependent variables must be transferred in order to compute the corrections $(L_j^{k-1}(\tilde{I}_{j,k}^{k-1}\tilde{w}^k(\tilde{x})), j = 2,3)$. Since only the pressure is stored, these corrections must be computed during the marching process.

It follows that in addition to the pressure on all grids, one has to save one correction term for each momentum equation on the coarser grids. Assuming N computational points on the finest grid, a simple-minded estimate gives $32N/7$ storage locations for the three-dimensional NS Multi-Grid solution, and $11N/7$ for the PNS marching MG solution. For the two-dimensional case the corresponding figures are $13N/3$ and $7N/3$.

<u>Interpolation</u>

Since the present marching scheme generates the velocity field from the pressure, only the correction to the pressure must be interpolated back to the fine grid.

5. GENERALIZATIONS

In order to generalize the preceding approach we note that the essence of the relaxation procedure is the replacement of the term $\Delta x \partial P/\partial x$ by the marching difference form:

$$-(2P_m^* + \bar{S}_m) \tag{5.1}$$

where $\bar{S}_m$ is already known. If (5.1) is differenced it will (using the definition of $\bar{S}_m$) give rise to

$$2(P_m^* - P_{m-1}^*) + \bar{S}_m - \bar{S}_{m-1} = -(P_{m+1}^{k-1} - 2P_m^* + P_{m-1}^k) .$$

Thus, the correct successive line over relaxation form is implicitly obtained for the second derivative of the pressure in the marching direction. (It should be emphasized again that the second order elliptic equation for the pressure is neither derived nor used in the algorithm itself.)

The implication of the present technique is: if it is known that the equations can be manipulated so that some variable will satisfy approximately a second order elliptic equation, we should use the replacement (5.1) for the derivative of that variable in the main flow direction. An efficient marching scheme will thus be generated.

The present version of the algorithm will be applied to the subsonic compressible multi-dimensional Navier-Stokes equations. Several particular cases will be examined.

The first step is the derivation of an elliptic equation starting with:

$$\vec{V}\cdot\nabla\vec{V} = -\frac{1}{\rho}\nabla p + \nu\nabla^2\vec{V} + \frac{\nu}{3}\nabla(\nabla\cdot\vec{V}) \tag{5.2}$$

where $\vec{V}$ is the velocity vector. In addition we will require the equation of state of a perfect gas

$$p = \rho RT \tag{5.3}$$

and the continuity equation in the form

$$\nabla\cdot\vec{V} = -\vec{V}\cdot\nabla \ln\rho \tag{5.4}$$

It follows from (5.3) that

$$\frac{1}{\rho}\nabla p = RT\nabla \ell n\rho + R\nabla T \tag{5.5}$$

and therefore

$$\nabla\cdot\frac{1}{\rho}\nabla p = RT\nabla^2 \ell n\rho + R\nabla^2 T + \text{l.o.t.} \tag{5.6}$$

(l.o.t. stands for lower order terms).

Also

$$\nabla\cdot(\vec{V}\cdot\nabla\vec{V}) = (\vec{V}\cdot\nabla)(\nabla\cdot\vec{V}) + \text{l.o.t.} = -(\vec{V}\cdot\nabla)^2 \ell n\rho + \text{l.o.t.} \tag{5.7}$$

Taking the divergence of (5.2) and using (5.6) and (5.7) we get

$$(\vec{V}\cdot\nabla)\ell n\rho = RT\nabla^2 \ell n\rho + R\nabla^2 T + \frac{4\nu}{3}\nabla^2(\nabla\cdot V) + \text{l.o.t.} \tag{5.8}$$

Several special cases follow:

1) Incompressible case. Here $\nabla\cdot\vec{V} = 0$, and we get directly (taking the divergence of (5.2)):

$$\nabla^2 p = \text{l.o.t.}\ , \tag{5.9}$$

thus the gradient term was differentiated once and the replacement (5.1) should apply in two or three dimensions. In the later case we have to compute simultaneously all the variables in the marching plane, m, and so we get a successive plane overrelaxation scheme. It is possible that an alternating direction scheme can be used to solve the coupled system in the m'th plane, but a multi-grid approach seems to be preferable. At the present time numerical results for the three dimensional case are not available.

2) Isothermal case. We get

$$(\vec{V}\cdot\nabla)^2 \ell n\rho = \frac{a^2}{\gamma}\nabla^2 \ell n\rho + \text{l.o.t.} + \text{v.t.}$$

where $a^2 = \gamma RT$ is the adiabatic speed of sound, γ is the ratio of specific heats, and v.t. is the viscous term to be discussed later.

3) Isentropic case. Here, $(\gamma-1)\nabla\ell n\rho = \nabla \ell n T$ and therefore:

$$(V\cdot\nabla)^2 \ell n\rho = a^2\nabla^2 \ell n\rho + \text{l.o.t.} + \text{v.t.}$$

4) Constant stagnation enthalpy. Here, $a_o^2 = a^2 + \frac{\gamma-1}{2}\vec{V}^2$ and we get

$$(V\cdot\nabla)^2 \ell n\rho + (\gamma-1)\nabla\cdot(\nabla\cdot V)\cdot\vec{V} = a^2\nabla^2 \ell n\rho + \text{l.o.t.} + \gamma\text{v.t.}$$

In all the compressible cases considered the prominent balance is:

$$(\vec{V}\cdot\nabla)^2 \ell n\rho \cong a^2\nabla^2 \ell n\rho \ .$$

Using local stream aligned coordinates s and n we find

$$(\vec{V}\cdot\nabla)^2 = V^2 \frac{\partial^2}{\partial s^2}$$

$$\nabla^2 = \frac{\partial^2}{\partial s^2} + \frac{\partial^2}{\partial n^2}$$

therefore $\tilde{p} \equiv \ell n\rho$ appears in the form:

$$(1-M^2)\frac{\partial^2 \tilde{p}}{\partial s^2} + \frac{\partial^2 \tilde{p}}{\partial n^2} = \text{other terms.}$$

After the parabolization of the viscous term, only the left hand side has a second derivative of $\tilde{p}$ in the streamwise direction. Specifically $\nu\nabla^2\nabla\cdot V$ is replaced by

$$\nu\frac{\partial^2}{\partial n^2}\nabla\cdot V = -\nu\frac{\partial^2}{\partial n^2}\vec{V}\cdot\vec{\nabla}\ell n\rho \ .$$

This term cannot become large since the pressure does not have large gradients in the boundary layer.

We argue that if our iteration is appropriate for the $\tilde{p}$ equation it will be a good scheme overall.

To get a successive line (or plane) over relaxation scheme, all we have to do is replace all the occurrences of $\frac{\partial \tilde{p}}{\partial x}$ with the marching form (5.1). All the properties of Section 4 will be the same as long as $M^2 < 1$.

In fact, better convergence can be expected as M^2 approaches 1 since the quantity q of (4.16) will become now $2 + (\Delta x^2 n^2/1-M^2)$. Only $\tilde{p}$ will have memory and must be globally saved and updated by the iteration procedure. $\tilde{p}$ will also transmit the downstream information and must be specified there.

For transonic flows a conservation form is preferred and it may be more convenient to work with ρ rather than $\ell n\rho$. An elliptic equation can be derived for ρ but care must be taken to transmit the downstream information via ρ. Upstream information should not be transmitted by ρ and ρ should not be specified at the inflow, otherwise the problem will be overspecified. Consider, for example, the term $\partial\rho u^2/\partial x$, it is discretized as

$(\rho_m U_m^2 - \rho_{m-1} U_{m-1}^2)$,

however at the station m we should compute the U_m velocities coupled with the ρ_{m-1} densities. The approach of Reddy and Rubin [12] where the pressure is specified both at inflow and at outflow is inconsistent unless one happens to know the right pressures before the computation. The inconsistency and consequent error can be easily demontrated by one dimensional examples.

6. RESULTS

In order to check the MG algorithm we choose the following analytical solution. It satisfies the continuity equation but gives rise to source terms in the momentum equations:

$$U = A + (x+y)^m; \quad V = -(x+y)^m; \quad P = -(E1+E2)(x+y)^m \tag{6.1}$$

where a and b from equations (3) are defined by:

$$a = E1 + F(x+y)^n; \quad b = E2 - F(x+y)^n \tag{6.2}$$

and $E1 = 1$; $E2 = .2$; $F = .2$; $A = 5$; $Re = 1000$; $m = 4$; $n = 2$. The coarsest grid consists of 4×4 intervals.

Figure 5 compares the MG convergence history of different relaxation schemes. In the MG solutions three levels were involved (M=3). The horizontal coordinate gives the number of Work Units (WU), where each work unit is equivalent to one global iteration on the finest grid. The vertical coordinate gives the logarithm of the dynamic residual ε. The dots show the solution of the equivalent Poisson equation (with the same solution for the pressure but with Dirichlet condition over all the boundaries). The linearized PNS equations were solved with and without the streamwise pressure gradient correction of [3]. The corresponding (17×17 points) single grid convergence history is plotted for comparison. The corrected discrete equations and the Poisson exhibit very similar convergence whereas the convergence of the unmodified equations is much worse. Upon increasing the number of grids in the unmodified equations the convergence deteriorates.

The Reynolds number independence of the scheme is demonstrated in Figure 6, where the convergence history is presented for Reynolds numbers 1, 10^3 and infinity.

In order to check the non-linear version of the code several test cases were run; the incompressible flow over a flat plate, the flow along an axisymmetric cylinder, entrance flow between two flat plates and the flow behind the trailing edge of a flat plate. In all cases good agreement was obtained with known solutions. The details will be presented elsewhere. Here we show (Figure 7) the convergence history for a flow over a flat plate with uniform upstream profile and Neumann condition for the pressure at the exit. While the number of levels is varied the finest grid remains the same and consists of 65×65 points. In Figure 8 there is a comparison between the present results for the flow near the trailing edge of a flat plate and the results of reference [11]. The skin friction coefficient CF is shown for $z < 1$ while the center line velocity UC is shown for $z > 1$. The trailing edge is at $z = 1$.

ACKNOWLEDGEMENT

This research was supported by the AFOSR under grant No. F49620-83-C-0064, by the Shiftung Volkswagenwerke, and by NASA under contract No. NAS1-17070 while the author was in residence at ICASE, NASA Langley Research Center.

REFERENCES

[1] Rubin, S.G., (1982), "Incompressible NS and PNS Solution Procedures and Computational Techniques", Von Karman Inst. Lecture Notes.

[2] Israeli, M., Reitman, V., Salomon, S. and Wolfshtein, M. (1981), "On the Marching Solution of the Elliptic Equations in Viscous Fluid Mechanics", Proc. of 2nd Int. Conf. on Numerical Methods in Laminar and Turbulent Flows, C. Taylor et al. editors, Venice.

[3] Israeli, M. and Lin, A. (1982), "Numerical Solution and Boundary Conditions for Boundary Layer Like Flows", 8th IDNMFD, Aachen, West Germany, Springer-Verlag, pp. 266-272.

[4] Reddy, D.R. and Rubin, S.G. (1984), Subsonic/Transonic, Viscous/Inviscid Relaxation Procedures for Strong Pressure Interactions. AIAA 17th Fluid Dynamics Conference, Snowmass, Colorado.

[5] Brandt, A. and Dinar, N., (1979), Multigrid Solution to Elliptic Flow Problems, Numerical Methods for PDE's, Academic Press, pp. 53-147.

[6] Israeli, M. (1982), NASA Lewis Seminar, July 1982.

[7] Rosenfeld, M. (1983), An Investigation of the Hierarchy of Approximations to the Multi-Dimensional Incompressible Navier-Stokes Equations, thesis proposal, Technion, Haifa.

[8] Rubin, S.G. and Reddy, D.R. (1983), "Analysis of Global Pressure Relaxation for Flows with Strong Interaction and Separation", Int. J. of Computers and Fluids, 11, 4, pp. 281-306.

[9] Rubin, S.G. and Lin, A. (1980), "Marching with the Parabolized Navier-Stokes Equations", Isr. J. of Tech., 18.

[10] Israeli, M. and Rosenfeld, M. (1983), "Marching Multigrid Solutions to the Parabolized Navier-Stokes (and Thin Layer) Equations", 5th GAMM Conf. on Num. Fluid Dynamics, Rome.

[11] Veldman, A.E.P., (1981), New, Quasi-Simultaneous Method to Calculate Interacting Boundary Layers, AIAA J., 19, pp. 79-85.

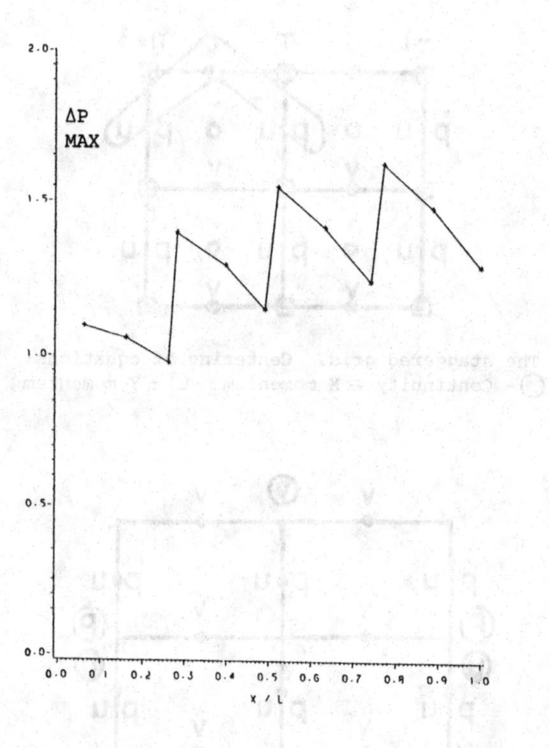

Figure 1. Pressure residue vs the global iteration number

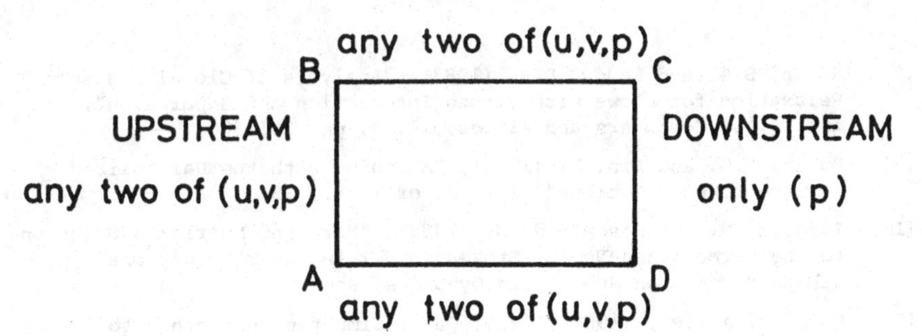

Figure 2. Example of permissible boundary conditions

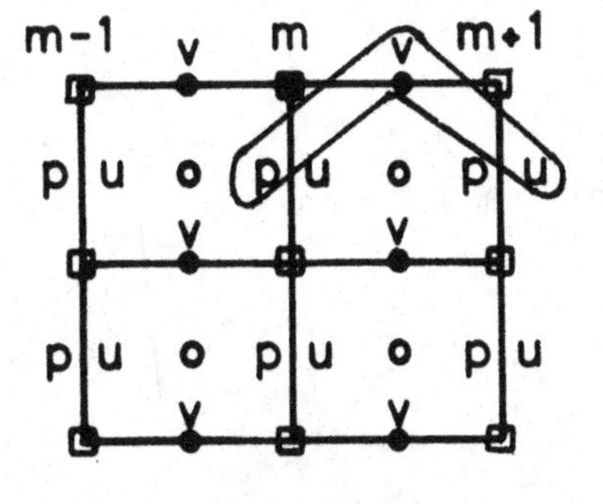

Figure 3. The staggered grid. Centering of equations:
○ - Continuity & X momentum □ - Y momentum

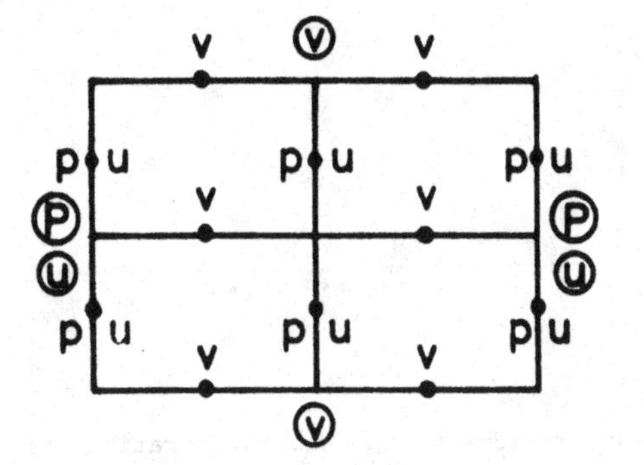

Figure 4. Relative placement of variables on two successive grids

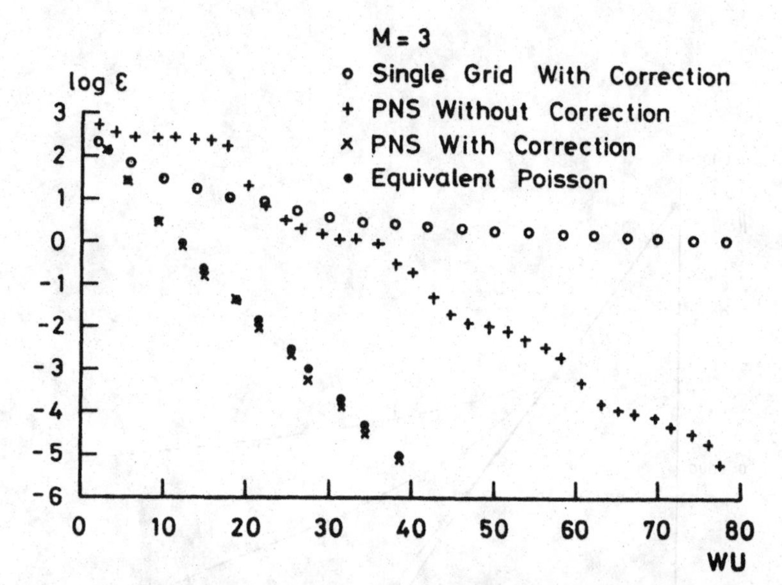

Figure 5. Convergence history for different relaxation schemes (M = 3)

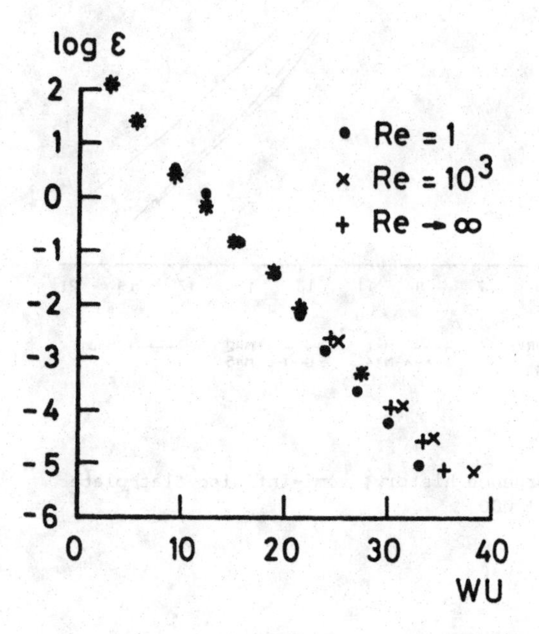

Figure 6. Convergence history for several Reynolds numbers (M = 3)

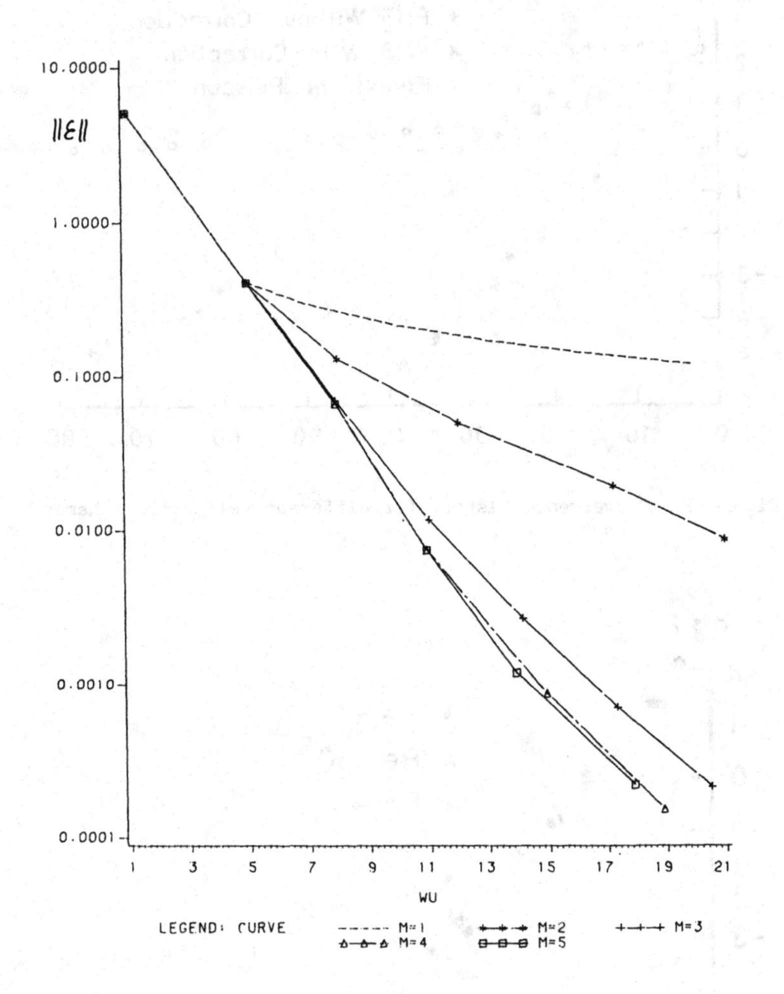

Figure 7. Convergence history, semi-infinite flat plate
RE = 10000

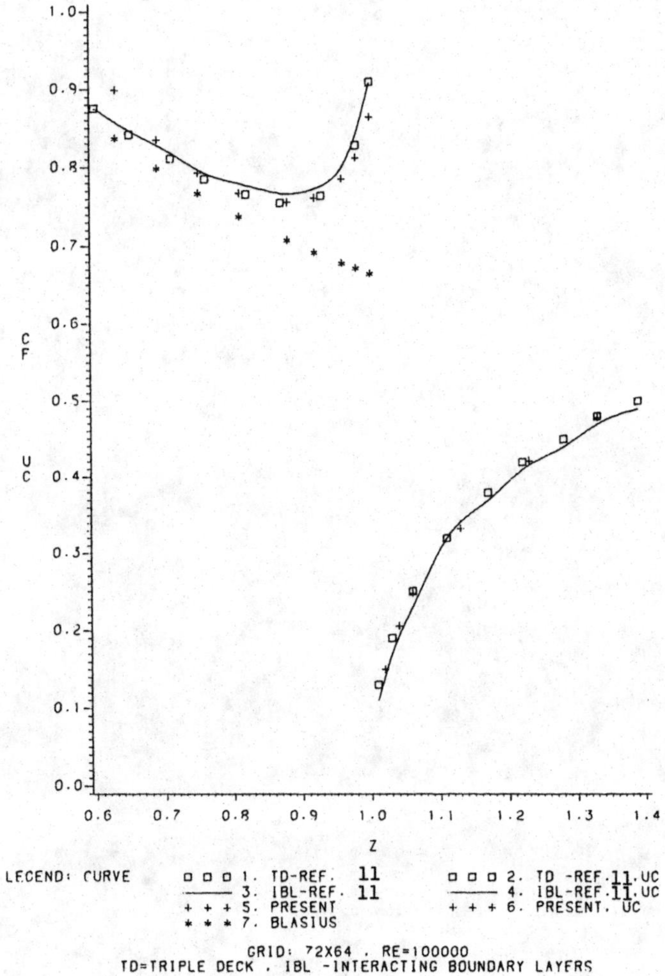

1.0
0.9
0.8
0.7
0.6
CF
0.5
UC 0.4
0.3
0.2
0.1
0.0
0.6 0.7 0.8 0.9 1.0 1.1 1.2 1.3 1.4
Z

LEGEND: CURVE
□ □ □ 1. TD-REF. 11
——— 3. IBL-REF. 11
+ + + 5. PRESENT
* * * 7. BLASIUS
□ □ □ 2. TD -REF.11.UC
——— 4. IBL-REF.11.UC
+ + + 6. PRESENT. UC

GRID: 72X64 . RE=100000
TD=TRIPLE DECK . IBL -INTERACTING BOUNDARY LAYERS

Figure 8. Flow Near the Trailing Edge of a Flat Plate.
CF - Skin friction
UC - Centerline velocity

Progress in Scientific Computing, Vol. 6
Proceedings of U.S.-Israel Workshop, 1984

Multigrid Solutions to Quasi-Elliptic Schemes

Achi Brandt and Shlomo Ta'asan

Department of Applied Mathematics
Weizmann Institute of Science, Rehovot, Israel
and
Institute for Computer Applications in Science and Engineering
NASA Langley Research Center, Hampton, Virginia, U.S.A.

Abstract. Quasi-elliptic schemes arise from central differencing or finite element discretization of elliptic systems with odd order derivatives on non-staggered grids. They are somewhat unstable and less accurate then corresponding staggered-grid schemes. When usual multigrid solvers are applied to them, the asymptotic algebraic convergence is necessarily slow. Nevertheless, it is shown by mode analyses and numerical experiments that the usual FMG algorithm is very efficient in solving quasi-elliptic equations to the level of truncation errors. Also, a new type of multigrid algorithm is presented, mode analyzed and tested, for which even the asymptotic algebraic convergence is fast. The essence of that algorithm is applicable to other kinds of problems, including highly indefinite ones.

Content

* This research is supported by the Air Force Aeronautical Laboratories, Air Force Systems Command, United States Air Force, under Grant AFOSR 84-0070.

1. Introduction

Quasi-elliptic schemes arise, for example, when central differencing is used to approximate odd-order derivatives in elliptic systems of partial differential equations, such as the Cauchy-Riemann, Stokes and Navier-Stokes systems. Usual finite element approximations to such systems also lead to quasi-elliptic schemes. Such schemes are in some sense unstable: Certain highly-oscillating components are amplified in the discretized solution much more than in the differential solution.

Instead of the quasi-elliptic schemes, other discretizations of the same system can usually be constructed which are h-elliptic, hence fully stable, and which are also more accurate than the quasi-elliptic schemes. Sometimes, however, these fully elliptic schemes are inconvenient to use. In case of elliptic systems with odd-order derivatives, for example, full ellipticity is obtained by grid staggering, i.e., by approximating different functions on different grids (cf. [3] and [8]). This is inconvenient, especially near curved boundaries. Also, the instability of quasi-elliptic approximations seldom really hurts, since the unstable components have very small amplitudes, which are still small even in the discrete solution. The inaccuracy is modest: The error in the quasi-elliptic solution is typically twice to four times larger than the error in an elliptic solution using the same grid size. Thus, quasi-elliptic schemes are often preferred and are widely used.

The instability of quasi-elliptic schemes does seem to hurt when multigrid solvers are applied: The asymptotic convergence turns out to be slow, and a simple mode analysis traces this slowness to the unstable modes. One approach, perhaps the best, to deal with this difficulty is simply to ignore it: The algebraic slowness does not matter because it occurs in modes whose amplitudes in the algebraic solution are erroneous anyway, bearing no relation to their amplitudes in the true differential solution. One should only take care not to initially admit large unstable amplitudes, and to average them out in case they must latter enter. We show, by mode analyses and numerical experiments, that the usual FMG algorithm is very effective in solving quasi-elliptic problems to truncation level (i.e., to the point where algebraic errors are dominated by discretization errors). Sometimes the FMG solution may even be better than the exact solution of the discrete equations, because the unstable components of the latter are slow to enter.

Although this is the easiest approach for obtaining fast *differential* convergence (convergence to the differential solution, through a sequence of grids), another algorithm is presented below which does provide fast *algebraic* convergence for quasi-elliptic schemes. This algorithm, based on multiple coarse-grid corrections, is interesting in its own right, since it is the simplest example of a new kind of algorithms for solving problems with highly-oscillating solutions, including highly indefinite problems (see [1, §3.2], [8] and a subsequent article). Smoothing rate analysis, for one quasi-elliptic example, suitably modified to account for the multiple coarse-grid corrections, shows that the new algorithm should *algebraically* be as efficient as usual multigrid cycles are for fully elliptic schemes. Numerical experiments exactly yield these expected convergence rates (see §8.6. Tests of such algorithms were also reported in [6].)

The significance of the present studies goes beyond elliptic PDE systems: Many non-elliptic systems, such as all subsonic steady-state flow problems, has determinants with at least one elliptic factor. Most discretizations of such systems provide quasi-elliptic approximation to that factor, leading to troubles and requiring cures similar to those reported herein.

Moreover, the techniques described in this article illustrate the following *general multigrid approaches to general non-elliptic problems*: (i) Differential, not algebraic convergence is sought, and usually easily obtained. Modified methods for apriori analyzing and aposteriori measuring such a convergence have been developed. (ii) With considerably more effort, fast *algebraic* convergence can also be obtained. (iii) The analysis of difference schemes, and the derivation of efficient smoothers, for any PDE system is based on the factors of the h-principal part of the operator determinant.

We thank Ruth Golubev for some of the calculations reported in §8.

2. Definitions and Examples

In the following L^h will represent a system of q real difference operators on q grid functions, where h, the meshsize of the grid, is for simplicity assumed to be uniform and the same in all directions. That is, L^h is a $q \times q$ matrix of real polynomials in $T_1, \ldots, T_d, T_1^{-1}, \ldots, T_d^{-1}$, where T_i are the grid translation operators, defined by

$$T_1^{\nu_1} \cdots T_d^{\nu_d} u(\underline{x}) = u(\underline{x} + \underline{\nu}\, h),$$

with $\underline{x} = (x_1, \ldots, x_d)$, $\underline{\nu} = (\nu_1, \ldots, \nu_d)$ and d being the dimension of the Euclidean space housing the grid. (In case of staggered grids there may appear non-integral powers of T_j and L^h will most usually be a matrix of polynomials in $T_j^{1/2}$ and $T_j^{-1/2}$, $j = 1, \ldots, d$).

Three common examples of difference operator are: (i) The five-point (compact) Laplacian

$$\Delta^h = \frac{1}{h^2}(T_{0,1} + T_{1,0} + T_{0,-1} + T_{-1,0} - 4T_{0,0}) = \frac{1}{h^2}\begin{bmatrix} & 1 & \\ 1 & -4 & 1 \\ & 1 & \end{bmatrix}, \qquad (2.1)$$

where $T_{\alpha,\beta} = T_1^\alpha T_2^\beta$ and the array on the left is the usual pictorial description of the weights of the operator. This is the simplest approximation to the two-dimensional Laplace operator $\Delta = \partial^2/\partial x_1^2 + \partial^2/\partial x_2^2$. (ii) The central non-staggered approximation to the Cauchy-Riemann operator

$$L_{CR}^h = \begin{pmatrix} \partial_1^c & \partial_2^c \\ -\partial_2^c & \partial_1^c \end{pmatrix}, \qquad (2.2)$$

where $\partial_i^c = \frac{1}{2h}(T_i - T_i^{-1})$. (iii) The central non-staggered approximation to the Stokes operator in two dimensions

$$L_S^h = \begin{pmatrix} -\Delta^h & 0 & \partial_1^c \\ 0 & -\Delta^h & \partial_2^c \\ \partial_1^c & \partial_2^c & 0 \end{pmatrix}. \qquad (2.3)$$

For simplicity we will deal in this article only with constant-coefficient operators L^h. In this case the *symbol* $\hat{L}^h(\underline{\theta})$ of L^h is defined by

$$L^h A e^{i\underline{\theta}\cdot\underline{x}/h} = \hat{L}^h(\underline{\theta}) A e^{i\underline{\theta}\cdot\underline{x}/h}, \qquad (|\underline{\theta}| \leq \pi)$$

for any q-vector A, where $\underline{\theta} = (\theta_1, \ldots, \theta_d)$, $\underline{\theta} \cdot \underline{x} = \theta_1 x_1 + \cdots + \theta_d x_d$ and $|\underline{\theta}| = \max(|\theta_1|, \ldots, |\theta_d|)$. Thus, $\hat{L}^h(\underline{\theta})$ is a $q \times q$ matrix of polynomials in $e^{\pm i\theta_j}$, $j = 1, \ldots, d$, obtained from L^h by replacing each T_j with $e^{i\theta_j}$.

Also for simplicity we will deal here only with ***homogeneous operators*** L^h, i.e., operators for which all terms in $\det L^h$ (the determinant of L^h) have the same power in h. (This means that L^h approximates a homogeneous differential operator L, i.e., $\det L$ is a homogeneous polynomial in $\partial/\partial x_1, \ldots, \partial/\partial x_d$. All examples above are homogeneous). For homogeneous difference operators, the general notion of ellipticity measure on a given scale (cf. [2, Sec. 3.1] or [3, Sec. 2.1]) is not needed, and we can use the following simpler definition.

Definition. The homogeneous difference operator L^h is ***elliptic of order*** $2m$ iff

$$|\det \hat{L}^h(\underline{\theta})| \geq C h^{-2m} \sum_{j=1}^{d} \theta_j^{2m} \quad \text{for all } |\underline{\theta}| \leq \pi, \tag{2.4}$$

where C is positive and independent of $\underline{\theta}$.

Ellipticity of differential operators is defined in the same way. (The parameter h is arbitrary then, and the range of $\underline{\theta}$ is unrestricted. It is thus more natural in the continuous case to replace $\underline{\theta}/h$ by another phase variable, $\underline{\omega} = \underline{\theta}/h$ say.) It is easy to see that both Δ and Δ^h are second-order elliptic. Generally, simplest central approximations to second-order scalar ($q = 1$) elliptic operators are themselves elliptic. But not all central approximations are. For example, the "skew Laplacian"

$$\Delta^{\times} = \frac{1}{2h^2}(T_{1,1} + T_{1,-1} + T_{-1,1} + T_{-1,-1} - 4T_{0,0}) = \frac{1}{2h^2}\begin{bmatrix} 1 & 0 & 1 \\ 0 & -4 & 0 \\ 1 & 0 & 1 \end{bmatrix} \tag{2.5}$$

or the "long Laplacian"

$$\Delta^{2h} = \frac{1}{4h^2}(T_{2,0} + T_{0,2} + T_{-2,0} + T_{0,-2} - 4T_{0,0}) \tag{2.6}$$

both approximating Δ, have the symbols

$$\hat{\Delta}^{\times}(\underline{\theta}) = \frac{1}{h^2}[(\cos\theta_1 - \cos\theta_2)^2 + \sin^2\theta_1 + \sin^2\theta_2]$$

$$\hat{\Delta}^{2h}(\underline{\theta}) = \frac{1}{h^2}(\sin^2\theta_1 + \sin^2\theta_2)$$

which clearly fail to satisfy (2.4). Indeed, $\hat{\Delta}^{\times}(\pi,\pi) = 0$ and $\hat{\Delta}^{2h}(\pi,0) = \hat{\Delta}^{2h}(0,\pi) = \hat{\Delta}^{2h}(\pi,\pi) = 0$. Whereas these examples seem somewhat artificial (although the skew Laplacian does naturally arise in various situations, e.g., in semi-implicit Lagrange codes [4, §IV] and for some kinds of finite elements [7]), non-elliptic operators are very common in approximations to elliptic ***systems*** ($q > 1$). The discrete Cauchy-Riemann (2.2) and Stokes (2.3) operators well represent this situation: They are the simplest (non-staggered) central approximations to elliptic operators, but $\det L^h_{CR} = \Delta^{2h}$ and $\det L^h_S = \Delta^h \Delta^{2h}$, hence they do not satisfy (2.4), their symbol vanishing wherever $\hat{\Delta}^{2h}$ does. Note that taking determinant commutes with passing to the symbol, hence ellipticity of L^h is equivalent to ellipticity of $\det L^h$, which in turn is equivalent to ellipticity of all factors of $\det L^h$.

Finite element discretizations of the same elliptic systems, with uniform non-staggered partitions, give rise to similarly non-elliptic difference operators. This is not usually recognized because finite element discretizations are seldom Fourier-analyzed as uniform-grid operators.

In all the above examples, even when L^h fails to satisfy (2.4), it still satisfies the weaker condition

$$|\det \hat{L}^h(\underline{\theta})| \geq Ch^{-2m} \sum_{j=1}^{d} \sin^{2m}(\theta_j), \quad \text{for all } |\underline{\theta}| \leq \pi, \tag{2.7}$$

where C is positive and independent of $\underline{\theta}$. The term ***quasi-elliptic*** was introduced in [8] to describe such operators.

Perhaps all reasonable approximations to homogeneous elliptic equations satisfy (2.7), but for the purpose of including some additional, not-so-reasonable approximations, we can extend the class of operators, and admit any homogeneous operators L^h for which $\det \hat{L}^h(\underline{\theta})$ vanishes only at a finite number of points. This class includes for example $\Delta^{2h\times} = (T_{2,2} + T_{2,-2} + T_{-2,2} + T_{-2,-2} - 4T_{0,0})/(8h^2)$, which satisfies neither (2.4) nor (2.7), but for which the methods described below are still applicable.

More generally, when inhomogeneous operators are also admitted, our methods will extend to any operator L^h with $O(1)$ "measure of quasi-ellipticity", defined by

$$E^{h,\alpha}(L^h) = \min_{\alpha \geq |\underline{\theta}| \geq |\underline{\theta}'|} |\det L^h(\underline{\theta})| / |\det L^h(\underline{\theta}')|, \tag{2.8}$$

for some reasonable $\alpha > 0$. $E^{h,\pi}$ is the usual measure of ellipticity E^h described in [3]. The methods here will in principle work for any positive α, although they will gradually deteriorate with the decrease of α for which $E^{h,\alpha}(L^h)$ is still $O(1)$.

For clarity, we discuss below only homogeneous operators, and the strict quasi-ellipticity (2.7) is assumed.

3. Instability and Inaccuracy

Quasi-elliptic operators do meet some general stability requirements even if they do not satisfy (2.4). For example, the skew Laplacian (2.5) is a positive type operator, hence satisfying the maximum principle. The associated matrix has a dominant diagonal. Nevertheless, in a certain sense such operators are not quite stable. Namely, since $\det \hat{L}^h(\underline{\theta}) = 0$ for some $\underline{\theta} \neq 0$, in an infinite space, or under periodic boundary conditions, there exists a highly-oscillating function $v^h(\underline{x}) = A\exp(i\,\underline{\theta}\cdot\underline{x}/h)$ which satisfies the homogeneous equation $L^h v^h(\underline{x}) \equiv 0$. Hence the solution, unlike the corresponding differential solution, is not unique-upto-an-additive-constant; it contains an undetermined highly-oscillating component. Similarly, in any bounded domain with any boundary conditions, functions $w^h(\underline{x})$ close to $v^h(\underline{x})$ (e.g., $w^h = \varphi_1 v^h + \varphi_2$, φ_j being smooth) exist which satisfy the boundary conditions and for which $L^h w^h$ is everywhere very small. Such w^h therefore forms an unstable mode: A small change in the equation can introduce a large change proportional to w^h. This is a kind of numerical instability, since a corresponding large change in the differential solution cannot occur.

This numerical instability need not hurt much: If the differential system is $LU = F$ and the discrete system is $L^h U^h = F^h$, all one has to do is to define $F^h = I^h F$, say, through an averaging operator I^h which liquidates the unstable modes, i.e. $\hat{I}^h(0) = 1$ and the ratio $\hat{I}^h(\underline{\theta})/\hat{L}^h(\underline{\theta})$ is uniformly bounded for all $|\underline{\theta}| \geq \mathcal{E} > 0$. For example, one can take $I^h = S^h I'^h$, where I'^h is any F averaging suitable for the fully elliptic case and S^h is like the solution averaging S^h described below. Even this is unnecessary in the usual, smooth case (in the same way that the above rule for I^h is frequently neglected for fully elliptic L^h), because the unstable modes, even when unduly magnified by the discretization, are usually still small.

Generally, the main disadvantage of quasi-elliptic operators is a certain loss of accuracy compared to corresponding truely elliptic operators, which is simply due to the larger differencing steps taken in certain terms of the quasi-elliptic scheme. In some cases this is particularly obvious, since the grid is locally decoupled into several subgrids which are not connected to each other by the quasi-elliptic operator. For example, the skew-Laplacian (2.5) introduces no coupling between red and black points (in the usual sense of checkerboard coloring, one color being associated with gridpoints where $(x_1 + x_2)/h$ is odd, the other with even). On each subgrid the discretization looks like the compact Laplacian (2.1) on a rotated grid with meshsize $h_1 = \sqrt{2}\,h$. Similarly, in case of (2.2), the grid is decoupled into 4 ***staggered*** subgrids, with meshsizes $2h$, on each of which the operator has good ellipticity (being in fact equivalent to the staggered-grid approximation described in

[3, §17.2] or [5, §5.2]). Thus, since the approximation is $O(h^2)$, the error in case of (2.5) is on the average twice larger, and in case of (2.2) four times larger, than the errors in corresponding fully elliptic approximations (assuming other discretization errors, related for example to the representation of right-hand sides or boundary conditions, behave similarly). In these cases, in other words, each of the subgrids can produce the resulting accuracy by itself, other subgrids only add work.

When derivatives are calculated from the solution, however, the approximating difference quotients may show much greater loss of accuracy, because they involve differences between values belonging to different subgrids. The error in ℓ-order derivatives will generally be $O(h^{-\ell})$ times the errors in the function itself. This excessive error can be avoided by taking differences only from one subgrid at a time, or, more generally, by using only difference operator D^h such that $\hat{D}^h(\underline{\theta})$ vanishes wherever $\hat{L}^h(\underline{\theta})$ does. In case of (2.3), for example, derivatives of the third unknown function (the pressure) should be approximated by long differences such as ∂_j^c, $\partial_j^c\partial_k^c$, etc.

The instability described above can also be removed, and the inaccuracy in derivatives proportionally reduced, by ***averaging the solution***, that is, by replacing the computed solution u^h by $S^h u^h$, where S^h is an averaging operator which removes all the unstable components. In other words, $\hat{S}^h(0) = 1$ and outside a neighborhood of $\underline{\theta} = 0$ the ratio $\hat{S}^h(\underline{\theta})/\hat{L}^h(\underline{\theta})$ should be uniformly bounded (wherever defined). For the quasi-elliptic L^h satisfying (2.7) there always exists such an averaging operator of the form

$$S^h = \Pi_{j=1}^d \left(\tfrac{1}{2}T_j^{1/2} + \tfrac{1}{2}T_j^{-1/2}\right)^{m_j}, \tag{3.1}$$

with integral $m_j \le m$. On the other hand, the averaging may further reduce the accuracy of the solution. With the averaging (3.1) the lost accuracy is $O(h^2)$. One can make that loss $O(h^{2s})$ by taking for example

$$S^h = \Pi_{j=1}^d \left(1 - \left(\tfrac{1}{2} - \tfrac{1}{4}T_j - \tfrac{1}{4}T_j^{-1}\right)^s\right)^{m_j}. \tag{3.2}$$

Another slight difficulty typical to quasi-elliptic approximations is the need to define extra boundary relations. This can satisfactorily be done by extrapolation (cf., e.g., §8.2).

In summary, although quasi-elliptic discretizations are in principle inferior to fully elliptic ones (obtainable for systems by grid staggering), they ***can*** be used. Since many programmers consider grid staggering a serious complication, especially near general boundaries, quasi-elliptic schemes and their fast solution become important.

4. Multigrid Troubles and Their Implications

Usual multigrid solvers yield ***poor asymptotic convergence rates*** when applied to quasi-elliptic schemes (see [4] and §8.4 below). The reason is simple: Slow to converge are the unstable modes, such as v^h or w^h above. They cannot significantly converge by coarse-grid corrections, since they are high-frequency modes, essentially invisible on coarser levels. Neither can they significantly converge by any type of relaxation, since an error like w^h shows a very small residual function $L^h w^h$ (compared with residuals shown by other modes with comparable amplitude) and the corrections introduced by any relaxation scheme are proportional to the size of the residuals (cf. [3, Sec. 1.1]). In particular, the amplification factor $\mu(\underline{\theta})$ of the error mode $\exp(i\,\underline{\theta}\cdot\underline{x}/h)$ per relaxation sweep must be 1 when $\hat{L}^h(\underline{\theta})=0$, and since the latter equality holds for some $|\underline{\theta}|=\pi$, the smoothing factor $\bar{\mu}=\max_{\pi/2\le|\underline{\theta}|\le\pi}|\mu(\underline{\theta})|$ cannot be smaller than 1.

The poor asymptotic rates are ***not a real trouble***, though. The modes slow to converge are exactly those unstable modes for which algebraic convergence is not really desired, their amplitudes in the algebraic solution being unrelated to their amplitudes in the differential solution. The only concern is that these amplitudes will remain suitably small.

This situation is ***typical to all problems which are not fully elliptic***, including most problems in fluid dynamics: Slow asymptotic convergence of suitable multigrid cycles occur exactly in those components where not much convergence is needed anyway. Whenever this situation arises, it is in a sense an absurd to try and fix the algorithm (although we show in Sec. 7 below how to do it), since one would then often end up investing most of his human and computer resources to obtain improvements which are meaningless in terms of solving the original ***differential*** equations.

Thus, the real objective of multigrid solvers should not be a fast ***algebraic convergence*** (convergence of the computed solution u^h to the exact discrete solution U^h), but fast ***differential convergence*** (convergence of u^h to the true differential solution U), using any sequence of meshsizes h and measured directly in terms of the decrease in $\|u^h-U\|$ as function of the overall computational work (cf. [3, §13]). This modified objective allows for simpler algorithms, but also calls for some modifications in our approach for analyzing algorithms, for apriori predicting and aposteriori measuring their performance. The next two sections will illustrate these modifications for the case of quasi-elliptic schemes.

5. Modified Mode Analysis

It was shown in Sec. 4 that in case of quasi-elliptic systems $\bar{\mu} \geq 1$, but that this bad smoothing factor is not relevant to our real objective. To analyze a given relaxation scheme, assume first that it is as efficient as needed for the ***differential*** convergence of the highest frequency modes (which should latter be checked by the 2-level FMG mode analysis mentioned below). The question then is what efficiency one should expect from the multigrid cycle (employing the given scheme on all levels) in reducing all other modes. As in the conventional smoothing analysis, our simplifying assumption here will be that relaxation on each level should efficiently treat all modes in only one segment of modes, and that the union of these segments should cover all relevant modes. Instead of assigning to grid h the conventional segment $\pi/2 \leq |\underline{\theta}| \leq \pi$, however, we can assign to it any segment of the form $\alpha/2 \leq \|\underline{\theta}\| \leq \alpha$, with any norm $\|\underline{\theta}\|$. That would automatically assign to grid $h/2$ the segment $\alpha/4 \leq \|\underline{\theta}\| \leq \alpha/2$, and so on. It means that we allow some of the highest frequency components on any intermediate level not to converge efficiently by relaxation on that level, as long as those components efficiently converge by the next-finer-level relaxation. This only leaves the highest frequency modes on the finest grid unaccounted for, which is exactly the segment where we do not seek simple algebraic convergence. Thus, the modified definition of the smoothing factor relevant for our purpose here is

$$\bar{\mu} = \min \max_{\alpha/2 \leq \|\underline{\theta}\| \leq \alpha} |\mu(\underline{\theta})|, \tag{5.1}$$

where $\mu(\underline{\theta})$ is the amplification factor of $\exp(i\,\underline{\theta} \cdot \underline{x}/h)$ per relaxation sweep, and the minimum can be taken over all $\alpha > 0$ and over all possible choices of the norm $\|\cdot\|$. (For a generalization of this definition to cases of semi coarsening, cf. [3, §12]).

In case of the skew Laplacian (2.5), for example, the lexicographically order Gauss-Seidel relaxation yields the amplification factor

$$\mu(\underline{\theta}) = e^{i\theta_1} \cos\theta_2 / (2 - e^{-i\theta_1} \cos\theta_2), \tag{5.2}$$

so that $\mu(\pi, \pi) = 1$ and the conventional smoothing factor is 1. But choosing $\alpha = \pi$ and $\|\underline{\theta}\| = \max(|\theta_1 + \theta_2|, |\theta_1 - \theta_2|)$ easily shows that the modified factor (5.1) yields $\bar{\mu} \leq .5$. The same can be shown for the long Laplacian (2.6), by taking $\alpha = \pi/2$ and $\|\underline{\theta}\| = |\underline{\theta}|$.

In case of systems ($q > 1$), the $q \times q$ amplification matrix μ of the mode $A \exp(i\,\underline{\theta} \cdot \underline{x}/h)$ depends both on $\underline{\theta}$ and on the q-vector A. The modified smoothing factor $\bar{\mu}$ is then defined by

$$\bar{\mu} = \min \max_{\alpha/2 \leq \|\underline{\theta}\| \leq \alpha} \|\mu A\| / \|A\|$$

where α is allowed to depend on both $\underline{\theta}/|\underline{\theta}|$ and $A/|A|$. With these definitions and suitable distributed Gauss Seidel (DGS) relaxation schemes (see e.g. [3, §18.6]) this again yields $\bar{\mu} \leq .5$, for both the Cauchy-Riemann (2.2) and the Stokes (2.3) operators. In all these cases, still better factors are obtained by four-color ordering, for which definitions (5.1) and (5.2) should further be extended (cf. (3.2) in [3]).

As for ***two-level analyses*** (cf. [3, §4.1] or[5, §4.6]), they always couple lowest with highest frequency modes. In non-elliptic cases some highest-frequency modes are not expected to converge fast. What the analysis should then tell us is how efficient is the entire multigrid algorithm in reducing the algebraic errors below the truncation errors. This can be done by a ***two-level FMG mode analysis***, which Fourier analyzes the N-FMG algorithm described below (usually for $N = 1$) by assuming exact solution of the coarse grid equations (both for obtaining the first approximation and in each of the N cycles) and by comparing for each mode the final algebraic error with the truncation error (see [3, §7.4]).

6. FMG Solution to Truncation Level

Since the multigrid cycling is inefficient in reducing unstable mode errors, the multigrid solver should take care not to start with an initial solution which contains large amplitudes of such errors. The overall initial error in unstable modes should better be smaller than the overall truncation error. This is easily obtained by taking a first approximation from a coarser grid, employing interpolation of suitable order. The usual "Full multigrid" (FMG; also called "nested iteration") algorithm can therefore be used, with slight modifications. The usual algorithm and its modifications are briefly described in the following. For a flowchart, and a detailed discussion of FMG algorithms and the order of the first interpolation, see Secs. 1.6 and 7 in [3]. For simplicity we describe here the Correction Scheme (CS) version of the algorithm, so the problems are assumed linear; it should be converted to Full Approximation Scheme (FAS) to treat nonlinear problems [3, §8].

6.1 Multigrid cycle. A sequence of grids is given with meshsizes h_k ($k = 1, 2, 3, \ldots$), where $h_{k+1} = h_k/2$. On the h_k grid the discrete equations have the form

$$L^k U^k = F^k \tag{6.1}$$

where L^k approximates L^{k+1}. Given u_0^k, an approximate solution to (6.1), the multigrid cycle MG for producing an improved approximation u_1^k

$$u_1^k \leftarrow MG(k, u_0^k, F^k) \tag{6.2}$$

is recursively defined as follows:

If $k = 1$ solve (6.1) by any direct or iterative method, yielding the final result u_1^k. Otherwise do (A) through (D).

(A) Perform ν_1 relaxation sweeps on (6.1), resulting in a new approximation $\bar{u}^k$.

(B) Starting with $u_0^{k-1} = 0$, make γ successive cycles

$$u_j^{k-1} \leftarrow MG(k-1, u_{j-1}^{k-1}, I_k^{k-1}(F^k - L^k \bar{u}^k)), \qquad (j = 1, \ldots, \gamma)$$

where I_k^{k-1} is a transfer ("reduction") of residuals from grid h_k to grid h_{k-1}. We have used the "full weighting"

$$\begin{aligned} I_k^{k-1} = {} & \tfrac{1}{4} T_{0,0} + \tfrac{1}{8}(T_{0,1} + T_{1,0} + T_{0,-1} + T_{-1,0}) \\ & + \tfrac{1}{16}(T_{1,1} + T_{1,-1} + T_{-1,1} + T_{-1,-1}). \end{aligned} \tag{6.3}$$

(C) Calculate $\tilde{u}^k = \bar{u}^k + I_{k-1}^k u_\gamma^{k-1}$, where I_{k-1}^k is a suitable interpolation ("prolongation") from grid h_{k-1} to grid h_k. For problems considered here, bilinear interpolation is used.

(D) Perform ν_2 relaxation sweeps on (6.1), starting with $\tilde{u}^k$ and yielding the final result u_1^k.

The cycle with $\gamma = 1$ is called ***V cycle*** or $V(\nu_1, \nu_2)$, and the one with $\gamma = 2$ is called ***W cycle*** or $W(\nu_1, \nu_2)$.

6.2 Full Multigrid (FMG). The N-FMG is an algorithm for calculating an approximate solution

$$u_N^k = FMG(k, F^k, N) \tag{6.4}$$

to equation (6.1), defined recursively as the following two successive steps.

(a) Calculating a first approximation u_0^k: If $k = 1$, put $u_0^k = 0$. Otherwise put

$$u_0^k = \Pi_{k-1}^k FMG(k-1, I_k^{k-1} F^k, N), \tag{6.5}$$

where Π_{k-1}^k is an interpolation of solutions from grid h_{k-1} to grid h_k, and I_k^{k-1} is some transfer ("averaging") from grid k to grid $k-1$, usually a full weighting of the type (6.3). The interpolation Π_{k-1}^k should usually be of order higher than that of the correction interpolation I_{k-1}^k mentioned above [3, §7.1]. In our experiments bicubic interpolation was used.

(b) Improve the first approximation by N successive MG cycles

$$u_j^k \leftarrow MG(k, u_{j-1}^k, F^k), \qquad (j = 1, \ldots, N)$$

as defined in Sec. 6.1.

6.3 Averaging. The algorithm above is the conventional one, and for equations with constant coefficients it requires no modifications. In case of quasi-elliptic equations with variable coefficients, and in particular in case of non-linear equations, it is not enough to guard against initial unstable errors, because such errors can also later be introduced due to interaction between modes. It is then better to explicitly reduce the unstable modes by averaging, such as (3.1) or (3.2). It may also then be important to replace $I_{k-1}^k u^{k-1}$ in step (C) above by $I_{k-1}^k S^{h_{k-1}} u^{k-1}$. In fact, experiments with non-staggered Navier-Stokes equations (cf. Sec. 8.2) gave slowly diverging MG cycles unless this averaging was used.

6.4 Measuring convergence. In various situations where algebraic convergence is not attempted, as in the present algorithm and double discretization [3, §10.2] and other algorithms, the question is raised how to measure convergence; how to know, in particular, that a solution to the truncation level (i.e., with algebraic errors dominated by discretization errors) has been obtained.

The answer is that solution to the truncation level is not really the important information when ***differential*** convergence is our objective (as it should most often be), because: (i) Solving to truncation level tells us nothing about the truncation error itself. We may for instance be doing good job in solving the algebraic system due to having chosen an easy-to-solve but badly-approximating discretization. (ii) A smaller differential error may often be obtained faster by switching to a finer grid before the equations on the present grid have been solved to truncation level.

The important information is the differential convergence itself, as function of computational work. This very information can directly be obtained from the N-FMG algorithm. Indeed, the sequence of approximations u_N^k, $(k = 1, 2, \ldots)$ is a sequence converging to the differential solution, hence the decrease in the sequence of differences $\delta_k = \|\bar{I}_k^{k-1} u_N^k - u_N^{k-1}\|$ exactly exhibit the speed of differential convergence, where the norm $\|\cdot\|$ used to measure δ_k can be chosen to exactly represent the sense in which convergence is sought. One only has to check that the smallness of δ_k is not governed by lack of change from u_0^k to u_N^k. It is enough for this purpose to check that the suitable residual norm $r_N^k = \|F^k - L^k u_N^k\|$ is considerably smaller than r_0^k. One can usually also verify that the algebraic errors are below truncation level, e.g., by confirming that r_N^k / r_0^k is considerably smaller than δ_k / δ_{k-1}.

7. Algorithm for Fast Algebraic Convergence

Although fast asymptotic algebraic convergence is not needed for fast differential convergence, it can still be produced by a more involved multigrid algorithm. This algorithm (also described in [6]) may be interesting in its own right, since it is the simplest example of a new kind of algorithms (first mentioned in [1, §3.2], and more fully in [8]) for solving problems with highly-oscillatory solutions, including highly indefinite problems.

7.1 Multiple coarse grid corrections. Let $\underline{\theta}^1, \underline{\theta}^2, \ldots, \underline{\theta}^\ell$ be all the components for which $\hat{L}^h$ vanishes, or, more generally, the centers of all neighborhoods in which $\hat{L}^h(\underline{\theta})$ is small. Usually $\underline{\theta}^1 = 0$. Then (by [3, §1.1], for example) there exists a relaxation scheme with fast convergence for all Fourier components except those close to some $\underline{\theta}^j$. The error after few such relaxation sweeps must therefore have the form

$$V^h(\underline{x}) = \sum_{j=1}^{\ell} V_j^h(\underline{x}) \exp(i\,\underline{\theta}^j \cdot \underline{x}/h), \tag{7.1}$$

where V_j^h are smooth functions. Whereas classical multigrid seeks to approximate V^h on a coarser grid, and the algorithm of Sec. 6 approximates V_1^h, the new algorithm will separately approximate each of the V_j^h, by successively employing ℓ different coarse-grid corrections.

Generally, denoting by H the coarser-grid meshsize ($H = 2h$), the equations for V_j^H, the coarse-grid approximation to V_j^h, should have the form

$$L_j^H V_j^H = I_{h,j}^H R^h \tag{7.2}$$

where $\hat{L}_j^H(\underline{\theta}) \approx \hat{L}_j^h(\underline{\theta}^j + \underline{\theta})$ for small $\underline{\theta}$, and $\hat{I}_{h,j}^H(\underline{\theta}^k) \approx \delta_{jk}$ (=0 except for $\delta_{jj} = 1$). The boundary conditions may couple V_j^H and V_k^H on any piece of boundary along which $\exp(i(\underline{\theta}^j - \underline{\theta}^k) \cdot \underline{x}/h)$ is a smooth function. There are various ways, variational ones and more direct ones, to derive L_j^H, $I_{h,j}^H$ and the boundary conditions. There also exist various ways for solving (7.2). In highly indefinite problems the latter leads to creating more components on grid $4h$, etc., so that on increasingly coarser grids the representation tends to a Fourier representation.

Here we give only the very simple example of solving for the skew Laplacian (2.5). (For a more general case, see [6].) In this case $\ell = 2$, $\underline{\theta}^1 = (0,0)$ and $\underline{\theta}^2 = (\pi, \pi)$, and one may simply take $L_1^H = L_2^H$ to be any H-approximation to the Laplace operator. In some situations, where the same mechanism creates both the fine grid and the coarse grid equations, these L_j^H may again be skew-Laplacians. As transfer operators one can use

$$I_{h,1}^H = \frac{1}{16}\begin{bmatrix}1 & 2 & 1\\ 2 & 4 & 2\\ 1 & 2 & 1\end{bmatrix} \quad \text{and} \quad I_{h,2}^H = \frac{1}{16}\begin{bmatrix}1 & -2 & 1\\ -2 & 4 & -2\\ 1 & -2 & 1\end{bmatrix}. \tag{7.3}$$

Considering the case that the fine-grid boundary conditions are Dirichlet conditions identically satisfied by any fine-grid approximation, the coarse-grid boundary conditions for both V_1^H and V_2^H are the homogeneous Dirichlet conditions. For solving the coarse-grid equations (7.2), the MG cycle of Sec. 6.1 can be used, ***even in the case that*** L_j^H ***are themselves quasi-elliptic***, because, for the purpose of accelerating the fine-grid algebraic convergence, equations (7.2) need to be solved each time only to their truncation level (i.e., only to the level of the error $V_j^H - V_j^h$). In case of similar equations but with non-constant coefficient, averaging as in Sec. 6.3 should better be used.

7.2 The modified algorithm. Given an approximate solution u_0^k to (6.1), the modified multigrid cycle MMG for producing the improved solution u_1^k

$$u_1^k \leftarrow MMG(k, u_0^k, F^k) \tag{7.4}$$

is defined ***non***-recursively as follows:

If $k = 1$ solve (6.1) by any direct or iterative method, yielding the final u_1^k. Otherwise, perform ν relaxation sweeps on (6.1), resulting in a new approximation $u^{k,0}$, and then, for $j = 1, 2, \ldots, \ell$, calculate

$$\begin{aligned} v^{k-1,j} &\leftarrow MG\left(k-1, 0, I_{h_k,j}^{2h_k}(F^k - L^k u^{k,j-1})\right) \\ u^{k,j} &= u^{k,j-1} + \exp(i\,\underline{\theta}^j \cdot \underline{x}/h)\, I_{k-1}^k v^{k-1,j} \end{aligned}$$

with $u_1^k = u^{k,\ell}$ being the final result. I_{k-1}^k again denotes linear interpolation. MG is the cycle defined in Sec. 6.1, with a choice of γ, ν_1, ν_2.

With this MMG cycle replacing the MG cycle, the modified FMG algorithm is defined in the same way as FMG in Sec. 6.2.

7.3 Modified smoothing analysis. The smoothing factor for the above MMG cycle, i.e., the ideal factor of convergence one can expect from such a cycle per relaxation sweep on the finest grid is defined by

$$\bar{\mu} = \max_{\pi/2 \le |\underline{\theta} - \underline{\theta}^j| \text{ for one } j,\ |\underline{\theta}| \le \pi} |\mu(\underline{\theta})|, \tag{7.5}$$

where $|\mu(\underline{\theta})|$ is the spectral radius of the amplification matrix (or the absolute value of the amplification factor, if $q = 1$). Note that for $\ell \geq 2^d$, the domain of $\underline{\theta}$ over which the maximum is taken may be empty. In such a situation convergence can in principle be obtained without any relaxation on the finest grid. This does not mean that the algorithm is more efficient than a conventional multigrid, because it employs at least ℓ times as many relaxation sweeps on each coarser grid.

A more precise ***two-level analysis*** can of course be made here in the conventional way [3, §4.1].

For the skew-Laplacian and the algorithm described above, the lexicographic Gauss-Seidel amplification factor (5.2) attains its maximum (7.5) at $(\pm\pi/2, 0)$ and at $(\pm\pi/2, \pi)$, yielding $\bar{\mu} = .447$.

8. Numerical Experiments

8.1 The skew Laplacian problem. Our main experimental studies were conducted with the skew Laplacian scheme (2.5) in the rectangle $\{0 \leq x_1 \leq 2, 0 \leq x_2 \leq 3\}$ with Dirichlet boundary conditions. These conditions and the right-hand side of the differential equation $\Delta U = F$ were chosen so that the solution U of the differential equations is known, to allow direct measurements of discretization errors. The sequence of grids have meshsizes $h_k = 2^{1-k}$ $(k = 1, 2, \ldots)$, each positioned so that the boundaries of Ω coincide with grid lines. On every level L^k is the skew Laplacian, and the relaxation is lexicographic Gauss-Seidel. The algorithms were those described in Secs. 6 and 7.

Table 1 shows the maximal differential error (maximal differences between computed and differential solutions) on various grids. In addition, columns headed by ∂ or ∂^c show maximal error in first derivatives, approximated either at grid midpoint by short difference quotients (the ∂ columns), or at gridpoints by ∂_j^c (the ∂^c columns). The upper part of the table gives these errors for the exact discrete solution, the lower part – for the solution obtained by a 1-FMG algorithm with $V(2,1)$ cycles. For grid 5 an additional result (5a) is sometimes given: It shows errors measured after the solution is $\underline{\text{a}}$veraged by $(T_1^{1/2} + T_1^{-1/2})/2$ (cf. (3.1)). The table compares skew-Laplacian with usual (compact) Laplacian (using the same meshsize and the same relaxation), and a case of smooth solution with a highly-oscillatory case. The latter is shown in order to emphasize how bad quasi-elliptic schemes ***can*** be. In practice such highly oscillatory components have very small amplitudes: If their amplitudes are bigger than $O(h^2)$ (here $h_5^2 = .001$), then second-order approximations cannot be obtained by ***any*** discretization. In the highly oscillating case it was of course necessary to use the full weighting (6.3) for I_k^{k-1} in (6.5); this was started with $k = 7$. In the smooth case, however, injection of F was used, in order to obtain a clearer picture, clean of F-averaging errors.

TABLE 1		$U = \sin(3x + 2y)$					$U = x(2-x)y(3-y)\cos\frac{\pi(x+y)}{h_5}$				
		$\Delta^\times$			Δ^h		$\Delta^\times$			Δ^h	
	grid		∂	∂^c		∂		∂	∂^c		∂
Exact	3	.1703	.410	.271	.0517	.108	2.25	3.93	3.37	.0152	.030
	4	.0417	.115	.092	.0129	.033	2.25	4.22	3.93	.0038	.008
	5	.0104	.031	.027	.0032	.009	608.	19494.	979.	.0009	.002
	5a	.0084	.027	.031			30.6	979.	38.1		
1-FMG	3	.1709	.436	.276	.0606	.151	2.25	3.93	3.37	.0198	.039
	4	.0418	.109	.092	.0169	.048	2.25	4.22	3.93	.0055	.012
	5	.0105	.030	.027	.0045	.014	21.5	752.	64.7	.0014	.003
	5a	.0085	.027	.031			2.13	64.7	6.41		

8.2 The Stokes and Navier-Stokes problems. We have also conducted experiments with the Stokes operator (2.3), described in detail in [3, §18.6] (with slight improvements, to be described in the new edition). The unknown grid functions of this operator are U^h, V^h and P^h – the discrete horizontal velocity, vertical velocity and pressure, respectively.

In the differential problem only velocities are normally given on the boundary. In the non-staggered discretization (2.3) some boundary conditions for P^h should be introduced (which is a disadvantage typical to many quasi-elliptic operators). For clarity of exposition we here avoid this issue by showing results for ***periodic boundary conditions*** (adjusting undetermined additive constants before measuring errors).

The exact treatment of boundary conditions is important only in measuring ***asymptotic*** convergence rates. It does not much affect results of 1-FMG. Therefore we will show such results also for the Dirichlet boundary conditions. In these experiments P^h at each boundary point is taken equal to the nearest interior value of P^h, and it changes whenever the latter does. This does not correspond to Neumann boundary conditions, but to coupling the four subgrids into which the P^h grid decouples. A partial relaxation sweep near Dirichlet boundaries is performed before each full relaxation sweep.

The relaxation employed is ***distributed Gauss-Seidel*** (DGS), a special case of a scheme for relaxing general PDE systems, explained in [3, §3.7]. Briefly, it is equivalent to writing $U^h = \varphi_1^h - \partial_1^c\varphi_3^h$, $V^h = \varphi_2^h - \partial_2^c\varphi_3^h$ and

$P^h = -\Delta^h \varphi_3^h$, and relaxing by usual Gauss-Seidel the resulting equations in φ_j^h. The changes in the latter imply changes in U^h, V^h and P^h, which define the actual changes performed by the DGS relaxation. The relaxation ordering is 4-colored, relaxing the four mentioned subgrids one at a time.

The domain for this problem is the square $\{0 \le x_j \le 2\pi\}$. The meshsizes are $h_k = 2^{-k}\pi$. The right-hand side and the boundary conditions are chosen so as to give the prescribed solution $U = V = P = \sin(\cos(x_1 + 2x_2))$, a periodic solution which includes many Fourier modes. The discrete right-hand sides were calculated by $F^{k-1} = I_k^{k-1} F^k$, using (6.3), starting at $k = 8$.

Some experiments were conducted with averaging (cf. Sec. 6.3). In the present case this means averaging of P^h only, since U^h and V^h vanish in the unstable modes. When used, this *P-averaging* employed (3.1), with $m_1 = m_2 = 2$, performed on P^h in any solution or correction just before interpolating it to a finer grid.

Also mentioned below are experiments with non-stagered ***incompressible Navier-Stokes (INS) equations***, with procedures similar to those for Stokes. For details see [3, §19]; the modification from staggered to non-staggered formulation and processing are the same as for the Stokes equations. Results are given for the Dirichlet problem (U and V given on the boundary, P on the boundary treated as above), for the case $U = V = P = 1 + .2\sin(\cos(x_1 + 2x_2))$. We have experimented with small and large Reynolds numbers, Re. In the latter case anisotropic artificial viscosity was used in relaxation, its magnitude being 1.4 times the viscosity introduced by upstream differencing. Central differencing without artificial viscosity was used for the fine-to-coarse residual calculations, allowing $O(h^2)$ solutions to be obtained. The large Re PDE problem is not elliptic (more precisely, it has small ellipticity measure), so its detailed discussion is beyond the scope here. Indeed, the present example is not fully typical for large Re, because it has no boundary layers and no gridline-streamline alignments.

Table 2 summarizes four numerical experiments: Three with the Stokes ($Re = 0$) problem (exact solutions for the periodic ("Per.") boundary conditions; 1-FMG solutions with $W(2,1)$ cycles for the same problem; and similar 1-FMG solutions for the Dirichlet ("Dir.") problem), and one experiment for "infinite" Re, i.e., with viscosity completely dominated by artificial viscosity. The latter experiment uses 2-FMG algorithm with $W(2,0)$ cycles, because double discretization (different artificial viscosities at different stages) is involved (cf. [3, §10.2]). For each experiment and

TABLE 2		Non-staggered Navier-Stokes: No P-averaging			Non-staggered Navier-Stokes: P-averaging			Staggered Nav.-Stokes	
	grid		∂	∂^c		∂	∂^c		∂
$Re = 0$	5	.00084	.0030	.0108	.00084	.0030	.0108	.00079	.0055
Per.		.00500	.0394	.0392	.00500	.0394	.0392	.00150	.0094
		.00542	.0395	.0395	.00542	.0395	.0395	.00513	.0212
Exact	6	.00024	.0007	.0026	.00024	.0007	.0026	.00018	.0013
Sol.		.00123	.0098	.0098	.00123	.0098	.0098	.00037	.0025
		.00135	.0100	.0108	.00135	.0100	.0108	.00127	.0054
$Re = 0$	5	.00090	.0036	.0113	.00097	.0031	.0113	.00080	.0052
Per.		.00661	.2086	.0555	.00978	.3452	.0682	.00163	.0215
		.00562	.0447	.0445	.00977	.0670	.0691	.00540	.0243
$W(2,1)$	6	.00024	.0008	.0027	.00025	.0008	.0027	.00018	.0013
1-FMG		.00136	.0536	.0146	.00216	.1346	.0181	.00036	.0036
		.00136	.0119	.0117	.00209	.0180	.0189	.00129	.0057
$Re = 0$	5	.00104	.0041	.0111	.00097	.0035	.0109	.00076	.0055
Dir.		.01285	.3715	.0851	.01480	.3946	.0763	.00198	.0176
		.00712	.0665	.0530	.00971	.0719	.0649	.00544	.0246
$W(2,1)$	6	.00027	.0011	.0028	.00026	.0008	.0027	.00017	.0013
1-FMG		.00337	.1586	.0451	.00371	.1776	.0264	.00047	.0038
		.00191	.0348	.0271	.00223	.0252	.0191	.00132	.0059
$Re = \infty$	5	.00272	.0433	.0253	.00215	.0250	.0097	.00168	.0180
Dir.		.01515	.4536	.2382	.00637	.2098	.0256	.00242	.0832
		.00957	.1945	.1594	.00273	.0168	.0147	.00106	.0076
$W(2,0)$	6	.00138	.0547	.0357	.00088	.0154	.0046	.00039	.0064
2-FMG		.01517	.8227	.4913	.00142	.0830	.0085	.00066	.0436
		.01051	.4062	.3413	.00074	.0048	.0033	.00038	.0016

each grid k, the three numbers shown in the first column are $\max(\|u^k - U\|, \|v^k - V\|)$, $\|p^k - P\|$ and $\|\bar{p}^k - P\|$, where (u^k, v^k, p^k) is the solution obtained for that grid, $\bar{p}^k = \frac{1}{4}\Pi_{j=1}^{2}(T_j^{1/2} + T_j^{-1/2})p^k$ and $\|\cdot\|$ is the discrete L_1 norm per unit area. The three numbers in the next column (headed by "∂") are $\max_{j=1,2}\max(\|\partial_j^k u^k - \partial_j U\|, \|\partial_j^k v^k - \partial_j V\|)$, $\|\partial_j^k p^k - \partial_j P\|$, and $\|\partial_j^k \bar{p}^k - \partial_j P\|$, where $\partial_j = \partial/\partial x_j$ and $\partial_j^k = (T_j^{1/2} - T_j^{-1/2})/h_k$. In the next column (headed by ∂^c), similar numbers are given, with the long difference quotient $\bar{\partial}_j^k = \partial_j^c = (T_j - T_j^{-1})/(2h_k)$ replacing ∂_j^k. The next 3 columns show similar sets of results for the case that P-averaging is used. The remaining 2 columns give for comparison results obtained on a *staggered* grid with the same meshsize, without P-averaging. (Using $\bar{p}^k$ for approximation, especially of derivatives, still may pay if $\nu_2 = 0$).

8.3 Accuracy and Stability. Tables 1 and 2 clearly show that the exact quasi-elliptic solutions ($\Delta^{\times}$ and the non-staggered Stokes, the latter mainly in terms of P) are several times less accurate than the corresponding fully elliptic ones (Δ^h and staggered Stokes, respectively), but they are still $O(h^2)$. Errors in the highly oscillating case, exhibiting instability, could of course all be reduced to $O(1)$ (or $O(h_s^{-1})$ in derivatives) by enough F-averaging (see §3). Averaging the solution (row 5a, or the $\bar{p}^k$ results), or taking suitable long difference quotients, cure the worst behavior too, but also somewhat further reduce the smooth-component accuracy, which nevertheless remains $O(h^2)$.

8.4 Poor asymptotic algebraic convergence. Denote by λ the asymptotic convergence factor per multigrid cycle, i.e., $\lambda = (r_\ell / r_m)^{1/(\ell - m)}$ for sufficiently large ℓ, m and $\ell - m$, where r_ℓ is any error (or residual) norm measured at any fixed stage of the ℓ-th cycle. As expected (see §4.1), the usual cycles $MG(k, \ldots)$ yielded poor λ for quasi-elliptic schemes:

In case of the Skew Laplacian and $V(2,1)$ cycles, our experiments exhibited $\lambda = .845$ and $\lambda = .96$ for levels $k = 4$ and $k = 5$, respectively. The convergence rate $\log 1/\lambda$ is clearly $O(h^2)$, as the rate of a simple Gauss-Seidel solver for the compact Laplacian Δ^h. Indeed, on each subgrid (red or black) the relaxation does look like Gauss-Seidel for Δ^{h_1}, and the coarse grid corrections are no help in case the black residuals cancel the red ones in the transfer to grid $k - 1$. For comparison: $V(2,1)$ cycles for the compact Laplacian Δ^h with lexicographic Gauss-Seidel yield $\lambda \approx .12$ on all grids.

Similarly, for the periodic Stokes problem and $W(2,1)$ cycles, $\lambda = .80$ and $\lambda = .945$ were obtained on levels 4 and 5, respectively, exhibiting again $O(h^2)$ rate. The rates were almost identically the same whether P averaging was used or not. For comparison, for staggered-grid Stokes discretizations the red-black DGS relaxation gives $\lambda = .30$ and $\lambda = .20$ for the $W(1,0)$ and the $W(2,0)$ cycles, respectively. These same excellent rates are obtained both for the periodic and the Dirichlet boundary conditions (provided some local relaxation near boundaries is added in the latter case). The same results are obtained for the Navier-Stokes problem with small Re. For large Re, divergence occur unless P-averaging is used (cf. §6.3).

8.5 FMG results. Despite the bad asymptotic convergence, Tables 1 and 2 clearly show that results obtained for the quasi-elliptic cases by short FMG algorithms are very good. In smooth cases they yield differential errors practically as small as in the exact discrete solutions. Moreover, in case of the unstable mode, the FMG results are visibly much better than the exact solution (precisely *because* the bad behavior is slow to enter). In case of non-linear equations (Table 2, $Re = \infty$) proper averaging (Sec. 6.3) is evidently necessary for good FMG results.

8.6 Asymptotic convergence with new algorithm. The $MMG(5, \ldots)$ cycle of §7.2 has been employed to solve the skew Laplacian problem with $\nu =$ 3 relaxation sweeps per cycle and with $V(2,1)$ used as the $MG(4, \ldots)$ inner cycle. For many cycles the convergence factor per cycle was steadily between .07 and .08, or a convergence factor of .425 per fine-grid relaxation, close to the value .447 expected by the smoothing mode analysis (§7.3).

References

[1] A. Brandt, Stages in developing multigrid solutions. *Numerical Methods for Engineering* (E. Absi, R. Glowinski, P. Lascaux, H. Veysseyre, eds.), Dunod, Paris, 1980, pp. 23-44.

[2] A. Brandt, Multigrid solvers for non-elliptic and singular-perturbation steady-state problems. The Weizmann Institute of Science, 1981.

[3] A. Brandt, Multigrid techniques: 1984 Guide with applications to fluid dynamics. Monograph. Available as GMD-Studie No. 85, from GMD-F1T, Postfach 1240, D-5205, St. Augustin 1, W. Germany.

[4] A. Brandt, J.E. Dendy, Jr. and H. Ruppel, The multigrid method for semi-implicit hydrodynamics codes, *J. Comp. Phys.* **34** (1980), pp. 348-370.

[5] A. Brandt and N. Dinar, Multigrid solutions to elliptic flow problems. *Numerical Methods for PDE* (S.V. Parter, ed.), Academic Press, 1979, pp. 53-147.

[6] J.E. Dendy, Jr., Multigrid semi-implicit hydrodynamics revisited. *Large Scale Scientific Computation* (S.V. Parter, ed.), Academic Press, 1984.

[7] V. Girault, Theory of finite difference methods on irregular networks, *SIAM J. Num. Anal.* **11** (1974), pp. 260-282.

[8] S. Ta'asan, Multigrid Methods for Highly Oscillatory Problems, Ph.D. Thesis, The Weizmann Institute of Science, Rehovot, Israel, 1984.

Progress in Scientific Computing, Vol. 6
Proceedings of U.S.-Israel Workshop, 1984

Secondary Instability of Free Shear Flows

Marc E. Brachet[1], Ralph W. Metcalfe[2],
Steven A. Orszag[3], James J. Riley[4]

Abstract

The three-dimensional stability of saturated two-dimensional vortical states of planar mixing layers and jets is studied by direct integration of the Navier-Stokes equations. Small-scale instabilities are shown to exist for spanwise scales at which classical linear modes are stable. These modes grow on convective time scales, extract their energy from the mean flow, and persist to moderately low Reynolds numbers. Their growth rates are comparable to the most rapidly growing inviscid instability and to the growth rates of two-dimensional subharmonic (pairing) modes. The three-dimensional modes do not appear to saturate in quasi-steady states. Indeed, they seem to lead directly to chaos. Results are presented for the resulting three-dimensional turbulent states.

[1] CNRS, Observatoire de Nice, 06-Nice, France
[2] Flow Research, Inc. Kent, WA 98032
[3] Princeton University, Princeton, NJ 08544
[4] University of Washington, Seattle, WA 98105

1. Introduction

Free shear flows like those of mixing layers and jets differ from wall-bounded flows in the sense that they are typically inflexional and, hence, are subject to inviscid instabilities. Thus, it may be thought that the process of transition to turbulence in free-shear flows would be inherently simple and amenable to analysis. Indeed, observations by Winant & Browand (1974), Brown & Roshko (1974), Wygnanski et al (1979), Ho & Huang (1982), Hussain (1984), and others show the central role played by two-dimensional dynamical processes through transitional regimes in these flows. While three-dimensional small scales are observed (Miksad 1972, Bernal et al 1979), they may not destroy the large-scale two-dimensional structure (Browand & Troutt 1980). In contrast, studies of wall-bounded flows have emphasized the central role of three-dimensional effects in their breakdown to turbulence.

In this paper, we investigate the nature of linear instabilities of saturated nonlinear two-dimensional flow states that arise from the primary inviscid instability of free shear flows. It is shown that these saturated two-dimensional states are subject to a class of strongly unstable three-dimensional modes that are present even at moderately low Reynolds numbers. It is possible that these three-dimensional instabilities can explain some of the initial stages of three-dimensional transition in free-shear flows. We find that the two-or three-dimensional character of these free-shear flows depend crucially on initial conditions as there is a close competition between the various modes of instability to be discussed below.

The approach followed here is similar to that used by Orszag & Patera (1980, 1981, 1983) in studies of secondary instabilities in wall-bounded flows. The parallel laminar flow is perturbed initially by a finite-amplitude two-dimensional disturbance that is allowed to evolve and to saturate in a quasi-steady state. The stability of this finite amplitude vortical state to both subharmonic (pairing) two-dimensional modes and smaller-scale three-dimensional modes is then studied by numerical solution of the full three-dimensional time-dependent Navier-Stokes equations. The character of the pairing instability was first explained theoretically by Kelly (1967) and numerically by Patnaik, Sherman & Corcos (1974) and Collins (1982) for stratified flows and by Riley & Metcalfe (1980) and Pierrehumbert & Widnall (1982) for unstratified flows; the present results confirm the strength of this kind of mode.

Pierrehumbert & Widnall (1982) have made a study of the linear two- and three-dimensional instabilities of a spatially periodic inviscid shear layer that is closely related to the present study. They consider the stability characteristics of the model family of two-dimensional vortex-modified mixing layers with velocity fields

$$u = \sinh z \,/(\cosh z - \rho \cos x)$$
$$w = -\rho \sin x/(\cosh z - \rho \cos x) \tag{1.1}$$

(Stuart 1967) for $0 \le \rho < 1$[+] and study subharmonic pairing instabilities and a new 'translative' three-dimensional instability. In contrast, we consider here both the linear and nonlinear stability characteristics of time-developing viscous shear layers. The three-dimensional secondary instability studied here is both the analog of the translative instability and the generalization of the instability analyzed by Orszag & Patera for wall-bounded flows.

[+] Note that for $\rho \ll 1$, the basic flow state (1.1) is of the form $\tanh z\,\hat{x} + \rho \mathrm{Re}[e^{ix}\tilde{v}(z)]$. This flow state is an inviscidly neutrally stable perturbation of the mixing layer $\tanh z\,\hat{x}$. At wavenumber 1, there are no primary two-dimensional instabilities that can compete with the subharmonic and secondary instabilities. In contrast, the results to be reported in Section 3 involve unstable primary perturbations to the mixing layer.

2. Numerical Methods

The Navier-Stokes equations are solved in the form

$$\frac{\partial \vec{v}}{\partial t} = \vec{v} \times \vec{\omega} - \nabla \pi + \nu \nabla^2 \vec{v} \tag{2.1}$$

$$\nabla \cdot \vec{v} = 0 \tag{2.2}$$

where $\vec{\omega} = \nabla \times \vec{v}$ is the vorticity and $\pi = p + 1/2\, v^2$ is the pressure head.
Periodic boundary conditions are applied in the streamwise, x, and spanwise, y, directions,

$$\vec{v}(x + \frac{4\pi}{\alpha}, y, z, t) = \vec{v}(x,y,z,t), \tag{2.3}$$

$$\vec{v}(x, y + 2\pi/\beta, z, t) = \vec{v}(x,y,z,t)$$

while the flow is assumed quiescent ($v \to U_{\pm} \hat{x}$, $U_{\pm}$ constants) as $z \to \pm \infty$. Note that the assumed periodicity length is $4\pi/\alpha$ to accommodate both the primary mode with x-wavenumber α and its subharmonic with x-wavenumber $\frac{1}{2}\alpha^{+}$.

$^{+}$ Pierrehumbert & Widnall (1982) point out that Floquet theory implies that the Navier-Stokes equations linearized about a flow periodic mix admit solutions of the more general form $v(x,y,z) = e^{i\gamma x} \bar{v}(x,y,z)$ where $\bar{v}$ is periodic in x with the same periodicity as the basic flow and γ is arbitrary. However, Pierrehumbert & Widnall consider only the subharmonic and primary cases. The analysis, which has not yet been done for more general γ, may yield important new results. Indeed, Busse (1979) points out the importance of these general γ modes in Benard convection. The present study is restricted to γ being a half-integer multiple of the primary wavenumber because our code is fully nonlinear with the periodicity condition (2.3).

The assumption of periodicity in the streamwise-x direction is unrealistic in a spatially growing mixing layer unless the modes being studied are localized in x and grow much more rapidly than the shear layer spreads. These latter approximations seem reasonably well justified for the three-dimensional modes studied here (see Sec. 3). However, future work using inflow-outflow boundary conditions in x should clarify the role of non-parallel effects in free-shear flows.

The dynamical equations are solved using pseudospectral methods in which the flow variables are expanded in the series

$$\vec{v}(x,y,z,t) = \sum_{|m|<\frac{1}{2}M} \sum_{|n|<\frac{1}{2}N} \sum_{p=0}^{P} \vec{u}(m,n,p,t)e^{im\alpha x}e^{in\beta y}T_p(Z) \tag{2.4}$$

where n and p are integers and m is a *half-integer* when pairing is allowed and a whole integer if pairing is excluded. Here $Z = f(z)$ is a transformed z-coordinate satisfying $Z = \pm 1$ when $z = \pm \infty$. Two choices of $f(z)$ have been studied, viz.

$$Z = \tanh \frac{z}{L} \qquad (|z|<\infty , |Z| < 1) \tag{2.5}$$

and

$$Z = \frac{z}{\sqrt{z^2 + L^2}} \qquad (|z|<\infty , |Z| < 1) \tag{2.6}$$

where L is a suitable scale factor. With these mappings, derivatives with respect to z are evaluated pseudospectrally using the relations

$$\frac{\partial \vec{v}}{\partial z} = \frac{1}{L} (1 - Z^2) \frac{\partial \vec{v}}{\partial Z} \tag{2.7}$$

$$\frac{\partial \vec{v}}{\partial z} = \frac{1}{L} \sqrt{1 - Z^2} \frac{\partial \vec{v}}{\partial Z} \tag{2.8}$$

for (2.5), (2.6), respectively.

Time stepping is done by a fractional step method in which the nonlinear terms are marched in time using a second-order

Adams-Bashforth scheme while pressure head and viscous effects are imposed implicitly using Crank-Nicolson differencing.

This scheme is globally second-order accurate in time, despite time splitting (Deville & Orszag 1983), because the various split operators commute in the case of quiescent boundary conditions at $z = \pm\infty$.

There is one further technical detail regarding the numerical method that should be discussed here. Various Poisson equations, like

$$\frac{d^2\Pi}{dz^2} - (m^2 + n^2)\Pi = g(z) \qquad (|z|<\infty) \qquad (2.9)$$

are solved by expansion in the eigenfunctions of d^2/dz^2:

$$\frac{d^2}{dz^2} e_k(z) = \lambda_k e_k(z) \qquad (|z|<\infty) \qquad (2.10)$$

Thus, if

$$g(z) = \sum_{k=0}^{P} g_k e_k(z)$$

Then

$$\Pi(z) = \sum_{k=0}^{P} \frac{g_k}{\lambda_k - (m^2+n^2)} e_k(z) \qquad (2.11)$$

We remark that this technique gives spectrally accurate solutions, despite the fact that the continuous version of the eigenvalue problem (2.10) has only a continuous, and hence singular, spectrum. Also, note that all the eigenvalues λ_k are real and non-positive; for both mappings (2.5) and (2.6), there are precisely three zero eigenvalues λ_1, λ_2, λ_3. One of these zero eigenmodes, λ_1, is physical, viz. $e_1(z) = 1$, but the other two are highly oscillatory and unphysical. Indeed, since the spectral (Chebyshev) derivative of $T_P(Z)$ vanishes except at $Z = \pm 1$, $e_2(Z) = T_P(Z)$ is a zero eigenfunction of d^2/dz^2; $T_P(Z_j) = (-1)^j$ at the Chebyshev collocation points $Z_j = \cos \pi j/P$. The third zero eigenmode oscillates and grows roughly like z. When $m = n = 0$, the incompressibility constraint (2.2) requires that this mode of the z-velocity field vanish identically so there is no difficulty with the zero pressure eigenvalues λ_1, λ_2, λ_3.

Comparisons of the behavior of linear Orr-Sommerfeld eigenmodes obtained using the mappings (2.5) and (2.6) show that (2.6) gives a superior representation of these modes unless L is fine tuned, which is not convenient in the nonlinear dynamical runs.[+] Some representative results are given in Table 1. Notice that as α increases, the optimal choice of map scale L decreases. Also, notice that the accuracy of the eigenvalue is much more sensitive to L for the hyperbolic tangent mapping (2.5) than for (2.6).

The nonlinear time-dependent Navier-Stokes code has been tested for the generalized Taylor-Green vortex flow (2.12) and for the behavior of linearized eigenfunctions, with satisfactory agreement being achieved with power series in t (Brachet et al 1983) and linear behavior, respectively.

[+] There is one case in which it seems that the hyperbolic tangent mapping (2.5) is more convenient than the algebraic mapping (2.6). This flow is the generalized Taylor-Green vortex flow that develops from the initial conditions

$$\begin{aligned} u(x,y,z,0) &= \sin x \cos y/\cosh^2 z \\ v(x,y,z,0) &= -\cos x \sin y/\cosh^2 z \\ w(x,y,z,0) &= 0 \end{aligned} \tag{2.12}$$

The evolution of this flow seems best studied, either by power series or initial value methods, using (2.5) with L = 1. The time evolution of this free shear flow is remarkably similar to that of the periodic Taylor-Green vortex (Brachet et al 1983).

Table 1. Growth Rates (Im c) of the Orr-Sommerfeld Eigenfunctions for the Mixing Layer $U_0(z) = \tanh z$ †

	x-wavenumber α								
	0.25			0.5			0.75		
	Number of Chebyshev Polynomials (P+1)								
L	17	33	65	17	33	65	17	33	65
	Hyperbolic Map (2.5)								
0.5	1.534	1.375	1.238	0.579	0.501	0.457	0.160	0.150	0.150
1	0.959	0.820	0.746	0.383	0.360	0.351	0.141	0.138	0.137
2	0.635	0.614	0.605	0.344	0.342	0.342	0.137	0.137	0.137
4	0.612	0.598	0.597	0.324	0.342	0.342	0.041	0.136	0.137
8	0.539	0.597	0.597	0.115	0.322	0.342	S	0.045	0.136
16	0.202	0.526	0.596	S*	S	0.321	S	S	0.046
	Algebraic Map (2.6)								
0.5	0.699	0.588	0.599	0.345	0.346	0.342	0.131	0.138	0.137
1	0.591	0.599	0.597	0.344	0.342	0.342	0.137	0.137	0.137
2	0.600	0.597	0.597	0.342	0.342	0.342	0.136	0.137	0.137
4	0.597	0.597	0.597	0.325	0.342	0.342	0.371	0.136	0.137
8	0.542	0.597	0.597	0.009	0.322	0.342	S	0.043	0.136

† Here the Reynolds number is $1/\nu = 100$ and the eigenvalue is the complex wave speed c for a temporal mode of the form $\psi(z)e^{i\alpha(x-ct)}$. For the most rapidly growing mode listed here, Re $c = 0$.

* S indicates that all modes are stable with the indicated parameter values

3. Results for Mixing Layers

In this Section, results are reported for the evolution of initial velocity fields of the form

$$\vec{v}(x,y,z,0) = U_0(z)\,\hat{x} + \mathrm{Re}[A_{10}\vec{v}_{10}(z)e^{i\alpha x} + A_{\frac{1}{2},0}\vec{v}_{\frac{1}{2},0}(z)e^{\frac{1}{2}i\alpha x} + A_{11}\vec{v}_{11}(z)e^{i\alpha x+i\beta y}] \qquad (3.1)$$

The laminar mean profile is assumed to be the mixing layer profile $U_0(z) = \tanh z$ and $\vec{v}_{ij}(z)$ is normalized so that $\max|v_{ij}(z)| = 1$. The initial functions $\vec{v}_{ij}(z)$ are normally chosen as the most unstable eigenfunctions of the linear Orr-Sommerfeld equation with the wavenumbers given in (3.1).* In this representation, A_{10} is the amplitude of the primary two-dimensional component, $A_{1/2,\,0}$ is the amplitude of its subharmonic or pairing mode, and A_{11} is the amplitude of the primary three-dimensional wave. In all cases, the initial conditions are chosen so that $A_{1/2,0}$ $A_{11} \ll A_{10}$; typically, $A_{10} = 0.25$. Also, the momentum thickness Reynolds number for the undisturbed flow is $R = 1/\nu$.

In the absence of subharmonic and three-dimensional perturbations ($A_{1/2,0} = A_{11} = 0$), the two-dimensionally perturbed flow quickly saturates to a quasi-steady state. In Figure 1, a plot is given of the time evolution of the two-dimensional disturbance energy $E_{10}(t)$ for various initial amplitudes A_{10}.

---

* The Reynolds numbers of the flows discussed below, while modest, are much greater than that of the onset of linear instability ($R_{crit} \approx 4$), so that even the linear modes are effectively inviscid. In this case, damped modes may lie only in the continuous spectrum (Drazin & Reid 1981) and so are singular. Whenever (3.1) calls for such a singular contribution to the initial condition (3.1), we choose instead the flow component $w_{nm} \equiv w_{10}$ of the primary mode (with u_{nm} and v_{nm} determined by incompressibility).

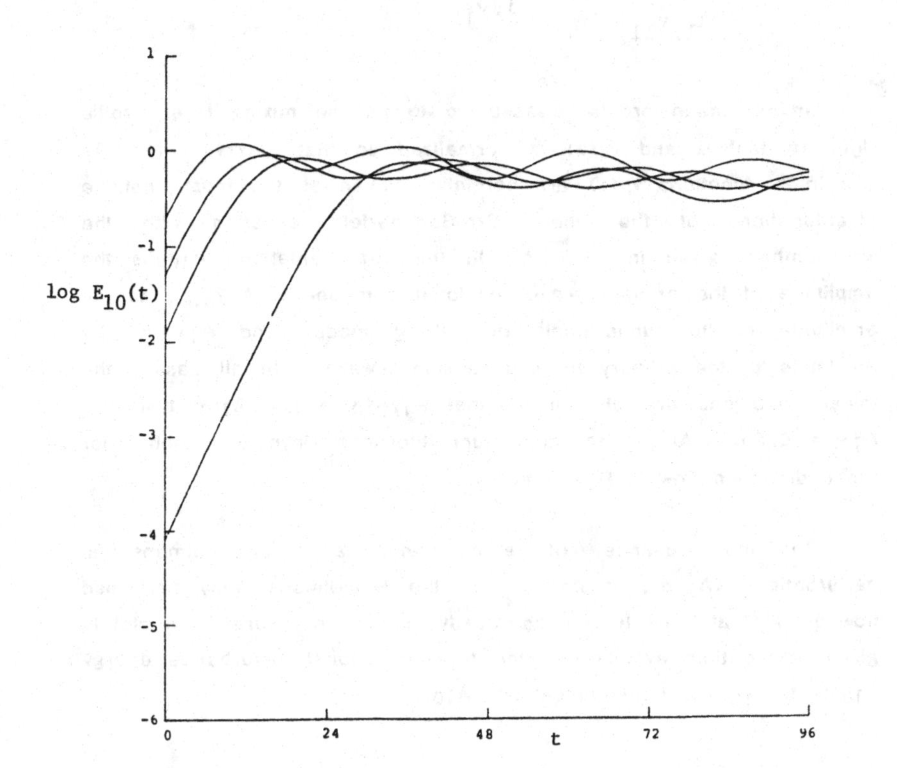

Figure 1.

A plot of $E_{10}(t)$ vs t for runs with $A_{\frac{1}{2},0} = A_{11} = 0$ and $A_{10} = 0.5,\ 0.25,\ 0.125,\ 0.01$. Here the Reynolds number is $R = 400$, $U_0(z) = \tanh z$, the spectral cutoffs in (2.4) are $M = 8$, $N = 1$, $P = 32$, (resolution $8 \times 1 \times 32$ with no pairing modes), the x-wavenumber is $\alpha = 0.4$, and the time step is $\Delta t = 0.02$. Note that the flow saturates into a vortical state nearly independent of the initial perturbation. Before such saturation occurs, the perturbation grows linearly like an Orr-Sommerfeld eigenfunction.

Here

$$E_{mn}(t) = \int_{-\infty}^{\infty} dz \, |\vec{v}_{mn}(z,t)|^2 \qquad (3.2)$$

where

$$\vec{v}_{mn}(z,t) = \sum_{p=0}^{P} \vec{u}(m,n,p,t) T_p(Z) \qquad (3.3)$$

and u is defined by (2.4). It is apparent that E_{10} saturates into a finite-amplitude vortical state on a time scale of order 10; indeed, the mean flow tanh z is inviscidly unstable to the perturbation A_{10} with maximum growth rate roughly 0.2 when $\alpha \approx 0.44$. [The range of inviscidly unstable wavenumbers for the tanh z profile is $0 < \alpha < 1$. Also, note that if we used a length scale in which the wavelength of the perturbation is of order unity (rather than our unity in which the shear layer thickness is order 1), saturation of E_{10} would occur on a time scale of order 1.] In Fig. 2, a plot is given of an instantaneous spanwise vorticity distribution in the developed two-dimensional flow.

Comparison of the energy evolution plotted in Figs. 3(a) and 3(b) shows that the initial phase of the subharmonic perturbation can affect its growth rate but not the eventual growth and saturation of the subharmonic. The present calculations differ from those of Riley & Metcalfe (1980) and Patnaik et al (1976) in that the initial disturbances are chosen to be computationally infinitesimal in our runs in contrast to their finite-amplitude initial perturbations. (Also Patnaik et al study stratified flows). While phase does affect the initial subharmonic growth rate, the perturbation eventually achieves its optimal growth rate during our long time runs. We conclude that the 'vortex shreddy' process found by Patnaik et al is a finite amplitude effect, not reproducable in the present long-time runs.

The saturated two-dimensional flow state discussed above can be unstable to subharmonic perturbations, $A_{1/2,0}$ in (3.1), for suitable α (Kelly 1967). In Figure 3, we plot the evolution of the

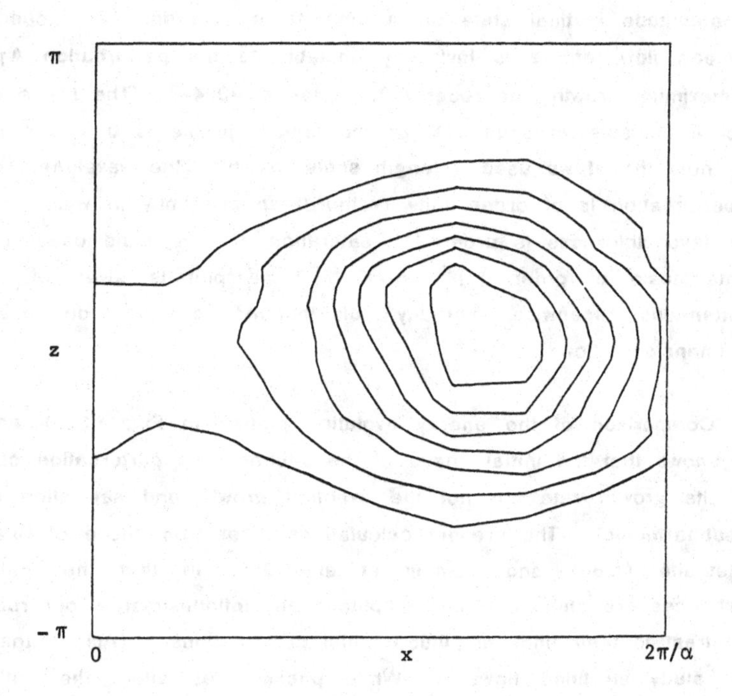

Figure 2.

A contour plot of spanwise (y) vorticity contours for the saturated flow state of the mixing layer at R = 400. The vortex prominent in this plot is nearly stationary.

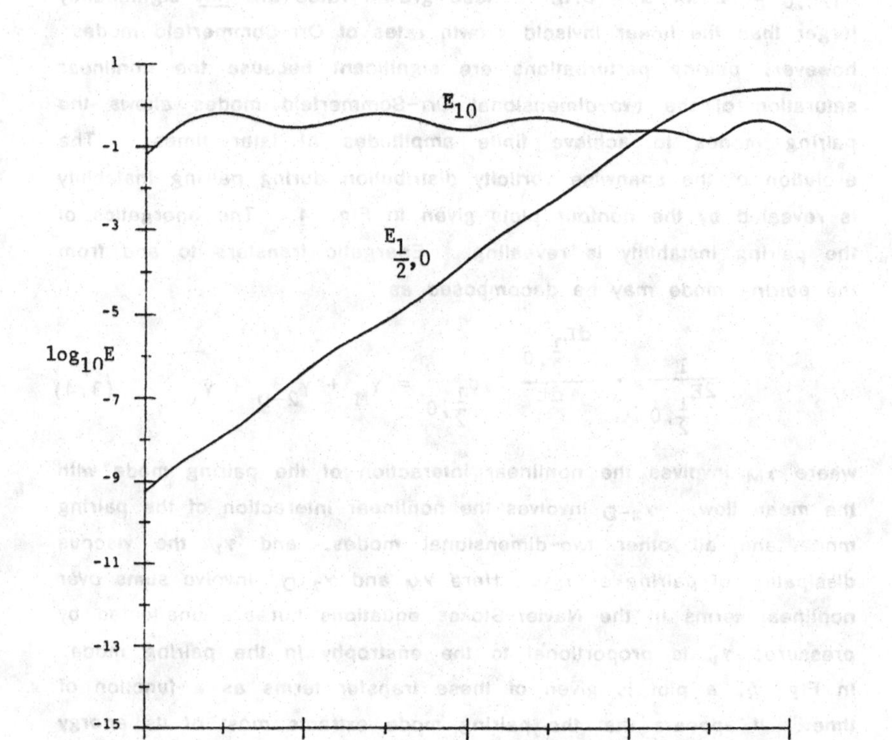

Figure 3.

Plots of the evolution of $E_{10}(t)$ and the two-dimensional pairing mode energy $E_{\frac{1}{2},0}(t)$ as functions of time. Here $R = 400$, $U_0(z) = \tanh z$, $A_{10} = 0.25$, $A_{\frac{1}{2},0} = 3 \times 10-4$, $M = 8$, $N = 1$, $P = 32$, (resolution $16 \times 1 \times 32$ with pairing modes) $\alpha = 0.5$ and $\Delta t = 0.02$. (a) and (b) differ by a 90° phase shift of initial subharmonic perturbation.

subharmonic perturbation energies $E_{1/2,0}(t)$ as well as the primary two-dimensional energy $E_{10}(t)$. Here we choose $a_{10} = 0.25$ and $a_{1/2,0} = 3 \times 10^{-4}$. In Figure 3(a), the primary and subharmonic perturbation vorticity are initially in phase; in Figure 3(b) they are initially out of phase. This subharmonic instability of the saturated two-dimensional vortical states is inviscid in character as its growth rate asymptotes to a finite limit as R increases. The growth rate $\sigma_{1/2,0} \approx 0.2$ for $\alpha = 0.8$. These growth rates are <u>not</u> significantly larger than the linear inviscid growth rates of Orr-Sommerfeld modes; however, pairing perturbations are significant because the nonlinear saturation of the two-dimensional Orr-Sommerfeld modes allows the pairing modes to achieve finite amplitudes at later times. The evolution of the spanwise vorticity distribution during pairing instability is revealed by the contour plots given in Fig. 4. The energetics of the pairing instability is revealing. Energetic transfers to and from the pairing mode may be decomposed as

$$\frac{1}{2E_{\frac{1}{2},0}} \cdot \frac{dE_{\frac{1}{2},0}}{dt} \equiv \sigma_{\frac{1}{2},0} = \gamma_M + \gamma_{2-D} + \gamma_\nu \qquad (3.4)$$

where γ_M involves the nonlinear interaction of the pairing mode with the mean flow. γ_{2-D} involves the nonlinear interaction of the pairing mode and all other two-dimensional modes, and γ_ν the viscous dissipation of pairing energy. Here γ_M and γ_{2-D} involve sums over nonlinear terms in the Navier-Stokes equations but are unaffected by pressure; γ_ν is proportional to the enstrophy in the pairing mode. In Fig. 5, a plot is given of these transfer terms as a function of time. It appears that the pairing mode extracts most of its energy from the mean flow and grows no faster than in the absense of the two-dimensional primary component. The important conclusions are that the presence of the saturated two-dimensional primary does not turn off the pairing mode and that the growth rate of this latter mode is of order that observed in the primary two-dimensional instability. These results imply that even a small pairing perturbation will quickly achieve finite amplitude after the primary mode saturates.

While these conclusions are in substantial agreement with those

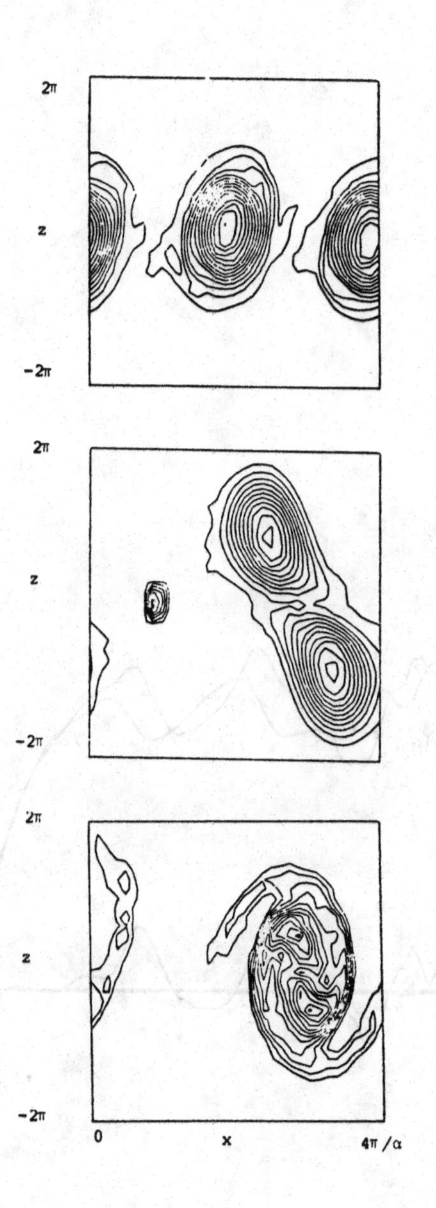

Figure 4.

Spanwise vorticity contours at $t = 48,\ 72,\ 96$ during a vortex pairing run with $R = 200$, $V_0(z) = \tanh z$, $A_{10} = 0.25$, $A_{\frac{1}{2},0} = 3 \times 10^{-5}$, $M = 16$, $N = 1$, $P = 32$, (resolution $32 \times 1 \times 32$) $\alpha = 0.43$, $\Delta t = 0.01$.

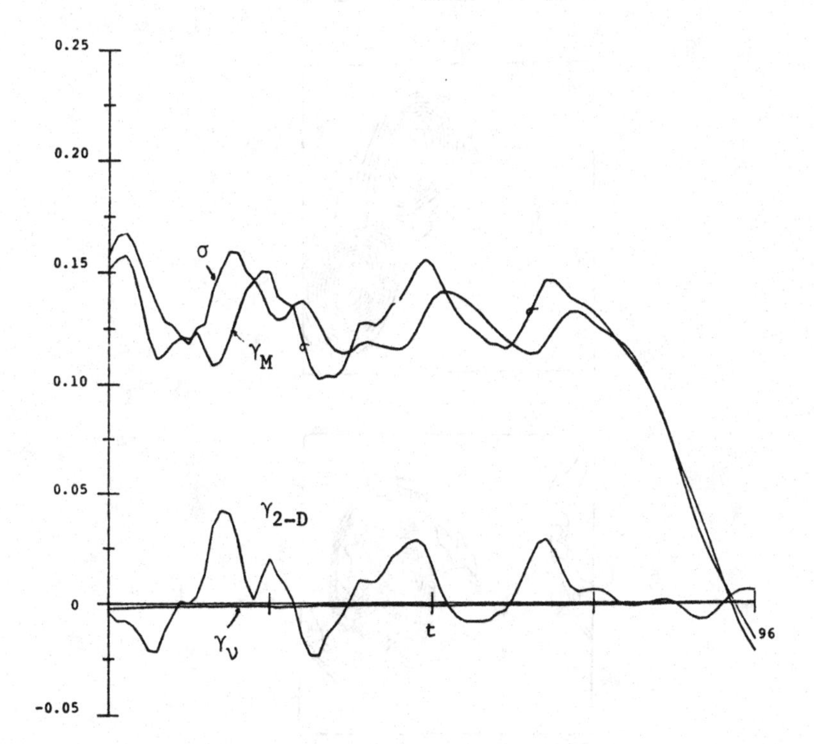

Figure 5.

A plot of the components γ_M, γ_{2-D}, γ_ν [see (3.4)] of the growth rate $\sigma_{\frac{1}{2},0}$ of pairing mode amplitude as functions of time for the same run as in Figure 4.

obtained by Kelly (1967) using perturbation theory, they differ in some important respects. First, we observe nothing very special about the 'resonant' wavenumber $\alpha \approx 0.44$ of maximum linear growth presumably because our study is a temporal, not sp atial, stability analysis. Second, we do not find that the growth rate of the pairing mode is significantly enhanced by the finite-amplitude primary mode (Pierrehumbert & Widnall 1982). On the contrary, the growth rate of the pairing mode at $1/2\alpha = 0.22$ seems to be slightly <u>less</u> when the primary achieves finite amplitude than for the parallel shear flow.

The saturated two-dimensional flow is also subject to three-dimensional instabilities. While the laminar mean flow is inviscidly unstable only for $\alpha^2 + \beta^2 < 1$, the finite-amplitude two-dimensional flow can be unstable for large β at high R. In Fig. 6, we plot the evolution of three-dimensional disturbance energy

$$E_{3D} = \sum_m E_{m1}(t) \tag{3.5}$$

for runs with initial conditions (3.1) with $A_{10} = 0.25$, $A_{1/2,0} = 0$, $A_{11} = 10^{-6}$ with $\alpha = 0.4$, $2 \leq \beta \leq 8$. For these parameter values, the mean flow tanh z is both viscously and inviscidly stable at these three dimensional scales. Nevertheless, the saturated two-dimensional disturbed flow is strongly unstable at these scales, with disturbances growing at roughly the same rate as the inviscid two-dimensional primary instability. Since the two-dimensional modes saturate, the three-dimensional modes can achieve finite amplitudes on convective time scales and thereby modify significantly the later evolution of the flow.

The growing three-dimensional wave is localized in space on top of the two-dimensional vortex motion. In Fig. 7, a plot is given of contours of the spanwise (three-dimensional) perturbation velocity $v(x,0,z,t)$, as well as a wind plot of the finite amplitude two-dimensional flow $(u(x,0,z,t), w(x,0,z,t))$. The structure of this three-dimensional mode is not dissimilar to that found in wall-bounded flow (see Orszag & Patera 1983 for a detailed discussion of these latter modes).

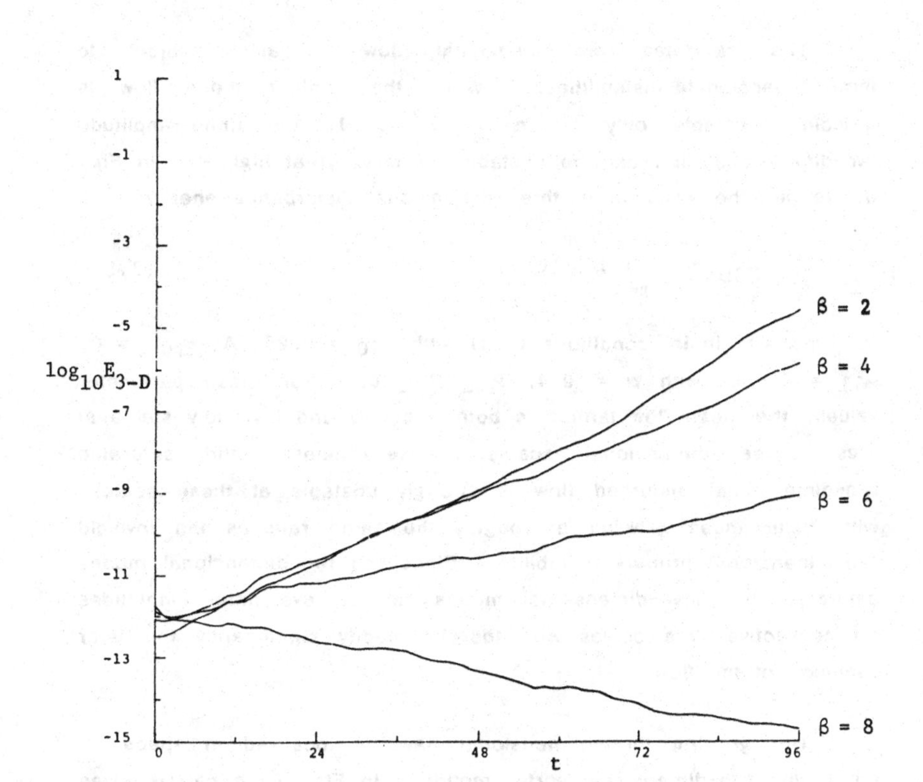

Figure 6.

A plot of the evolution of the three-dimensional disturbance energy $E_{3-D}(t)$ vs t for runs with R = 400, $U_0(z)$ = tanh z, M = 8, N = 4, P = 32 (resolution 8 x 4 x 32) α = 0.4, Δt = 0.02, A_{10} = 0.25, A_{11} = 10^{-6}, and β = 2, 4, 6, 8.

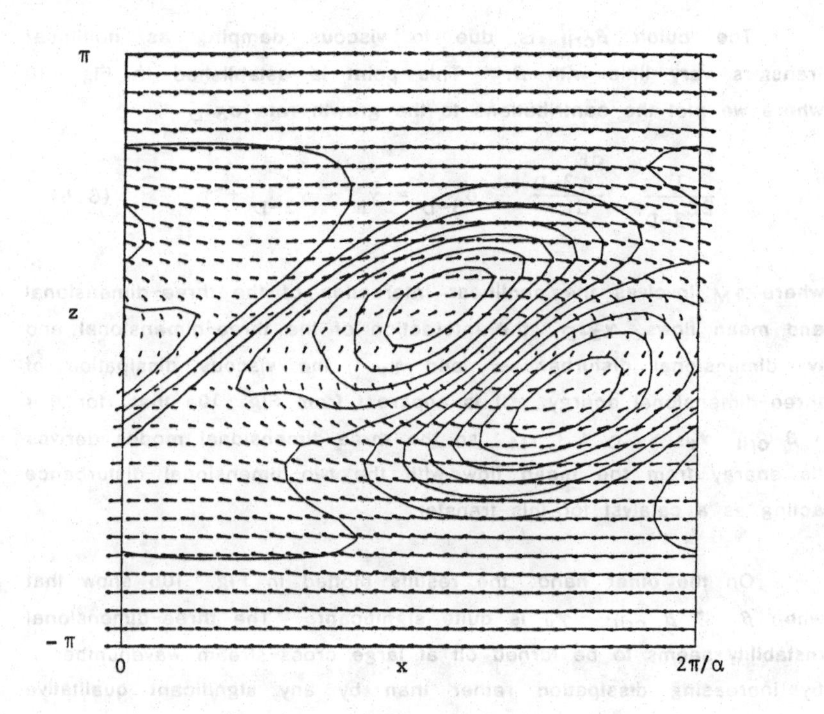

Figure 7.

Contour plot of the perturbation three-dimensional velocity component $v(x,0,z,t)$, in the plane $y = 0$ superimposed on a wind (vector) plot of the finite amplitude vortex flow $(u(x,0,z,t), w(x,0,z,t))$ whose vorticity contours are plotted in Figure 2. Here $R = 200$, $\alpha = 0.8$, $\beta = 0.8$, $U_0(z) = \tanh z$, $M = N = P = 32$ (resolution $32 \times 32 \times 32$ with no pairing modes) and the contours are plotted at $t = 24$.

In Fig. 8, a plot of the average three-dimensional growth rate σ_{3-D} vs. β is given for various R when $\alpha = 0.4$. It is apparent from the results plotted in Fig. 8 that, as R increases, σ_{3-D} approaches a finite limit for fixed β (so the secondary instability discussed is inviscid in character) and that the instability turns off for $\beta > \beta_{crit} \sim 1/3\sqrt{R}$.

The cutoff β_{crit} is due to viscous damping as nonlinear transfers vary little with β. This point is established in Fig. 10 where we plot the contributions to the growth rate σ_{3-D},

$$\frac{1}{2E_{3-D}} \cdot \frac{dE_{3-D}}{dt} \equiv \sigma_{3-D} = \gamma_M + \gamma_{2-D} + \gamma_\nu \qquad (3.6)$$

where γ_M involves the nonlinear interaction of the three-dimensional and mean flows, γ_{2-D} the interaction of the three-dimensional and two-dimensional disturbances, and γ_ν the viscous dissipation of three-dimensional energy. It is apparent from Fig. 10a that, for $\beta << \beta_{crit}$, γ_ν, $\gamma_{2-D} << \gamma_M$ so the three-dimensional modes derives its energy from the mean flow with the two-dimensional disturbance acting as a catalyst for this transfer.

On the other hand, the results plotted in Fig. 10b show that when $\beta \sim \beta_{crit}$, γ_ν is quite significant. The three-dimensional instability seems to be turned off at large cross-stream wavenumber β by increasing dissipation rather than by any significant qualitative change in nonlinear transfers from the mean and two-dimensional components.

The flows that develop from the three-dimensional secondary instability do not saturate in ordered states like those of the primary two-dimensional and pairing instabilities. Instead, the three-dimensional modes seem to lead to chaos and, finally, turbulence. In Figs. 11, contour plots of spanwise vorticity and wind plots in three-dimensional mixing layer runs (at resolution M = N = P = 32) are given at t = 36 after the three-dimensional fluctuations

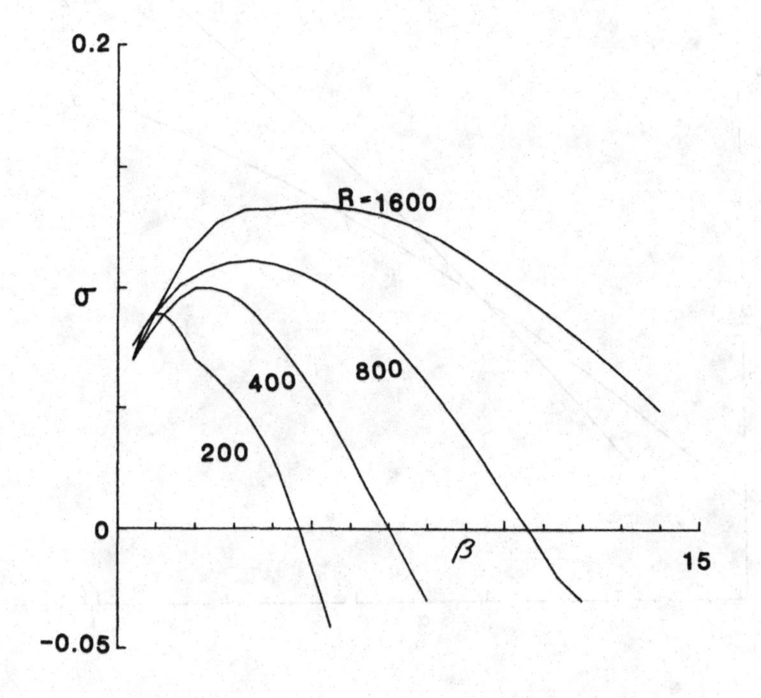

Figure 8.

A plot of the computed three-dimensional growth rate σ_{3-D} [see (3.6)] as a function of spanwise wavenumber β for various Reynolds numbers for $U_0(z) = \tanh z$. Here $\alpha = 0.4$.

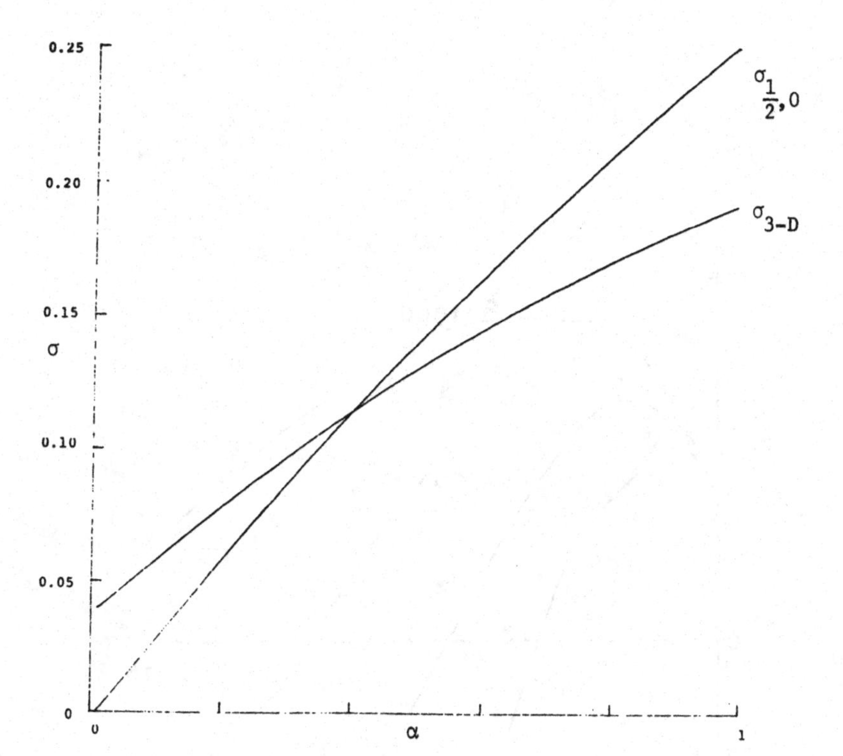

Figure 9.

A plot of the computed pairing growth rate $\sigma_{\frac{1}{2},0}$ and three-dimensional growth rate σ_{3-D} as a function of α at $R = 400$, $\beta = 0.8$ with $U_0(z) = \tanh z$. (Note that the wavenumber of the pairing mode is $\frac{1}{2}\alpha$)

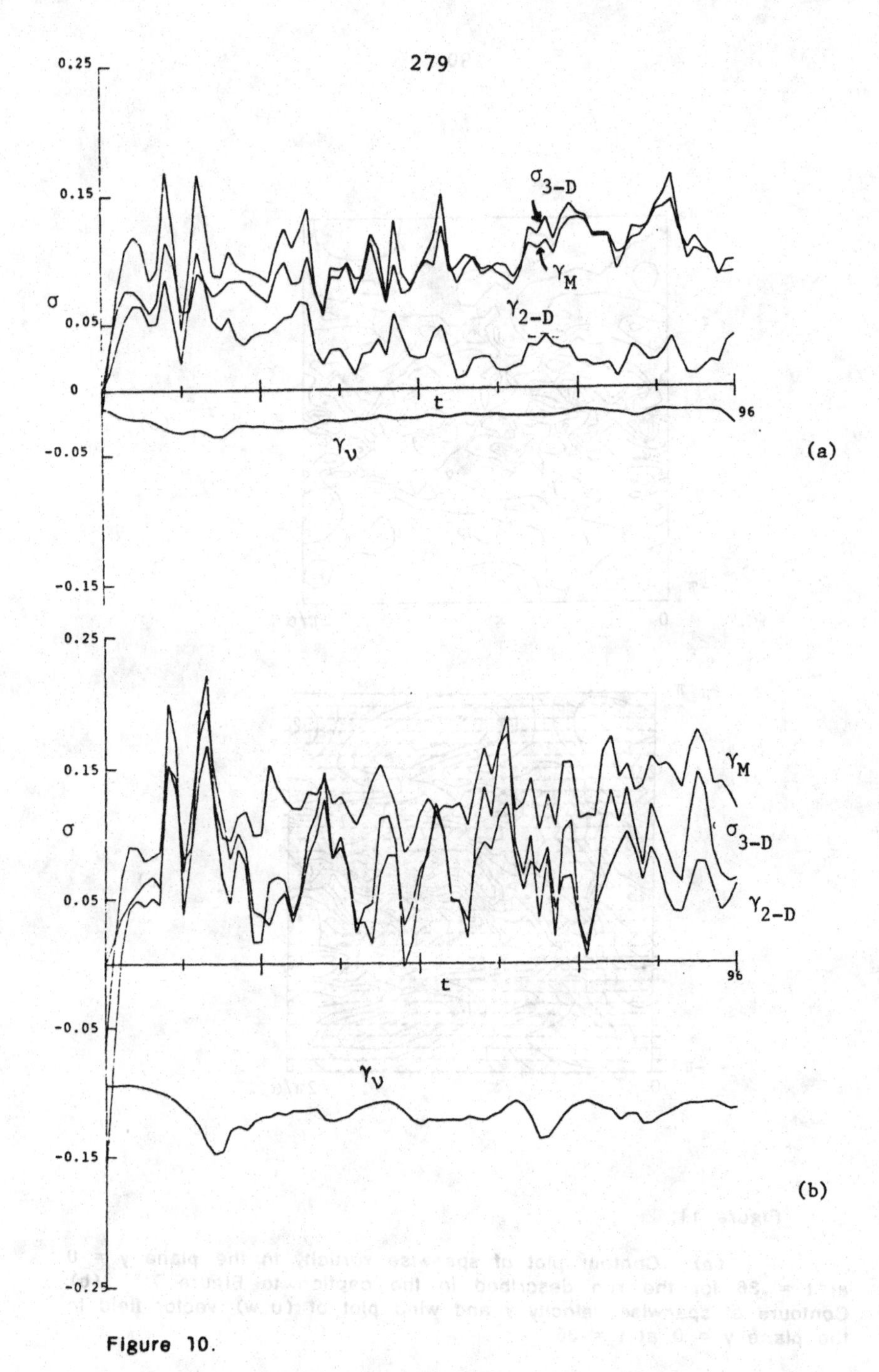

Figure 10.

A plot of the components γ_M, γ_{2-D}, γ_ν [see (3.6)] of the three dimensional growth rate σ_{3-D} as functions of time for R=400, α = 0.4, A_{10} = 0.25, A_{11} = 10^{-6}. (a) β = 4 (b) β = 6

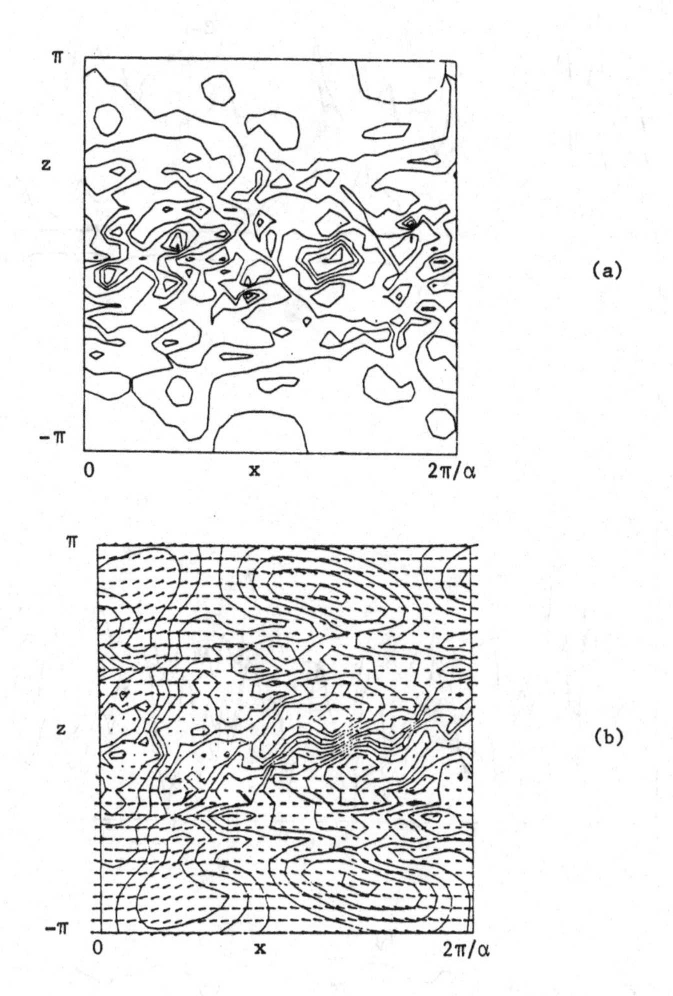

Figure 11.

(a) Contour plot of spanwise vorticity in the plane $y = 0$ at $t = 36$ for the run described in the caption to Figure 7. (b) Contours of spanwise velocity v and wind plot of (u, w) vector field in the plane $y = 0$ at $t = 36$.

become comparable to the two-dimensional amplitudes. The order apparent in Fig. 7 is partially obliterated by the three-dimensional excitations apparent in Fig. 11, but two-dimensional structure is still significant.

The nature of the competition between two-dimensional pairing and three-dimensional instability is further illustrated by the results plotted in Figs. 12, 13. In both figures, results of runs with $R = 400$, $\alpha = 0.4$, $\beta = 0.2$ are plotted. In Fig. 12, the initial conditions are chosen so that the pairing mode perturbation is much larger than that of the three-dimensional perturbation; it seems that the pairing process slightly inhibits the three-dimensional instability. In Fig. 13, the evolution of the instabilities are plotted when the initial three-dimensional perturbation is much larger than the pairing mode. In this case, it seems that the pairing instability is nearly unaffected by the three-dimensional instability before finite amplitudes are reached; when the three-dimensional mode becomes finite amplitude, the flow is chaotic so a higher resolution three-dimensional code should be used to study the energetics.

5. Discussion

The principal result of this paper is the demonstration that small-scale three-dimensional instabilities like those previously studied by Orszag & Patera (1980, 1981, 1982), Pierrehumbert & Widnall (1982) exist in viscous free shear flows and that these instabilities persist to moderately low Reynolds number. It is possible that these modes are responsible for the initial development of three dimensionality in these shear flows. The dynamics of the three-dimensional instability is qualitatively the same as that of the three-dimensional instabilities studied by us in wall-bounded shear flows. In particular, the instability does not appear to be similar to the Görtler instability in curved channels, as the instability has significant streamwise variation along the two-dimensional eddy. While the instability shares some features of a classical inflectional instability, including phase locking with the primary vortex, inflectional instability is preferentially

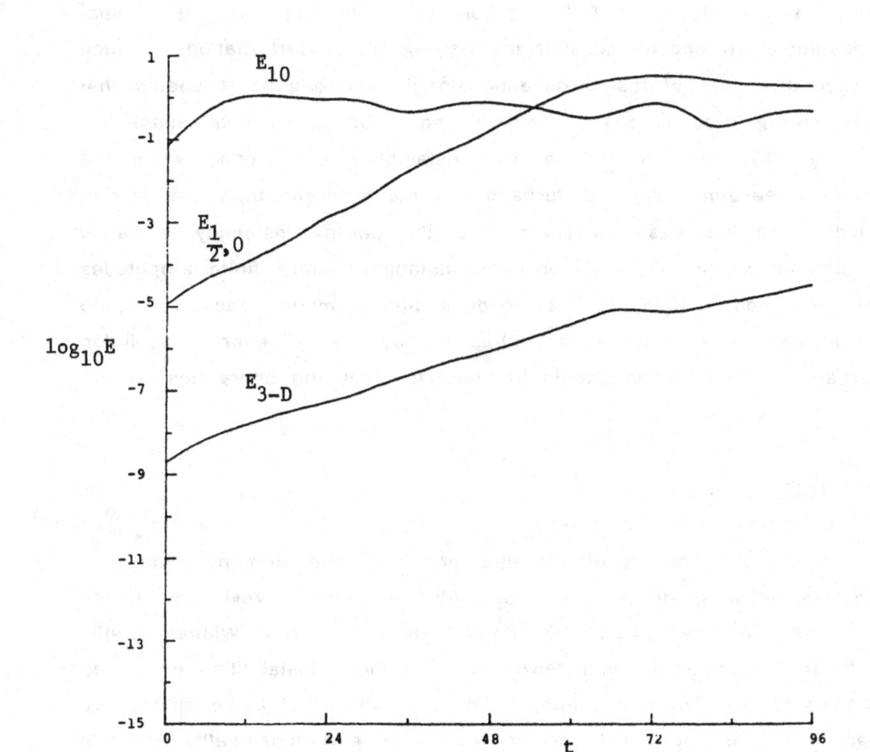

Figure 12.

A plot of the evolution of the energies E_{10}, $E_{\frac{1}{2},0}$, E_{3-D} vs t for a run with R = 400, $U_0(z)$ = tanh z, α = 0.4, β = 0.2, M = 8, N = 4, P = 32, and initial conditions A_{10} = 0.25, $A_{\frac{1}{2},0}$ = 4 x 10^{-3}, A_{11} = 3.3 x 10^{-5}. The subharmonic pairing mode dominates the three-dimensional mode.

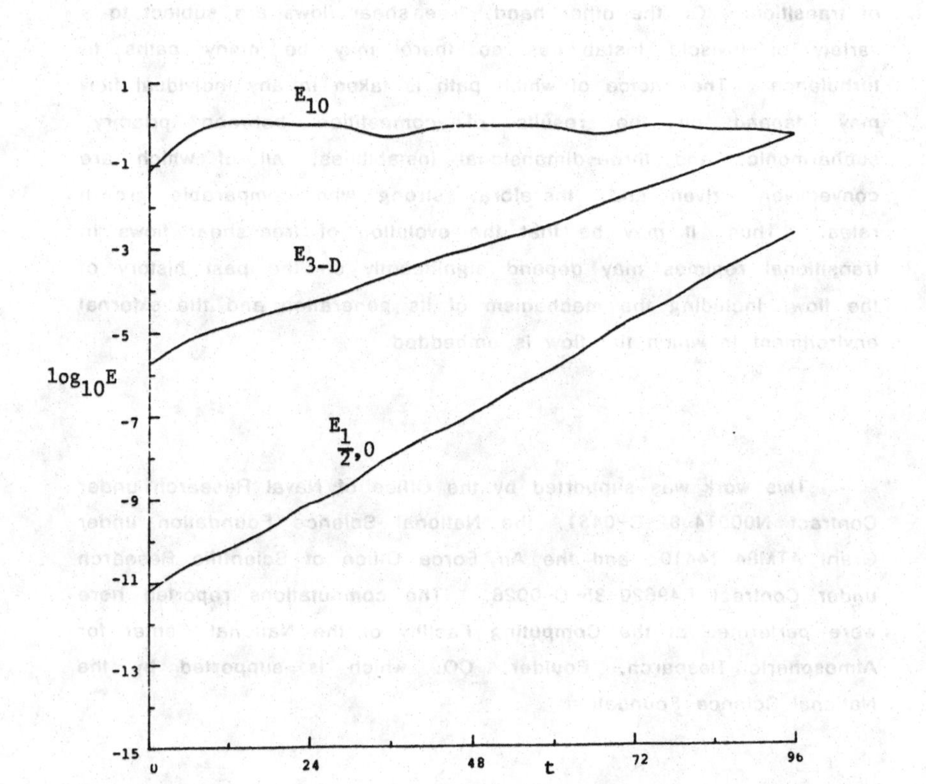

Figure 13.

Same as Figure 12, except that the initial conditions are A_{10} = 0.25, $A_{\frac{1}{2},0}$ = 3 x 10 $^{-6}$, A_{11} = 10^{-3}. The three-dimensional mode initially dominates the pairing mode.

two-dimensional but the present instability is not.

It seems that the mechanics of transition in the free shear flows studied here may, in a sense, be rather more complicated than in the case of wall-bounded shear flows. In the latter case, linear instabilities are often viscously driven and, therefore, weak, so they can not be directly responsible for the rapid distortions characteristic of transition. On the other hand, free shear flows are subject to a variety of inviscid instabilities so there may be many paths to turbulence. The choice of which path is taken in any individual flow may depend on the results of competition between priamry, subharmonic, and three-dimensional instabilities, all of which are convectively driven and, therefore, strong with comparable growth rates. Thus, it may be that the evolution of free-shear flows in transitional regimes may depend significantly on the past history of the flow, including the mechanism of its generation and the external environment in which the flow is embedded.

This work was supported by the Office of Naval Research under Contract N00014-82-C-0451, the National Science Foundation under Grant ATM84-14410, and the Air Force Office of Scientific Research under Contract F49620-85-C-0026. The computations reported here were performed at the Computing Facility of the National Center for Atmospheric Research, Boulder, CO, which is supported by the National Science Foundation.

References

Bernal, L.P., Breidenthal, R.E., Brown, G.L., Konrad, J.H., & Roshko, A. 1979 On the development of three dimensional small scales in turbulent mixing layers. In Proc. 2nd Int. Symp. on Turbulent Shear Flows, Imperial College, London.

Brachet, M.E., Meiron, D. I., Orszag, S.A., Nickel, B.G., Morf, R.H. & Frisch, U. 1983 Small Scale structure of the Taylor-Green vortex. J. Fluid Mech., 130,411-452.

Browand, F.K. & Troutt, T., 1980 A note on spanwise structure in the two-dimensional mixing layer. J. Fluid Mech. 97, 771.

Brown, G.L. & Roshko, A. 1974 On density effects and large structures in turbulent mixing layers. J. Fluid Mech. 64, 775.

Collins, D.A. 1982 A numerical study of the stability of a stratified mixing layer. Ph.D. Thesis, Department of Mathematics, McGill Univ., Montreal.

Deville, M. O., Israeli, M. & Orszag, S.A. 1983 Splitting methods for incompressible flow problems. To be published.

Drazin, P.G. & Reid, W.H. 1981 Hydrodynamic Stability, Cambridge University Press.

Ho, C.M. & Huang, L.S. 1982 Subharmonics and vortex merging in mixing layers. J. Fluid Mech. 119, 443-473.

Miksad, R.W. 1972 Experiments on the nonlinear stages of free shear layer transition. J. Fluid Mech. 56, 645.

Orszag, S.A. & Patera, A.T. 1980 Subcritical transition to turbulence in plane channel flows. Phys. Rev. Lett. 45, 989.

Orszag, S.A. & Patera, A.T. 1981 Subcritical transition to turbulence in planar shear flows. In Transition and Turbulence (ed. R.E. Meyer). Academic, New York.

Orszag, S.A. & Patera A.T. 1983 Secondary instability of wall-bounded shear flows. J. Fluid Mech., in press.

Patnaik, P.C., Sherman, F.S. & Corcos, G.M. 1976 A numerical simulation of Kelvin-Helmholtz waves of finite amplitude. J. Fluid Mech. 73, 215.

Pierrehumbert, R.T. & Widnalll, S.E. 1982 The two- and three-dimensional instabilities of a spatially periodic shear layer. J. Fluid Mech. 114, 59.

Riley, J.J. & Metcalfe, R.W. 1980 Direct numerical simulation of a perturbed, turbulent mixing layer. AIAA Paper No. 80-0274.

Stuart, J.T. 1967 On finite amplitude oscillations in laminar mixing layers . J. Fluid Mech. 29, 417.

Winant, C.D. & Browand, F.K. 1974 Vortex pairing: the mechanism of turbulent mixing-layer growth at moderate Reynolds number. J. Fluid Mech. 63, 237.

Wygnanski, I. Oster, D., Fiedler, H. & Dziomba, B. 1979 On the perservance of quasi-two-dimensional eddy-structure in a turbulent mixing layer. J. Fluid Mech. 93, 325.

Progress in Scientific Computing, Vol. 6
Proceedings of U.S.-Israel Workshop, 1984

TURBULENT FLOW SIMULATION: FUTURE NEEDS

by

Joel H.Ferziger
Department of Mechanical Engineering
Stanford University
Stanford, CA

I. Introduction

A classification of approaches to the simulation of turbulent flows according to the kind of averaging employed was given by Kline et al. (1978). This paper, concerns three types of methods: ones based on time or ensemble-averaged equations, large eddy simulation, and full simulation. Other methods are important but will not be dealt with.

Full turbulence simulation (FTS) solves the Navier-Stokes equations without modeling and is applicable only to low Reynolds number flows. Large eddy simulation (LES) employs models for the small scales while simulating the large scales. Its advantage over full simulation is its ability to handle a wider variety of flows and higher Reynolds numbers; its principal disadvantage is its reliance on models. Full simulation should be used wherever possible and large eddy simulation when necessary.

Methods based on one-point averages of the Navier-Stokes equations are the best choice for predicting turbulent flows in complex geometries at present. However, the turbulence models used in these methods need to be carefully monitored. They are not as robust as one would like; however, they are often blamed for errors due to numerical methods. FTS and LES cannot be applied directly to simulation of engineering flows and, for the foreseeable future, will be useful in engineering principally for their application to testing models used in the one-point average equations.

This paper will explore the accomplishments of these methods and the problems that need to be solved to extend their usefulness. Needs include better models and numerical methods for both compressible and incompressible flows. Because the author's experience is mainly with incompressible flows, they will receive the majority of the attention in this paper. Except for very high Mach number flows, the same turbulence models can be used for both types of flows. However, numerical methods for incompressible flows are not as well established as those for compressible flows. Also, because they are expected to deal with more geometric complexity, numerical methods used for the time-averaged equations differ from those used for FTS and LES.

II. Reynolds Average Turbulence Models

Let us begin by examining why turbulence models are needed. In the Reynolds average approach, the mean velocity field is defined by ensemble averaging, time averaging, or averaging over a spatial coordinate in which the mean flow is homogeneous. The velocity can be regarded as a combination of a mean velocity (U) and turbulent fluctuations (u') so that $u = U + u'$; this leads to the Reynolds averaged Navier-Stokes equations:

$$\frac{\partial U_i}{\partial t} + \frac{\partial}{\partial x_j} U_i U_j = - \frac{\partial p}{\partial x_i} + \nu \frac{\partial^2 U_i}{\partial x_j} - \frac{\partial}{\partial x_j} \overline{u_i' u_j'} \qquad (2.1)$$

The mean-turbulent decomposition is a convenience whose physical significance needs to be questioned.

The terms $\overline{u_i' u_j'}$ in Eqs. (2.1), which result from the nonlinearity of the unaveraged equations, are called the Reynolds stresses; they represent the effect of the turbulence on the mean field. Unknowns outnumber equations in (2.1); closure is the function of turbulence modeling. Several kinds of models are used.

Turbulence converts kinetic energy of the mean flow into thermal energy (dissipation) and increases the rate of transport of mass, momentum, and energy normal to the streamlines (diffusion). In laminar flows, both of these phenomena are consequences of viscosity, so the effect of turbulence on the mean flow is usually represented as an increased viscosity:

$$\overline{u'_i u'_j} = - \nu_T \left(\frac{\partial U_i}{\partial x_j} + \frac{\partial U_j}{\partial x_i} \right) \quad (2.2)$$

where ν_T is an eddy viscosity.

The eddy viscosity can be determined either algebraically in terms of the mean velocity or via dynamic equations. Good correlations are known for the eddy viscosity in two-dimensional flows. However, the proportionality assumed in Eq. (2.2) should be regarded with caution in the general case.

A simple description of turbulence characterizes it by its kinetic energy, $q^2/2$, and an average length scale, L, in terms of which the eddy viscosity can be written:

$$\nu_T = C_\mu qL \quad (2.3)$$

Many current models are based on this assumption.

In mixing length models, q and L are determined directly by the properties of the mean velocity field. For free shear flows, the turbulence length scale is a fraction of the width of the shear layer. Near a solid boundary, the distance to the surface is a better choice. To complete the model, an expression for the turbulent velocity, q, is required. It is usually approximated by $q = LdU/dy$ (cf. Tennekes and Lumley, 1975).

The simplicity of mixing length models allows them to account for extra effects such as pressure gradients, curvature, and transpiration through a wall with relative ease. Kays and Crawford (1978) constructed a model which does an excellent job for boundary layers; Cebeci and Bradshaw (1981) gave an alternative model. The chief deficiency of mixing length models is their inability to handle separation and reattachment.

Two-equation models employ the Boussinesq eddy viscosity relation (2.2) but the evolution of the eddy viscosity is governed by a pair of partial differential equations. One PDE determines the turbulent kinetic energy (TKE) and, from it, the velocity scale. An exact TKE equation can be derived from the Navier-Stokes equations; it contains terms representing convection, production of new turbulent energy, dissipation, and diffusion. Of these, the convection term is exact, the production can be computed from the model itself, the method of obtaining the dissipation is explained below, and the diffusion is usually modeled by a gradient transport model. We also need the length scale and the dissipation. Turbulence theory suggests that these are related by $L = q^3/\epsilon$, which leads to the idea of using a PDE for the dissipation; this is the basis of the most popular two-equation models. Other possibilities have had less success.

Two-equation models (particularly the k-ϵ model) simulate a variety of flows with reasonable accuracy, including flows for which mixing length models have trouble. However, caution is needed. Although rarely a disaster, two-equation models are not always accurate enough for engineering purposes. Well-tuned mixing length models may be more accurate in particular flows. The deficiencies of two-equation models include use of the eddy viscosity relationship, which prohibits prediction of counter-gradient fluxes, unrealistic large and abrupt changes in the Reynolds stresses when a strain is newly applied to a flow, and lack of treatment of nonequilibrium effects.

Reynolds stress models are designed to avoid these problems by using evolutionary PDEs for the Reynolds stresses themselves. Details will not be given here. They have been applied to relatively few engineering flows to date, with mixed success. In the 1980-81 Stanford meeting (Kline et al, 1981), Reynolds stress models fared no better than the models discussed above.

Algebraic stress models (Rodi, 1976) approximate the Reynolds stress PDEs by algebraic equations. PDEs for the turbulence kinetic energy and dissipation are retained. The resulting model is only slightly more expensive to use than the two-equation model. However, it is not clear that it performs better than the other models.

Current turbulence models can be used with confidence only to simulate flows which are not too different from those used to calibrate the model. In the author's opinion, one should think of turbulence models as sophisticated engineering correlations. A single universal model applicable to all flows may not be attainable, because turbulence structure differs from flow to flow for simple flows so one model may not be able to represent all of the physics. Instead, a different model can be developed for each flow. In complex flows, one can use a model appropriate to each zone. In the transitions between zones, a blending model is required. Zonal models should be both simpler and easier to construct than universal models.

III. Numerical Methods

A. Conservation

Incompressible flow computation offers mathematical difficulties absent in compressible flow. There is no separate energy conservation equation; the momentum equation conserves both momentum and kinetic energy. Lack of a time derivative in the continuity equation is another manifestation of this problem. As a result, conservative methods for the incompressible equations are difficult to construct.

Global conservation laws for mass and momentum state that the total amount of the conserved quantity in the macroscopic volume is changed only by flows through the surface of that volume; for the kinetic energy, there is a volumetric viscous dissipation term. A discretization scheme is called conservative if has a macroscopic conservation property. Numerical methods which conserve momentum are easy to obtain, but energy conservative ones are more difficult. Finding a scheme that has the desired conservation properties can be tedious.

The issue of whether a numerical scheme needs to be conservative or not has been controversial for both compressible and incompressible flows. Conservative methods are generally better behaved than nonconservative ones and tend to be more stable. On the other hand, some nonconservative schemes produce excellent results.

One method of guaranteeing conservation is to use a staggered grid which locates each component of the velocity and the pressure at different grid points. The control volumes for each variable are also different; conservation is readily obtained. As difference schemes on the staggered grid are more accurate than those for the regular grids, staggered grids are used almost exclusively for incompressible flows.

Although central differencing is natural on the staggered grid, most numerical methods are unstable when used with it, so upwind methods have been used for nearly all high Reynolds number flow simulations. In the last few years, the first-order upwind method has been supplanted by second-order schemes such as Raithby's (1976) skewed upwind method and Leonard's (1977) QUICK method.

B. Solution Strategies for Steady Flows

For stationary problems in incompressible flows, a number of strategies are used. The system of algebraic equations resulting from finite differencing the steady Navier-Stokes equations is large, nonlinear, and sparse. Iterative techniques are the methods of choice for such systems. As the cost is only mildly sensitive to the quality of the initial guess, a uniform initial flow is usually assumed.

In elliptic flows, the velocity in the principal flow direction may change sign; such flows are also referred to as recirculating or separated flows. Solution of the unsteady incompressible Navier-Stokes equations for these flows is made difficult by the lack of a time derivative in the continuity equation or an explicit equation for the pressure. For flows in which velocity reversal does not occur, the equations may be approximated by parabolic ones, and solved by marching methods. In the general case, three major solution strategies are employed: time-like methods, artificial compressibility methods, and direct methods. These are briefly described below.

A common strategy for stationary problems is to treat them as the steady limit of unsteady problems. Relaxation methods do not need to be time-accurate, so use of the largest possible time step is desirable. However, fully implicit methods are costly, making linearization a necessity. One common method updates the velocity via the momentum equations. The velocity field is then made divergence free by requiring the pressure to satisfy a Poisson-like equation.

Except in simple geometries for which fast, direct solvers are available, the Poisson equation must be solved by an iterative method. However, there is no point in computing the pressure accurately until the steady state is reached, so the usual strategy is to iterate the Poisson equation only once at each step.

The most commonly used method of this type is the SIMPLE algorithm (Caretto et al., 1972). This method begins with guesses at the pressure and velocity fields, p^n and u^n. A new velocity field is computed using a linearized implicit approximation to the staggered-grid finite difference momentum equations. The equations are solved line by line using a block tridiagonal algorithm. For efficiency reasons, only one sweep is made. The result is u^*, a first approximation to the velocity at the new time step. Next, the momentum equations are linearized about p^n and u^* and simplified by eliminating the contributions from the velocities at the neighboring points. Requiring the divergence of these equations to be zero, we obtain a Poisson equation for a pressure correction; it is given a single line iteration. The resulting pressure correction is then used to update the pressure and the velocity. This completes one iteration.

It is difficult to analyze these methods. Most of what is known about them is based on experience. It requires upwind differencing to converge at high Reynolds numbers. It can handle a wide range of flows. Its rate of convergence is nearly Reynolds number independent; typically, 200-400 iterations are required, which is many more than that required by methods for simple elliptic equations. A number of explanations of the slow convergence have been given; these include the asymmetry and the lack of diagonal dominance of the matrices, nonlinearity, and the mixed elliptic-parabolic nature of the equations. While all of these contribute, none provides a complete answer. Understanding of the behavior of numerical methods for incompressible flows is still in the future.

A variant of time-like methods, fractional step methods treat the differential equation as a composite of simpler equations. They offer flexibility in the selection of algorithms and bear a strong resemblance to splitting methods. The incompressible Navier-Stokes equations can be regarded as a composite of convective, pressure gradient, and viscous equations, each of which may be time-advanced independently or in combination with the others. A great variety of methods is possible.

An example is the method of Kim and Moin (1980), which consists of three fractional steps. First, velocities are computed by treating the convective terms explicitly and viscous diffusion in the x_1-direction implicitly. The second step treats the x_2-direction diffusion implicitly and the convection terms explicitly. Then the pressure is constructed so as to remove the divergence of the velocity; it is determined by a Poisson equation.

The state of the art is more advanced for numerical methods for compressible flows than it is for incompressible flows. A method of applying compressible flow methods to the incompressible case was developed by Chorin (1967). Improvements, mainly the use of improved compressible flow methods, have been made by Steger and Kutler (1977) and by Chang and Kwak (1984).

These methods introduce an unphysical time-derivative term into the continuity equation. For time accuracy, the added term must have a small coefficient. However, the added term has no effect on the converged result, so its coefficient is unimportant for steady flows. The term added to the continuity equation is usually proportional to the time derivative of the pressure:

$$\frac{1}{\beta}\frac{\partial p}{\partial t} = -\rho\frac{\partial u_i}{\partial x_i} \tag{3.1}$$

where β is the artificial compressibility. This equation is solved together with the momentum equations; they form a hyperbolic set which can be solved with methods used for the compressible equations. The effective Mach number is much less than unity, making the problem stiff and implicit methods mandatory. The number of iterations is not sensitive to the magnitude of the compressibility parameter and is comparable to the number of iterations required by time-like methods.

An alternative to iterative methods is to regard the finite-differenced Navier-Stokes equations as a large set of nonlinear equations to be solved directly by methods such as Newton-Raphson. These methods converge in a small number of iterations, making them very attractive. In this approach, the equations are linearized about some initial guess or previous iterate to produce a system of linear algebraic equations. The number of equations is 3N, where N is the total number of mesh points used. Even if a sparse matrix storage scheme is used, this matrix will not fit in the central memory of most computers.

Although it is possible to solve the linearized equations with iterative methods, Vanka and Leaf (1983) pointed out that this approach leads to the difficulty this method is trying to avoid--the slow convergence of iterative methods in these applications. They chose a sparse matrix solver; an alternative, suggested by Vanka (1983) is to use domain decomposition. Experience with these methods is limited. Vanka reports overall computation times of one-fifth of that required by SIMPLE for a few test cases. The method seems promising but further experience with it is required.

The partial differential equations of the k-ε model are stiff, because the production and dissipation terms are large and approximately equal. As their difference determines the rate of change of the kinetic energy, the equations are very sensitive to small changes in either quantity, i.e., they are stiff. Consequently, solution methods for the Reynolds average equations using the two-equation model need to be constructed with more care than is needed for methods for laminar flows.

IV. Full and Large Eddy Simulation

A. Overview of Numerical Methods

As noted earlier, full and large eddy simulation are three-dimensional, time-dependent methods of simulating turbulent flows. A principal aim of LES and FTS is to simulate the structures peculiar to a flow, but it is difficult to determine whether this goal has been accomplished. Reviews of LES and FTS have been given by the present author (Ferziger, 1983) and Rogallo and Moin (1984). We shall concentrate on the numerical problems common to both types of simulations. As these methods need to be time-accurate, explicit methods are used whenever there is no overwhelming factor requiring implicit treatment.

Homogeneous flows are those in which every point in the flow is statistically equivalent to every other. In these flows, it is sufficient to study a representative fluid mass large enough to contain all significant scales of motion of the turbulence. The surroundings are assumed identical to the considered piece of fluid (this is correct in a statistical sense), and periodic boundary conditions are applied. Periodic boundary conditions provide two important advantages: first, the difficult issue of specifying conditions at inflow boundaries is avoided; second, spectral methods, which are more accurate than finite difference approximations, become the natural method of computing spatial derivatives.

Inhomogeneous flows, in which the statistical properties of the turbulence vary with location, are more difficult to simulate for a number of reasons. The range of Reynolds numbers that can simulated is more limited. Flows that have been simulated to date are inhomogeneous in only one direction; work on flows with inhomogeneities in two directions is just beginning. The homogeneous directions are treated as in homogeneous flows, so the inhomogeneous directions are those on which we must concentrate. By definition, inhomogenous flows have boundaries on which periodic boundary conditions cannot be applied. This introduces several difficulties:

1. In directions in which the flow is unbounded, the computational domain needs to be extended to infinity.

2. At inflow boundaries, conditions that represent the large, three-dimensional, time-dependent structures of a single realization of a turbulent flow must be constructed.

3. At solid boundaries, no-slip velocity boundary conditions are required. However, the pressure boundary conditions are not known.

4. At outflow boundaries, conditions which allow the eddies to flow out smoothly are required. A convective condition of this kind was recently found by Lowery (1985).

5. Initial conditions are more difficult to generate than in homogeneous flows, and more time is required for the flow to develop.

6. In all of these flows, the eddy sizes vary greatly through the flow.

7. It may be necessary to deal with complex geometry.

More will be said about these items below.

The simulated flows can be divided into time-developing and statistically steady flows. The first class includes homogeneous flows and most free shear flows. Statistically steady flows that have been simulated include channel and annular flows and some free shear flows and boundary layers.

B. Initial Conditions

The optimum initial condition for any full or large eddy simulation is a snapshot of a single realization of the flow, but the required data are almost never available. Most simulations use artificial initial conditions. As a result, the flow must be allowed to develop for some time, which may be quite long, before the statistically steady state is reached.

In homogeneous flows, the turbulence length scales increase with time and eventually become larger than the considered region. Periodic boundary conditions are then inappropriate. Thus, homogeneous flow simulations are unphysical in their early stages due to the initial conditions and in the later later stages due to the boundary conditions. This may leave a relatively short time span in which the results are realistic; this span increases with the number of grid points.

For time-developing free shear flows, simulations of transitional and fully developed flows have been made. The latter suffer the same problems as homogeneous flows; the accurately simulated time span is short. In transitional flows, the initial conditions consist of a laminar free shear flow and perturbations; authors choose various perturbations depending on their interests.

For statistically steady flows, particularly channel and annular flows, a very long initial development time is required unless the initial conditions are accurate. The initial conditions used in these flows are typically superpositions of a mean profile, finite amplitude eigenmodes of the Orr-Sommerfeld equation, and noise. An alternative is to take the initial conditions from the results of another simulation. Long periods are also required to obtain stable time averages after the flow has developed. For these reasons and the need for fine resolution near the wall, these flows require longer run times than any of the others simulated to date.

As more complicated flows are considered, both the difficulty of constructing accurate initial conditions and their influence on the run time required will increase. At present, the only known method of generating these conditions is to use results from related simulations. This has not yet been done but should be tried in the near future.

C. Numerical Treatment of Spatial Derivatives and Boundary Conditions

As noted above, in any direction in which the flow is homogeneous, periodic boundary conditions and spectral methods of computing the spatial derivatives are employed. For homogeneous flows which include mean strain, coordinate systems which deform with the mean strain are used. This causes the considered region to become thin in some coordinate direction and invalidates the simulation sooner than would be the case on a non-deforming grid. For sheared turbulence, it is possible to regrid periodically, although at the cost of some aliasing error. For strained turbulence, no method of regridding has yet been found.

Simulations of free shear flows have used either no-stress conditions some finite distance from the center of the flow or a mapping which brings the point at infinity to a finite distance. Cain et al. (1981) developed a spectral method appropriate to such mappings.

Most simulations of channel flow have used use finite difference methods with nonuniform grids in the inhomogeneous direction. Recently, spectral methods based on Chebychev polynomials were developed by Moser and Moin (1984). In this approach, the velocity is expanded in divergence-free vector functions, eliminating the pressure and one velocity variable and thereby reducing the memory requirements of the method.

In nearly all simulations of inhomogeneous flows, artificial boundaries separate the considered region from the surroundings. Their location may have an important effect on the quality of the results. This is as true for LES as for any other flow simulation method. It has not been a problem for the flows considered to date but may become an important consideration when more complex flows are simulated.

At solid walls, no-slip velocity boundary conditions are well established. However, the boundary conditions to be applied to the pressure (if any) are not established. Indeed, Moser and Moin (1984) suggested that the pressure may have non-analytic behavior near the wall. Inclusion of no-slip wall boundary conditions in LES requires that a large fraction of the grid points be placed in close proximity to the wall or that many spectral modes be used.

To avoid the waste associated with putting so many points near the wall, Deardorff (1970), Schumann (1973), and others used artificial boundary conditions. They place the first grid point outside the region in which viscous effects are important. Schumann's boundary condition is a linear relationship between the fluctuating components of the stress and the velocity. Experimental evidence indicates that such a relationship is more nearly correct if it contains a time lag but further investigation is needed.

Inflow boundary conditions are extremely difficult to generate and have only been used in FTS or LES of transition flows. Inappropriate inflow conditions may contaminate the flow for a considerable distance downstream of the boundary on which they are applied. For fully developed flows, they must contain structures which can only be obtained from another simulation. Generating such conditions is an important issue to be faced in the near future. The problem is not as serious for transitional flows, for which initial conditions can be constructed from the Orr-Sommerfeld equation or simply by introducing noise.

All turbulent flows have small regions in which energy is concentrated in the small scales. In full simulations, the grid size must be chosen so that all of the scales are resolved, meaning that the grid size is determined by the smallest eddy in the flow. Naturally, this results in a great deal of waste. Use of adaptive grids, especially grids which can be created and destroyed as needed (Berger, 19982), could be a valuable tool in reducing this waste. The application of adaptive grid methods to full simulation of turbulent flows should therefore receive high priority.

D. Flows to be Considered in the Future

The future of development of supercomputers will play an important role in determining the future of LES. Significantly faster machines with substantially larger fast memories should lead to major changes in the kinds of problems that are attacked with FTS and LES in the next five years. Some possibilities are laid out in this section.

For homogeneous turbulence, it may be possible to use grids as large as $128 \times 128 \times 128$ or $256 \times 256 \times 256$, allowing simulation of Reynolds numbers which permit inertial subranges in the spectra. It may become possible to produce simulations of sheared and strained turbulence with long enough accurately simulated time spans to answer questions that have long perplexed turbulence theorists.

It should become possible to simulate transitional free shear flows through the fully developed state. This will permit study of the effect of various perturbations on transition. Prediction of sound generation by turbulent flows, which can be done crudely at present, will be possible. It may even become possible to study the impingement of free shear flows on solid walls.

For wall-bounded flows, the new computers will bring with them the ability to simulate fully turbulent channel flows at moderate Reynolds numbers, temporally developing boundary layers, and, possibly, spatially developing boundary layers. It will also be possible to simulate transition without need of a model. The addition of other phenomena like unsteadiness and heat transfer will be feasible.

It should be possible to simulate flow in a rectangular channel by full or large eddy simulation in the near future. This flow contains a secondary flow in the corner which is still not completely understood or modeled. Simulation of this flow should help in understanding and modeling the corner flows that occur in many engineering applications. Simulation of this flow has all of the difficulties encountered in channel flow but requires implicit treatment of more terms and thus the development of new numerical methods.

Flows which are inherently three-dimensional and time-dependent should be the first engineering flows simulated. For these, simulations could as easily be large eddy simulations as ensemble-average calculations. The choice depends on the questions one is trying to answer. Operationally, the major difference between the two approaches lies in the length scale appearing in the turbulence model; the same code can be used in both approaches with appropriate model changes.

There are many free shear flows of technological interest, including combusting flows. For these, needs include the development of accurate boundary conditions for the computational inflow and outflow surfaces and a method of dealing with the rapid growth of length scale in free shear flows.

Many of the same issues arise in wall-bounded flows. Simulation of spatially developing wall-bounded flows requires methods of dealing with in- and out-flow boundaries. Another problem is that of simulating a sufficiently large part of the flow on a small enough grid to capture the important eddies in the flow. This may require removing the layers closest to the wall from consideration by means of artificial boundary conditions; improved conditions of this type need to be developed. LES applications in wall-bounded flows will also be expanded to allow many 'extra effects' including heat transfer, blowing, suction, curvature, and rotation.

The new supercomputers should make it possible to begin to simulate turbulent combusting flows. Applications in meteorology and oceanography should also be possible.

V. Conclusions

Although a number of methods of computing steady, incompressible, turbulent flows are available, they are relatively slow and leave room for improvement. The turbulence models used in these methods are fairly good but in need of constant review. Zonal models may be necessary. Full and large eddy simulations can assist in model development.

Large eddy simulation has made important strides. To further extend the capability of LES, a number of developments need to take place. These include the development of better models for the unresolved scales, better methods of deriving initial conditions, and better treatment of the boundary conditions. A number of directions in which these developments may take place have been laid out in this paper.

The coming improvements in computers should lead to growth of LES as a tool for turbulence research and as a top-line tool for applications.

References

Berger, M., "Adaptive Mesh Refinement for Hyperbolic Partial Differential Equations," Report STAN-CS-82-924, Computer Science Dept., Stanford Univ., 1982.

Cain, A. B., Reynolds, W. C., and Ferziger, J. H., "Simulation of the Transition and Early Turbulent Regions of a Free Shear Flow," Report TF-14, Dept. of Mech. Engr., Stanford Univ., 1981.

Caretto, L. S., Gosman, A. D., Patankar, S. V., and Spalding, D. B., "Two Calculation Procedures for Steady, Three-Dimensional Flows with Recirculation," Proc. Third Intl. Conf. Num. Meth. Fluid Dyn., Paris, 1972.

Cebeci, T., and Bradshaw, P., Momentum Transfer in Boundary Layers, McGraw-Hill, New York, 1981.

Chang, J. L. C., and Kwak, D., "On the Method of Pseudo-Compressibility for Numerically Solving Incompressible Flows," AIAA paper 84-0252, 1984.

Chorin, A. J., "A Numerical Method for Solving Incompressible Viscous Flow Problems," J. Comp. Phys., Vol. 2, 12, 1967.

Ferziger, J. H., "Higher Level Simulations of Turbulent Flow," in Computational Methods for Turbulent, Transonic, and Viscous Flows (J.-A. Essers, ed.), Hemisphere, 1983.

Kays, W. M., and Crawford, M. E., Convective Heat and Mass Transfer, (second ed.), McGraw-Hill, New York, 1978.

Kim, J., and Moin, P., "On the Numerical Solution of Time-Dependent Fluid Flows Involving Solid Boundaries," J. Comp. Phys., Vol. 35, 301, 1980.

Kline, S. J., Ferziger, J. H., and Johnston, J. P., "Calculation of Turbulent Shear Flows: Status and Ten-Year Outlook," ASME J. Fluids Engrg., Vol. 100, 3, 1978.

Kline, S. J, Lilley, G. M., and Cantwell, B. J., Proceedings of the 1980-81 AFOSR-HTTM-Stanford Conference on Complex Turbulent Flows, Dept. of Mech. Engr., Stanford Univ., Stanford, CA, 1981.

Leonard, B. P., "Upstream Parabolic Interpolation," Proc. Second GAMM Conf. on Num. Meth. in Fluid Mech., Cologne, 1977.

Lowery, P. S., private communication.

Moser, R. D., and Moin, P., "Direct Simulation of Turbulent Flow in a Curved Channel," Rept. TF-20, Dept. of Mech. Engr., Stanford Univ., 1984.

Raithby, G. D., "Skew Upstream Differencing Schemes for Problems Involving Fluid Flow," Comp. Meth. Appl. Mech. Engrg., Vol. 9, 75, 1976.

Rodi, W., "A New Algebraic Relation for Calculating the Reynolds Stress," ZAMM, T219, 1976.

Rogallo, R. S., and Moin, P., "Numerical Simulation of Turbulent Flows," Ann. Revs. Fluid Mechanics, Vol. 16, 99, 1984.

Schumann, U., "Ein Untersuchung über der Berechnung der Turbulent Stromungen im Platten- und Rinspalt-Kanelen," dissertation, Karlsruhe, 1973.

Steger, J. L., and Kutler, P., "Implicit Finite Difference Procedures for the Computation of Vortex Wakes," AIAA J., Vol. 15, 581, 1977.

Tennekes, H., and Lumley, J. L., A First Course in Turbulence, MIT Press, 1972.

Vanka, S. P., and Leaf, G. K., "Fully Coupled Solution of Pressure-Linked Fluid Flow Equations," Rept. ANL-83-73, Argonne Natl. Lab., 1983.

Vanka, S. P., "Fully Coupled Calculation of Fluid Flows with Limited Use of Computer Storage," Rept. ANL-83-87, Argonne Nat'l. Lab., 1983.

Progress in Scientific Computing, Vol. 6
Proceedings of U.S.-Israel Workshop, 1984

NUMERICAL CALCULATION OF THE REYNOLDS STRESS AND TURBULENT HEAT FLUXES

Micha Wolfshtein
Technion, Haifa, Israel

1. Introduction

1.1 Purpose of the Paper

1.1.1 The problem considered

This paper is concerned with the numerical calculation of the turbulent momentum and heat fluxes. Direct simulation of the unsteady turbulent flow field is still very expensive. Therefore we shall concentrate here only on the much cheaper solutions of the mean flow equations. The calculation is composed of two parts: Formulation of the problem, and the numerical solution. The formulation problem may be viewed as the specification of a closure to the governing equations for turbulent flows. In recent years the term "closure" was often replaced by "turbulence models," and this is how we shall refer to such methods. A good closure should offer adequate predictive capabilities to the engineering profession at a reasonable expenditure of computer resources. The choice of a formulation influences the numerical schemes and the programming techniques to be used as the stability and efficiency of different algorithms is not identical for laminar and turbulent flows. All these problems are discussed in the paper with special reference to these models in which the Reynolds stresses and turbulent heat fluxes are governed by differential equation.

1.1.2 The relevance of the problem

Turbulent flows are so abundant in nature and in engineering that turbulent flow may be considered the "natural" form of fluid flow. However, this situation does not reflect on the ratio of research on turbulent flows to that performed on other types of flows. A possible reason for this odd situation is the difficulties which are encountered when solutions of turbulent flows are attempted.

Although it is usually assumed that the Navier-Stokes equations are applicable to turbulent flows, it is very difficult to solve these equations for a turbulent flow. The spread of scales of the motion is extreme, the boundary conditions are not simple to define, and the equations are not amenable to analytical methods. This is why turbulence models are employed. Such models may require very few assumptions (e.g. the sub-grid-scale models used for large eddy simulations) or depend on very strong assumptions (as say the integral techniques used in boundary layers).

Unfortunately, the price and complexity of the solution are inversely proportional to the number of assumptions and their severity. At the present time the more realistic models tax the computer resources too heavily for most engineering calculations, even when the biggest computers are used. It is therefore of interest to develop efficient solvers for various turbulence models in order to allow the more complex ones to be used in engineering.

1.1.3 What is a turbulence model?

Turbulence modeling is based on the assumption that the real turbulent flow field may be substituted by an imaginary field of some mathematically defined continuous functions, which can be used to obtain a "sufficiently good" description of the turbulent flow field. These functions usually represent physical quantities measured in the flow field. Many examples in the contemporary literature show that turbulence modeling is considered by a large number of researchers as an acceptable compromise. The models are not prohibitively expensive, yet they often yield sufficiently reliable answers and offer a degree of universality sufficient to justify their usage in comparison to either cheaper, less general, methods or to more expensive, but potentially more reliable, methods.

Within this group of methods we shall concentrate on models for the Reynolds stresses and turbulent heat fluxes. Although these models are rather complex, there is some evidence to suggest that the calculation of the Reynolds stresses and the heat fluxes is not a prohibitively expensive task, although it calls for the development of special numerical techniques. Some of these numerical problems will be presented, as well as special numerical schemes designed to answer these problems.

1.1.4 Turbulence modeling as a part of CFD

We wish to comment here on the relation of turbulence modeling to

Computational Fluid Dynamics (CFD). This branch of numerical mathematics is of significant practical importance, and is expected to supply answers to complicated problems which are difficult or expensive to obtain otherwise. As many of these problems are turbulent, it appears that the scope of CFD is greatly reduced if adequate methods for the calculation of turbulent flows are not available. On the other hand, calculation of turbulent flows is so complicated and difficult that it cannot be undertaken without the utilization of very powerful numerical schemes. It is not surprising that under these circumstances the two subjects are very closely related, and no serious attempt to calculate turbulent flows is made nowadays without a coupled numerical effort. The two subjects are so strongly interwoven that the calculation of turbulent flow is considered by many as a part of CFD.

1.1.5 Discussion of the mathematical problems

The numerical calculation of the Reynolds stress (or the turbulent heat fluxes) is very difficult due to a number of reasons. First, we are dealing with a large system of equations. Typically we have to solve the continuity, momentum, Reynolds stress, and turbulent scale for isothermal problems (totaling 8 equations in the 2D case and 11 in the 3D case). When non-isothermal problems are considered, the number rises further as the energy, turbulent heat fluxes, and temperature fluctuation equations have to be solved (bringing the number of equations to 12 in the 2D case and 16 in the 3D case).

The second problem is concerned with the coupling of the equations. Typically all the equations of the turbulence model are very strongly coupled to one another. While some of the coupling is related to the nonlinear convection terms, the most difficult part is related to strong linear coupling through algebraic source terms. This problem is aggravated by the very large values which these sources reach when the turbulence is in the relatively common situation of local equilibrium.

Another problem is that of the very steep profiles of the turbulent quantities in various parts of the flow field. Unfortunately different quantities reach the maximal steepness at different locations, thus making the problem less amenable to coordinate stretching.

1.2 Outline of the History of the Subject

1.2.1 Early contributions

The first attempts to use turbulence models were made by Kolmogorov (1942) and Prandtl (1945). In both these papers the turbulence was characterized by the intensity and scale (Kolmogorov actually used the frequency). The intensity was defined as the turbulent energy, which was calculated using the corresponding differential equation. The scale was assumed to be known. Both these models utilized an eddy viscosity. This assumption, which is restrictive at least in principle, was removed by Chou (1945), who derived the Reynolds stress equation and tried to identify mechanisms and models in this equation. A few years later Rotta (1951) extended Chou's Reynolds stress model, and derived an equation for the macro-scale. Typically all these efforts were not continued because the equations were very complex and analytical solutions could be obtained only for very few simple cases. Thus, the development stopped and was not to start again until high speed digital computers became available in the 1960's. When this happened, fast development started, and turbulence modeling became a very active branch of CFD.

1.2.2 Reynolds stress versus two equation models

Even in the early days a distinction could have been made between the work of Prandtl and Kolmogorov, who used the turbulence energy and scale to calculate an eddy viscosity, and that of Chou and Rotta who calculated the Reynolds stresses directly. A major difference is the number of equations and the computational load, which makes the simpler two equation models a more attractive possibility. However, the eddy viscosity hypothesis contradicts many experimental findings and is likely to be adequate only when a single component of the Reynolds stress is important. Therefore, such formulations can work very well for boundary layers, but their applicability to recirculating flows or swirling flows is limited. This difficulty is removed when a Reynolds stress model is used. The price is a larger numerical load. For instance, in an isothermal problem we need 5 or 6 equations for a two-dimensional or a three-dimensional case, respectively, while the corresponding numbers for a Reynolds stress model are 8 or 11. Typical formulations were proposed by Naot et al (1974) and Launder et al (1975). The former is used here.

1.2.3 Turbulent heat flux models

Modeling of the turbulent heat flux equations is similar in principle to that of the Reynolds stresses, although the heat fluxes are components of a vector and as such easier to model. Launder (1976) assumed that the modeling principles are similar to those of the Reynolds stresses and formulated the problem. Lumley (1975) formulated the heat flux equations using some of their tensor properties. The present author suggested elsewhere to employ a quasi-isotropic model for the two point temperature-velocity correlation. The results are similar in principle to those of Launder, although not identical. As this formulation appears to be more consistent with the governing equations, it was chosen here.

1.2.4 Algebraic stress modeling

If a Reynolds stress model is used, the number of equations increases and they become very complex. The system may be simplified by replacing the partial differential equations for the Reynolds stresses by algebraic equations. This may be done by assuming a local equilibrium, which means that all spatial derivatives of the Reynolds stresses may be neglected. A better approximation was suggested by Rodi (1979), namely to relate the convection and diffusion of the Reynolds stresses to those of the turbulence energy. Such a practice may be applied to the heat fluxes as well. The resulting equations are much simpler, but they are not as widely applicable. Still the improved computational efficiency makes them very popular for complex flow fields. We shall not consider such models here. Indeed, the major goal of this paper is to identify algorithms which allow for better structuring of programs and more efficient solution of the Reynolds stress equations.

1.3 Definition of the Problem

The governing equations for a stationary flow problem are the continuity, 3 momentum equations, 6 Reynolds stress equations, and a scale equation, totaling 11 equations for a 3-dimensional problem and 8 for a 2-dimensional case. When heat transfer is calculated as well, the energy equation, the 3 turbulent heat flux equations, and the temperature fluctuation equation should be added, bringing the total number of equations to 16 for the 3-dimensional case and 12 for the 2-dimensional case. These equations are coupled and nonlinear. The Reynolds stress and turbulent heat fluxes equations tend to become rather stiff when the turbulence approaches local equilibrium.

Another difficulty arises from the boundary conditions. The formulation of the turbulence model near solid walls is not always possible. It is therefore very common to remove the region adjacent to the wall from the computational domain. This means that boundary conditions should be specified at an artificial boundary inside the flow field. Such a formulation is usually dependent on universal empirical wall laws. Fortunately such laws may be defined.

2. The Governing Equations

2.1 The Reynolds Stress Equations

The Reynolds stresses, u_{ij}, are governed by partial differential equations which are obtained from the time dependent Navier-Stokes equations. The equations require some modeling, which is usually related to the tensorial properties of the Reynolds stresses. The resulting equations are:

$$\frac{D}{Dt} u_{ij} = V_{ij} + T_{ij} + \pi_{ij} + P_{ij} + L_{ij} + G_{ij}$$

$$V_{ij} = u_{ij,\ell\ell} \qquad T_{ij} = -(\overline{u_i u_j u_\ell})_{,\ell}$$

$$\pi_{ij} = -u_{i\ell} u_{j,\ell} - u_{j\ell} u_{i,\ell}$$

$$\begin{aligned} P_{ij} + L_{ij} &= -\alpha(\pi_{ij} - \pi_{\ell\ell}\delta_{ij}/3) \\ &\quad + \beta(D_{ij} - D_{\ell\ell}\delta_{ij}/3) + 4/3\,\gamma\, k\, E_{ij} \\ &\quad + \frac{\varepsilon}{k}(\lambda u_{ij} + 2/3\, k\,\delta_{ij}) \end{aligned} \tag{1}$$

$$D_{ij} = -u_{ik} u_{k,j} - u_{jk} u_{k,i} \qquad E_{ij} = (u_{i,j} + u_{j,i})/2$$

$$G_{ij} = -\beta_T(g_i \psi_i + g_j \psi_i)$$

where α, β, γ, λ are empirical constants and β_T is the coefficient of thermal expansion.

2.2 The Turbulent Heat Flux Equations

The turbulent heat flux vector ψ_i is governed by a partial differential equation. The exact form of the equation can be derived by manipulation of the momentum and energy equation. The modeling of this equation yields the following equation:

$$\frac{D}{Dt}\psi_i = T_{T_i} + \hat{G}_{T_i} + \frac{2}{3} C_t \frac{\varepsilon}{k}\psi_i + \Gamma_i + \hat{\pi}_i$$

$$\hat{\pi}_i = \left(\left(\frac{2}{3}\zeta - 1\right) u_{i,j} - \frac{2}{3}\xi u_{j,i}\right)\psi_j - u_{ij}\, T_{ij} \tag{2}$$

$$\hat{G}_{T_i} = -\frac{2}{3}\beta_T\, g_i\, \overline{\tau^2}$$

$$T_{T_i} = -(\overline{u_i u_j \tau})_{,j} \qquad \frac{2}{3} C_t = -\left(C_T + (1+\sigma)/C_G^2\right)$$

where C_T, C_G, ξ, ζ are empirical constants.

2.3 The Temperature Fluctuating Equation

The temperature fluctuations squared, $\overline{\tau^2}$, are required when free convection problems are considered. The equation is derived by manipulation of the Navier-Stokes equations and the energy equation. The modeling of this equation is not difficult, and the final result is:

$$\frac{D}{Dt}\overline{\tau^2} = -(\overline{u_j \tau^2})_{,j} - 2T_{,j}\psi_j - 2\frac{\nu}{\sigma}\frac{\overline{\tau^2}}{\lambda_\eta^2} + \frac{\nu}{\sigma}(\overline{\tau^2})_{,jj} \tag{3}$$

2.4 Boundary Conditions

2.4.1 Wall functions

The Reynolds stress equations are not valid in the region adjacent to solid walls. The reasons for this are the highly non-isotropic character of this region and the low level of the turbulence in the viscous sublayer. The most common cure to this situation is the use of the so-called wall functions. Essentially we define a narrow strip very near to a wall where the distribution of the Reynolds stresses is not obtained by a solution of the partial differential equations, but by substitution of a given distribution which is obtained either from experiments or from some other theory. This practice has another advantage as well. The steepest gradients usually occur near solid walls. The elimination of this region from the numerical solution allows us to use coarser meshes and to reduce the requirements of the computer resources.

The most common forms of wall functions are those derived from the logarithmic law of the wall. These are highly reliable formulations of the distribution of the mean velocity and Reynolds stresses near solid walls, which correlate very well with experimental data for a very wide range of flows. It is therefore advantageous to move the boundary from the wall to the logarithmic region. The major difficulty here is that the logarithmic region is very thin and its exact location is not known in advance. Still, it is usually possible to define a new boundary which is always inside the logarithmic region.

The logarithmic law formulation always includes the friction velocity as its major parameter. This quantity is not known in advance and should be eliminated. This can be done by using the velocity distribution as well as the velocity gradient. Unfortunately this elimination cannot be explicitly performed due to the logarithmic formulation. An explicit formulation can be obtained if the logarithmic law is replaced by a power law approximation (e.g. the 1/7th power law).

2.4.2 Some examples

Consider the logarithmic law

$$\frac{ku}{v^*} = \ell n \left(\frac{Eyv^*}{\nu}\right) \tag{4}$$

and its derivative

$$\frac{du}{dy} = \frac{v^*}{ky} \quad . \tag{5}$$

Elimination of v^* between these two equations results in a cumbersome implicit expression for the relation of u and du/dy. Although this relation may be used, it is easier to utilize the much simpler power law

$$u = a\ y^{(1/n)} \quad . \tag{6}$$

This relation yields the following expression for the boundary condition

$$\frac{du}{dy} = \frac{u}{ny} \ . \tag{7}$$

The Reynolds stresses are uniform in a logarithmic boundary layer. As we do not know their actual value, their boundary condition should be specified as a zero gradient one.

3. The Numerical Scheme

3.1 Assessment of the Problem

3.1.1 Stiffness

The problem to be solved may be described as

$$f_t = (X + Y + A)f + S \tag{8}$$

where X and Y are second order ordinary differential operators, A is an algebraic linear operator, and S is an arbitrary source term. As f is a vector (4 - 11 elements long), X, Y, and A are matrix operators. When the turbulence is in local equilibrium, A becomes very large, and sometimes it even dominates the equation. When this happens, the equation becomes stiff and special techniques are required for the solution.

3.1.2 Computer memory

Solution of the numerical problem described above is done by marching through a set of elliptic solutions. The numerical solution requires usually two levels of all the variables. For a 2D parabolic problem we need at least $2\cdot(8-11)\cdot n^2$ words of memory, while the requirements for a 3D problem are at least $2\cdot(11-16)\cdot n^3$, where n is the number of mesh points in each direction. These numbers are quite significant, and the ability to use fine meshes to get high accuracy is rather limited. Therefore it is necessary to use high order techniques to obtain higher accuracy with relatively coarse meshes.

3.1.3 CPU time

The solution of the Reynolds stress equations involves a large amount of computations for the calculation of the coefficients of the finite difference equations. This number is directly related to the number of mesh points. Therefore, these calculations require a very large portion of the computer time. A high order solution cuts this number, and is usually worthwhile even if the high order numerical scheme requires more operations than a low order scheme.

3.2 Numerical Details

For improved stability and efficiency, staggered meshes and primitive variables are recommended. The solution at each step is obtained in two levels. First we have to solve the momentum equations using an approximate pressure field to get an approximate velocity field. This field does not satisfy the continuity equation. Second we have to add a velocity potential which is used to ensure that the pressure field is so chosen as to make the related velocity field satisfy the continuity equation. The potential satisfies the Laplace equation, with the following boundary conditions:

$$\nabla^2 \phi = - \nabla \cdot V_P \tag{9}$$

$$\left(\frac{\partial \phi}{\partial \eta}\right)_B = 0 \ . \tag{10}$$

This formulation is a two level method. An approximate velocity field is predicted, and then the velocity field is modified by the addition of the potential velocity. Consistency with original equations and boundary conditions requires that the boundary conditions for the velocity components are given by

$$V_{P,B} + \nabla \phi_B = V_B \ . \tag{11}$$

3.3 The Three Level Splitting

Consider the differential equation (8), $f_t = (X + Y + A)f$. As explained above, A may become the dominant term in equation (8). The density of the matrix A causes stability problems or requires complex programming. These problems are alleviated if the following three level splitting is used :

$$\left(I - \frac{kA}{2}\right) f^* = (A+X+Y) f^n \quad , \quad \left(I - \frac{kX}{2}\right) f^{**} = f^* \quad ,$$

$$\left(I - \frac{kY}{2}\right) f^{***} = f^{**} \quad , \quad f^{n+1} = f^n + k f^{***} \tag{12}$$

where k is the time step. This scheme is second order accurate, and stable. Moreover, as A is an algebraic operator, the first step involves only a solution of a small system of equations at each mesh point. This system is dense, but its size not large (between 4 x 4 and 11 x 11). Even the larger size does not tax the computation too heavily. The second and third steps are uncoupled and therefore only a solution of a tri-diagonal matrix is required at each row or column of the finite difference mesh. Usually the operators in the above ADI scheme are difference operators. However, if they are retained in their differential form then standard ODE solvers may be used for the solution.

3.4 Fourth Order Compact Scheme.

3.4.1 Ordinary differential equations

3.4.1.1 The scheme

Let us consider the following equation:

$$f'' + p f' + q f = r \ . \tag{13}$$

It is easy to show that this equation can be approximated to the fourth order by $a\, f_{i-1} + b\, f_i + c\, f_{i+1} = A\, r_{i-1} + B\, r_i + C\, r_{i+1}$

where

$$a = \frac{(A+B+C)}{h^2} + \frac{\left(-3A\, p_{i-1} - B\, p_i + C\, p_{i+1}\right)}{2\,h} + A\, q_{i-1} \quad (14)$$

$$b = \frac{-2(A+B+C)}{h^2} + (A\, p_{i-1} - C\, p_{i+1}) + B\, q_i$$

$$c = \frac{(A+B+C)}{h^2} + \frac{(-A\, p_{i-1} + B\, p_i + 3C\, p_{i+1})}{2h} + C\, q_{i+1}$$

$$A = 6 + (2p_{i+1} - 5p_i)h - p_i\, p_{i+1}\, h^2$$

$$B = 60 + 16(p_{i+1} - p_{i-1})h - 4p_{i-1}\, p_{i+1}\, h^2$$

$$C = 6 + (5p_i - 2p_{i-1})h - p_i\, p_{i-1}\, h^2 \;.$$

This is a tri-diagonal matrix formulation, and the problem can be solved easily and efficiently using standard algorithms.

3.4.1.2 Boundary conditions

The general boundary condition to an ordinary differential equation of the second order can be written as

$$sf' + qf = w. \quad (15)$$

On the other hand, the differential equation (13) may be written as

$$f'' = r - qf - pf' \;.$$

Differentiation of the differential equation yields

$$f''' = r' - pr + (p^2 - p' - q)f' + (pq - q')f \quad (16)$$

$$f'''' = r'' - pr' + (p^2 - 2p' - q)r$$
$$+ (-p'' + 3pp' + 2pq - p^3 - 2q')f'$$
$$+ (2p'q + pq' - p^2q' + q^2 - q'')f \;. \quad (17)$$

Using these definitions of the derivatives, the variable f may be expanded near the bottom boundary, to give

$$f_2 = f_1(1+F) + f'_1\, G + E \quad (18)$$

where E, F and G are calculated on the boundary using the following formulas:

$$E = \frac{hr}{2} + \frac{h(r' - pr)}{6} + \frac{h\left(r'' - pr' + (p - 2p' - q)r\right)}{24} \quad (19a)$$

$$F = h - \frac{hq}{2} + \frac{h(pq - q')}{6} + h(2p'q + pq' - pq + q - q'') \tag{19b}$$

$$G = h - \frac{hp}{2} + \frac{h(p - p' - q)}{6} + h(-p'' + 3p\,p' + 2pq - p - 2q')\,. \tag{19c}$$

Simple elimination of f' between the boundary condition (15) and the Taylor series expansion (18) gives

$$(Gt - s - sF)f_1 + sf_2 = sE + wG \quad . \tag{20}$$

The corresponding equation for the upper boundary is

$$(Gt - s - sF)f_{N+1} + sf_N = sE + wG \tag{21}$$

where N+1 is the number of mesh points.

The functions p, q, and r are given on the boundaries. Their first and second derivatives are calculated by one-sided second-order 3-point and 4-point formulas respectively. Examination of the equations shows that a second order calculation of the derivatives is sufficient to guarantee fourth order accuracy of the solution. The extension to nonlinear problems is not difficult, but iterations are required.

3.4.2 Partial differential equations

3.4.2.1 ADI formulations with differential operators

We have already discussed the equation

$$f_t = (X + Y + A)f + S \tag{22}$$

where X and Y are ordinary differential operators in the x and y directions respectively, A is an algebraic operator, and S is a source term. As any ODE solver can be applied to the second and third stages, we can choose any efficient solver, including the fourth order algorithm described above.

The application of the algorithm is straightforward, and no preparatory work is required. However, this is not enough to guarantee fourth order accuracy. Indeed, the right hand side of the equations contains some derivatives in the lateral direction which should be computed to a fourth order if such accuracy is required.

Fourth order accuracy is obtained when the right hand side of the equation for the first ADI step is computed to fourth order as well. This involves a fourth order evaluation of first and second derivatives. Using Pade approximations, this is not a difficult task. On the boundary a fourth order one-sided analog of the Pade approximation is used. These steps are enough to warrant fourth order accuracy. Altogether the numerical load is only slightly increased. While the previous solution required the solution of 6(n-1) ODE's, the present one requires an addition of four tridiagonal matrices. These equations have constant coefficients. Therefore, it is possible to invert them once and obtain further solutions (for new time steps or iterations) by matrix multiplication.

4. Examples

4.1 Nonlinear ODE's

The fourth order ODE solver was tested on a one-dimensional Burger's equation. A typical case of $Re = 2$ converged to an accuracy of 10^{-7} with 5 mesh points and 5 iterations.

4.2 Poisson Equation

The fourth order ADI scheme as outlined above was used to solve a Poisson equation

$$f_{xx} + f_{yy} = \frac{a\left(2e^{ax}e^{ay} - e^{ax} - e^{ay}\right)}{e^{a} - 1} \tag{23}$$

with $a = 10$. The solution of this equation is

$$f = \frac{\left(e^{ax} - 1\right)\left(e^{ay} - 1\right)}{e^{a} - 1} \tag{24}$$

The solution converged as $4.71/N$, where N is the number of mesh points in the x and y direction. About 100 iterations were required to obtain convergence with a 20×20 mesh, and with a maximum error of the order of 10^{-5}.

4.3 Turbulent Duct Flow

The second order ADI scheme was used to solve the parabolic three-dimensional Reynolds stress equations in a square duct. The solution on an IBM 370/168 computer with a 20×20 mesh required 2 - 4 seconds CPU time per integration step. Fair agreement with experiments was obtained.

5. Conclusions

The main conclusions of this paper are:

(i) The Reynolds stress and turbulent heat flux equations can be solved at an acceptable price and without serious programming difficulties.

(ii) High order methods should be used to obtain accurate solutions with coarse meshes. This resolves the problems of CPU time and computer storage.

(iii) The formulation and programming of the Reynolds stress and turbulent heat flux equations are made much easier if the solvers can be simply used, without having to formulate the high order finite difference formulas (these should be automatically performed inside the program).

6. References

Chou, P.Y., (1945), On the velocity correlations and the solutions of equations of turbulent fluctuation, Quart.J.Appl.Math., vol 3, pp 38-54.

Launder, B.E., (1976), Heat and mass transport, Chapter 6 in TURBULENCE, ed. by P. Bradshaw, Topics in Applied Physics, vol 12, Springer Verlag.

Lumley, J.L., (1975), Prediction methods for turbulent flows, von Karman Inst. Lecture Series, 76.

Kolmogorov, A.N., (1942), Equation of a turbulent motion of an incompressible fluid, Izv. Akad. SSR Ser. Phys., vol VII, 1-2 (in Russian).

Naot, D., A. Shavit and M. Wolfshtein, (1974), Warme und Stoff Ubertraggung, vol 7, pp 151-161.

Prandtl, L., (1945), Uber ein neus Formelsystem fur die ausgebildete Turbulence, Nachrichten von der Akad. der Wissenschaften in Gottingen, van der Loeck & Ruprecht, pp 6-19.

Rodi, W., (1979), Turbulence models and their application to hydraulics, I.A.H.R. section on Fundamentals of Division II: Experimental and Mathematical Fluid Dynamics, Rotterdamsweg 185, 2600 MH Delft, the Netherlands.

Rotta, J., (1951), Statistical theory of non-homogeneous turbulence, Physic, vol 129, p 547 (in German).

Progress in Scientific Computing, Vol. 6
Proceedings of U.S.-Israel Workshop, 1984

NUMERICAL INVESTIGATION OF ANALYTICITY PROPERTIES OF HYDRODYNAMIC EQUATIONS USING SPECTRAL METHODS

P. L. Sulem

1. THE METHOD

Proving all time regularity or occurrence of a singularity for non-linear partial differential equations (PDE) like those arising in fluid mechanics may be a mathematical challenge. For example, existence in the large of classical solutions to the three-dimensional Navier-Stokes (NS) equation for an incompressible fluid, remains unproved, except for small Reynolds numbers (large viscosities) [21] and [38]. For the three-dimensional Euler (inviscid NS) equation, only local results are available [19], [11], [5], [4], and [37]. It is in particular unknown whether this flow develops a singularity in a finite time [39], [13], and [29]. The situation is much better in two-dimensions: the conservation of vorticity (Curl of velocity) insures existence in the large of a classical solution even at zero viscosity [40], [18], [11], and [32]. However, when this conservation is broken by the coupling of the velocity field with another field like the magnetic field in magnetohydrodynamics or the temperature in stratified flows, only local existence has been proved in the absence of dissipation [30].

The aim of this paper is to show on several examples, that like in other fields of non-linear dynamics [41], the computer, used in a heuristic mode, can shed some light on the regularity properties of non-linear PDE's. High resolutions are generally required, and super-computers are thus needed to study equations with two or three spatial dimensions.

A simple method has recently been implemented to equations with periodic boundary conditions [33]. This assumption which strongly simplifies the analysis, seems reasonable when the interest is in the internal dynamics rather than in the effects of external geometrical constraints. Assuming analytic initial conditions, the method consists of studying the time evolution of the width of the spatial analyticity strip of the solution. This quantity is easily estimated from the large wavenumber behavior of the Fourier transform of the solution. Indeed, if a function $v(\vec{z})$ with $\vec{z} \in C^n/2\pi Z^n$, is analytic in the strip $S_\rho = \{|\mathrm{Im}\,\vec{z}| < \rho\}$ and continuous in the closure of S_ρ, the Fourier coefficients satisfy

$$|\hat{v}_k| \leq M \exp\{-|\vec{k}|\rho\} \tag{1}$$

with $M = \sup_{S_\rho} v$. Reciprocally, if (1) holds, the function is analytic in the strip S_ρ. A more precise estimate is available for a function of one variable whose analytic continuation has poles or branch points [15]. The leading term of the asymptotic behavior of the Fourier modes behaves like an exponential with a logarithmic decrement equal to the width of the analyticity strip. A possible algebraic prefactor reflects the nature of the singularity the closest to the real domain. Oscillatory behavior may be produced by the dominance of several complex singularities (e.g. in problems with symmetries).

Spectral (or pseudo-spectral) methods [26], [17] based on Fourier mode expansions, are very well adapted to these investigations, provided that the spatial resolution be large enough. More precisely, the width of the analyticity strip of the solution must be large compared to the mesh size. This condition is required for the uncertainty on the position of the closest complex singularity to be much smaller than its distance to the real domain. As the numerical integration proceeds in time, two situations may occur. If the width of the analyticity strip is bounded away from zero, the computation can in principle be carried on for an arbitrary long time, provided that the mesh size be small enough. If it decreases indefinitely, the calculation has to be interrupted when the width of the analyticity strip becomes comparable to the mesh size. From the available numerical data, one can try to characterize quantitatively the time evolution of the analyticity strip. If the spatial resolution is high enough, this evolution may be followed up to times large enough to be considered as asymptotic, and thus extrapolated to $t \to \infty$ or $t \to t_*$, depending on the situation.

The method was first tested on the one-dimensional Burgers equation [33], the analyticity structure of which is well known [12]. Other one-dimensional problems which have been considered include non linear

Schrodinger equations [33], [35] and the Kelvin Helmholtz interface instability [20]. In the following sections, we shall report on results obtained in the case of multi-dimensional evolution problems arising in fluid mechanics. For such problems, instead of dealing with the Fourier transform $\hat{v}_k$ of the solution which is a vector field depending on a multi-dimensional wavevector, it may be more convenient in practice to consider the angle average "energy spectrum"

$$E(k) = \sum_{k \leq |\vec{k}'| < k+1} |\hat{v}_{k'}|^2 \tag{2}$$

on which the analyticity of the solution in S_ρ is reflected by an exponential decay with a logarithmic decrement 2ρ.

2. THE TWO-DIMENSIONAL EULER EQUATION

Analyticity in the large for the two-dimensional Euler equation was proved in [3]. Like the existence of classical solutions, it is based on the conservation of vorticity, here along the analytic continuation of the fluid trajectories. It is shown that if the (periodic) initial condition $u_0(\vec{x})$ can be continued as an analytic function in a strip $|\mathrm{Im}\, \vec{z}_0| < d_\rho$, then, for any real time, the solution remains analytic in some strip $|\mathrm{Im}\, \vec{z}| < D(t)$ where $D(t)$ does not go to zero faster than an exponential of an exponential.

A numerical simulation of the two-dimensional Euler equation with $(256)^2$ Fourier modes is presented in [33]. The temporal scheme is a stabilized leap-frog. The initial velocity field is that considered in [23]:

$$\vec{u}(-\sin y - \sin 2y\ ,\ \sin x + \sin 2x) \tag{3}$$

Fig. 1 taken from [33] shows the angle-average spectrum (2) in lin-log scales up to $t = 2.2$. The solid lines are least-square fits to the logarithm of the energy spectrum assuming

$$E(k,t) = C(t)\, k^{-\alpha(t)} \exp\{-\beta(t)k\}\ . \tag{4}$$

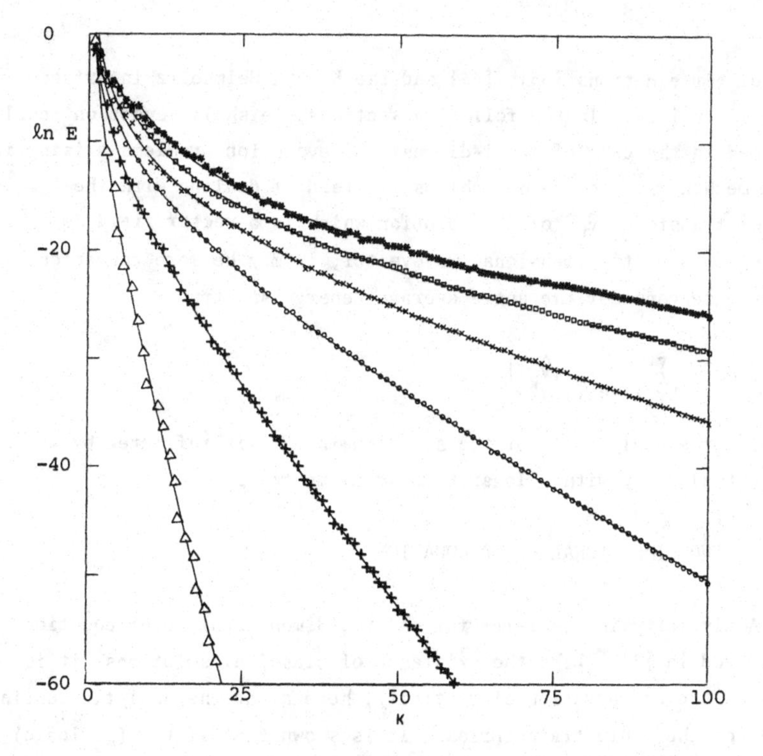

Figure 1. Angle-averaged spectrum of the two-dimensional Euler equation in linear-log scales at equally spaced times between $t = 0.2$ and 2.2 in steps of 0.4 for initial condition (3).

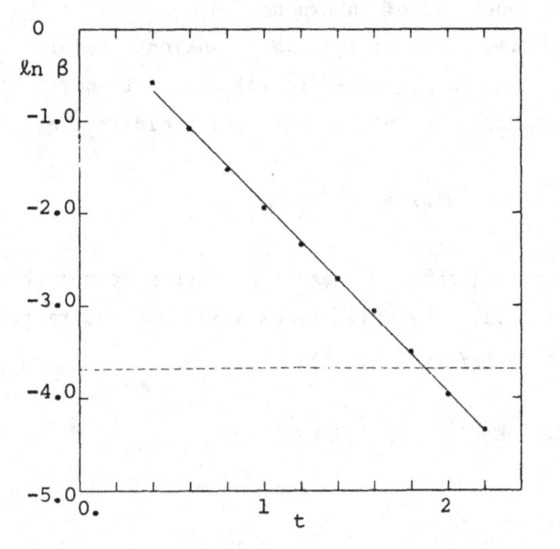

Figure 2. Logarithmic-decrement versus time of the two-dimensional Euler equation. The dashed line shows the mesh size ($2\pi/256$).

The time evolution of the logarithmic decrement $\beta(t)$ is shown in figure 2 (taken from [33]) for $t \lesssim 2.2$. Clearly, $\beta(t)$ decays exponentially. This exponential is almost insensitive to the precise choice of the fitting range of wavenumbers. Figure 2 corresponds to a fit in the range $8 \lesssim k \lesssim 100$. Similar numerical experiments have been done with other (even random) initial conditions and highest resolutions have been used [7], [8]. An exponential decay of the logarithmic decrement of the energy spectrum has always been observed, suggesting that the width of the analyticity strip of the two-dimensional Euler equation decays in fact exponentially in time. The dynamics is characterized by the formation of layers of high vorticity gradient which have been predicted by Saffman [28] and probably reflect the existence of nearby complex singularities. This quasi-one dimensionality of the small scale structures, combined with the incompressibility makes the non-linear interactions weaker than predicted by the usual estimates on the Euler equation which does not take these geometrical features into account.

3. THE NON-DISSIPATIVE TWO-DIMENSIONAL MHD-EQUATIONS

The analysis was also applied to the non-dissipative magnetohydrodynamic equations in two-dimensions which read

$$\left\{\begin{array}{l} \dfrac{\partial v}{\partial t} + v\nabla v = -\nabla p + b\nabla b \\ \dfrac{\partial b}{\partial t} + v\nabla b = b\nabla v \\ \nabla . v = 0 \quad ; \nabla . b = 0 \end{array}\right. \tag{5}$$

Only local existence, regularity and analyticity have been proved for these equations [30]. Numerical simulations were performed in [16] for different kinds of initial conditions:

(i) Initial condition with neutral x-points. In these points, the magnetic field vanishes and the lines of magnetic potential have an hyperbolic structure. To be specific, we concentrate here on the results obtained with the Orszag-Tang (OT) vortex [27] corresponding to a stream function

$$\psi_0(x,y) = 2\cos x + 2\cos y \tag{6a}$$

and a magnetic potential

$$a_0(x,y) = 2\cos x + \cos 2y \,. \tag{6b}$$

Similar results were obtained with random initial conditions. Figure 3 taken from [16] displays the magnetic energy spectrum (the kinetic energy spectrum is very similar), computed with a resolution of $(512)^2$ Fourier modes. Again a very good fit is obtained with (4).

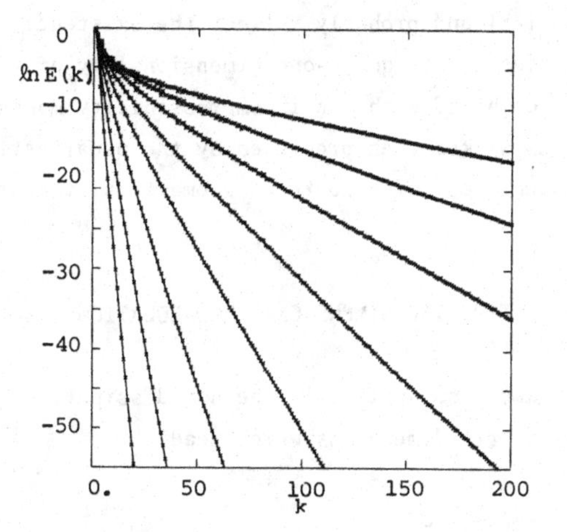

Figure 3. Magnetic energy spectrum for the O-T vortex, resolution 512^2. Time t from 0.1 (steepest) to 0.8 (shallowest) in steps of 0.1.

Figure 4 shows that the logarithmic decrement of the energy spectrum decreases exponentially in time. An asymptotic analysis near the neutral point indicates that this law of decay is not a transient but proceeds forever [34]. In this case, the solution remains analytic in the large and the spontaneous appearances of real singularity in the two-dimensional MHD equations must be ruled out. Nevertheless, smaller and smaller scales of motion develop in time.

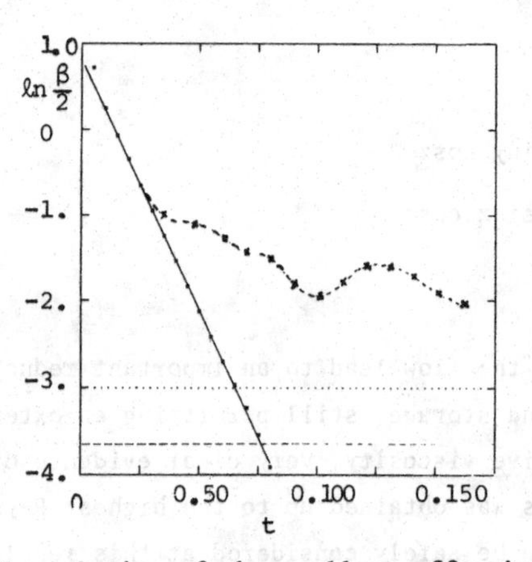

Figure 4. Temporal evolution of the smallest effectively excited scale (β = log decrement of the energy spectra). O-T vortex, linear-log coordinates. Dotted and dashed lines are the respective limits of reliability for resolutions 256^2 and 512^2 (β = twice the mesh). The crosses forming an undulating structure correspond to the case where a uniform magnetic field of strength 4 in the x_1 direction is added to the O-T vortex. Data up to $t \approx 0.25$ are undistinguishable from the pure O-T case.

(ii) A qualitatively different dynamics is observed with flows without neutral points. The simplest way to insure this, is to add to the O-T vortex, a uniform background magnetic field of sufficient strength B_0. For short time, the logarithmic decrement of the energy spectra is indistinguishable from the pure O-T, but when integration is continued, it appears to stay bounded away from zero, indicating that the width of the analyticity strip of the solution has a positive lower bound. It is likely that this regularization effect is due to the dispersive action of the Alfven waves which, in the case of no neutral points, are present through all the flow. Regularization by Alfven wave was proved in [31] for a one-dimensional model which generalized the Burgers equation to MHD.

4. THE NAVIER STOKES AND EULER EQUATIONS IN THREE-DIMENSIONS

Numerical simulations at resolutions up to $(256)^2$ Fourier modes were performed in [6] in the case of the Taylor Green vortex [36], [23], [9], and [10] corresponding to the initial conditions:

$$\begin{cases} u = \cos x \sin y \cos z \\ v = -\cos y \sin x \cos z \\ w = 0 \end{cases} \tag{7}$$

The symmetries of the flow lead to an important reduction to the amount of calculations and storage, still permitting an extended range of scales. At positive viscosity, very clear evidence of spatial analyticity for all times was obtained up to the highest Reynolds number $R = 3000$ which can be safely considered at this resolution. The inviscid problem seems more delicate to investigate. The energy spectrum is again well fitted with (4). It appears that in time interval $t < 2$, where the computation is reliable, the logarithmic decrement of the energy spectrum decays exponentially in time. It is however unclear whether this behavior may be considered as asymptotic. Quoting the authors, there are indirect evidences that this is only a transient and that "more violent vortex stretching takes place at later time, possibly leading to a real singularity at a finite time". Mathematically, only local regularity has been proved. It can be viewed as the consequence of an abstract Cauchy Kovalewska theorem whose hypotheses are satisfied by the Euler equation [1]. Using more specifically the hyperbolic character of the equation, it has been shown that the solution remains analytic as long as it is continuously differentiable [2]. This result which also holds for MHD equations [30], strenghtens the interest of studying the analyticity properties of these equations.

5. TEMPORAL ANALYTICITY

An alternative method to investigate the regularity properties of non-dissipative hydrodynamic equations was developed in [23] and [6]. It consists basically in studying the (temporal) analytic structure of (spatial) Sobolev norms of the solution.

The solutions of the Euler or non-dissipative MHD equations are known to be locally analytic in time [1] and [2]. This insures that the solution can thus be computed as a temporal series around the origin. Because of the quadratic non-linearity of the equations, then, for initial conditions which are linear combinations of sines and cosines, the coefficients of the series are trigonometric polynomials. The accuracy is limited by the rounding error only. Time expansions of various Sobolev norms of the solution can thus be obtained. The analytic structure of these norms is then determined by the behavior of the successive coefficients. The analysis is relatively simple and gives very precise results when the singularity the closest to the origin is real [22]. When it is not the case, it is much more delicate to implement because analytic continuation is required outside the circle of convergence of the Taylor Series, by mean of Padè approximants or Weierstrass continuation. The two-dimensional Euler and non-dissipative MHD equations were investigated in [24], and evidences of analyticity in the large were obtained. The situation seems more delicate for the three-dimensional inviscid Taylor Green vortex where the results are still inconclusive [6]. As noticed in [14], the presence of non-isolated singularities (natural frontiers) could make analytic continuations very delicate.

REFERENCES

[1] Baouendi, M. and Goulaouic C: Problèmes de Cauchy Pseudo-Differentiels Analytiques. Sèminaire Goulaouic - Schwartz (1975-1976). Ecole Polytechnique, Paris.

[2] Bardos C. and Bènachour S: An. Sc. Norm. Sup. Pisa, IV, 4(1977), 648.

[3] Bardos C., Bènachour S. and Zerner M.: C. R. Ac. Sc. Paris, A 282, (1976), 995.

[4] Bardos C. and Frisch U.: C. R. Acad. Sc. Paris 281, A 775 (1975).

[5] Bourguignon, J.P. and Brèzis H.: J. Funct. Anal. 15 (1974), 341.

[6] Brachet, M.E., Meiron, D., Orszag, S.O., Nickel B., Morf, R. and Frisch U.: J. Fluid Mech. 130, (1983), 411.

[7] Brachet, M.E. and Sulem, P.L.: Direct Numerical Simulation of Two-dimensional Turbulence, MHD Flows and Turbulence. (Proc. 4th Beersheva Sem. on MHD Flows and Turbulence, 1984). Progress in Astronautics and Aeronautics. In Press.

[8] Brachet, M.E. and Sulem, P.L.: Free Decay of High Reynolds Number Turbulence. Proc. 9th Int. Conf. Numer. Math. in Fluid Dyn. (Saclay 1984). Springer Lecture Notes in Physics. In Press.

[9] Chorin, A.J.: Comm. Pure Appl. Math. 34, (1981), 853.

[10] Chorin, A.J.: Comm. Math. Phys. 83, (1982), 517.

[11] Ebin, D.G. and Marsden: J. Ann. Math. 92, (1970), 102.

[12] Fournier,J.D. and Frisch, U.: J. Mèc. Thèor. Appl. 2, (1983), 699.

[13] Frisch, U.: Fully Developed Turbulence and Singularities. Les Houches Summer School 1981. G. Iooss, R.H.G. Helleman and R. Stora eds. North Holland, (1983).

[14] Frisch, U.: The Analytic Structure of Turbulent Flows. Proc. 6th Kyoto Summer Institute. Springer Lecture Notes, (1984).

[15] Frisch, U. and Morf, R.: Phys. Rev. A23, (1981), 2673.

[16] Frisch, U., Pouquet A., Sulem, P.L. and Meneguzzi, M.: J. Méc. Pure Appl. Special Issue on Two-Dimensional Turbulence, (1981), 191.

[17] Gottlieb, D. and Orszag, S.A.: Numerical Analysis of Spectral Methods: Theory and Applications. SIAM. (1977).

[18] Kato, T.: Arch. Ration. Mech. Anal. 25, (1967), 188.

[19] Kato, T.: Funct. Anal. 9, (1972), 296.

[20] Krasny, R.: A Numerical Study of Kelvin-Helmholtz Instability by the Point Vortex Method. Ph.D. Thesis. Lawrence Berkeley Laboratory. University of California, (1983).

[21] Lions, J.C.: Quelques Methodes de Resolution des Problèmes aux Limites Non-linéaires. Dunod-Gauthier-Villard.

[22] Meiron, D.I., Baker, G.R. and Orszag, S.A.: J. Fluid Mech., 114, (1982), 283.

[23] Morf, R.H., Orszag, S.A. and Frisch U.: Phys. Rev. Lett. 44 (1980) 572.

[24] Morf, R.H., Orszag, S.A., Meiron, D.I., Frisch, U. and Meneguzzi, M.: Proc. 7th Int. Conf. Numer. Meth. in Fluid Dyn. Lecture Notes in Physics 141, Springer Verlag, (1981), 292.

[25] Orszag, S.A.: Statistical Theory of Turbulence. 1973. Les Houches Summer School. R. Balian and J.L. Peube, eds. Gordon and Breach, (1977), 235.

[26] Orszag, S.A. and Patterson, G.S.: Phys. Rev. Lett. 28, (1972), 76.

[27] Orszag, S.A. and Tang, C.M.: J. Fluid Mech. 90, (1979), 129.

[28] Saffman, R.G.: Studies in Appl. Math. 50, (1971), 377.

[29] Saffman, R.G.: J. Fluid Mech. 106, (1981), 49.

[30] Sulem, C.: C.R. Acad. Sci. Paris 285, A365, (1977).

[31] Sulem, C., Fournier, J.D., Frisch, U. and Sulem, P.L.: C.R. Acad. Sci. Paris 288, A571, (1979).

[32] Sulem, C. and Sulem, P.L.: J. Mec. Théor. Appl. Special Issue on Two-Dimensional Turbulence. (1983), 217.

[33] Sulem C., Sulem, P.L. and Frisch, H.: J. Comp. Phys. 50, (1983), 138.

[34] Sulem, P.L., Frisch U., Pouquet, A. and Meneguzzi, M.: On the Exponential Flattening of Current Sheets Near Neutral X-Points in Two Dimensional Ideal MHD Flow. Submitted to J. Plasma Phys., (1984).

[35] Sulem, P.L., Sulem, C. and Patera, A.: Numerical Simulation of Singular Solutions to the Two-Dimensional Cubic Schrodinger Equation. Comm. Pure Appl. Math. In Press, (1984).

[36] Taylor, G.I. and Green, A.E.: Proc. Roy. Soc., London, A158, (1937), 499.

[37] Temam, R.: In Turbulence and Navier Stokes Equation, (Orsay, 1975). Lecture Notes in Mathematics 565, Springer Verlag, (1976), 184.

[38] Temam, R.: Navier Stokes Equation, North Holland, (1977).

[39] Von Neumann, J.: Recent Theories of Turbulence, (A Report to ONR 1949). Collected Works 6, (1949 - 1963), 437.

[40] Wolibner, W.: Math. Z. 37, (1933), 668.

[41] Zabusky, N.: J. Comp. Phys. 43, (1981), 195.

Tel-Aviv University, Israel and

CNRS, Observatoire de Nice, France.

Progress in Scientific Computing, Vol. 6
Proceedings of U.S.-Israel Workshop, 1984

Order and Disorder in the Kuramoto-Sivashinsky Equation

by Daniel Michelson

1. Introduction. The Kuramoto-Sivashinsky equation

$$u_t + u_{xxxx} + u_{xx} + \frac{1}{2} u_x^2 = 0, \quad -\infty < x < \infty, \quad t > 0 \qquad (1.1)$$

has attracted for the last decade a considerable attention ([3]-[6], [8], [12], [14]-[17]). It was originally derived by Kuramoto and Tsuzuki [3] in the context of a reaction diffusion system, and by Sivashinsky [14] in the context of flame front propagation. In the latter case $u(x,t)$ represents the perturbation of a flame front which propagates in a fuel - oxygen mixture. Numerical experiments ([4], [6], [8], [17]) have shown that eg. (1.1) when solved on a sufficiently large interval $-\ell < x < \ell$ with periodic boundary conditions tends to a turbulent state as $t \to \infty$. The solution $u(x,t)$ has the form

$$u(x,t) = -c_0^2 t + v(x,t) \qquad (1.2)$$

where $c_0^2 \approx 1.2$ is a universal constant independent of the initial condition, while the mean value of $v(x,t)$ is close to zero. For a fixed t the function $v(x,t)$, although irregular, has an appearance of a quasi-periodic wave with a characteristic wave length $\ell_0 = 2\pi/\omega_0$, $\omega_0 = \sqrt{2}/2$. Note that the frequency ω_0 is maximally amplified by the linear terms in (1.1). Formula (1.2) suggests that one should look for steady solutions of (1.1), namely solutions of the form

$$u(x,t) = -c^2 t + v(x) \qquad (1.3)$$

Clearly $v(x)$ satisfies the O.D.E.

$$\frac{d^4 v}{dx^4} + \frac{d^2 v}{dx^2} = c^2 - \frac{1}{2} \left(\frac{dv}{dx}\right)^2 \qquad (1.4)$$

or a third order equation for the slope $y = \frac{dv}{dx}$

$$\frac{d^3y}{dx^3} + \frac{dy}{dx} = c^2 - \frac{1}{2}y^2, \quad -\infty < x < \infty \tag{1.5}$$

The main objective of this work is to study the set of bounded solutions of (1.5) and its dependence on the parameter c.

2. Topological properties of the set of bounded solutions

First we shall show that for c in a closed interval $c \in [0,c_*]$ the set of all bounded solutions of (1.5) is uniformly bounded. Indeed, let $y_n(x)$ be a sequence of bounded solutions corresponding to a sequence of parameters $c_n \in [0,c_*]$ such that $\sup_x |y_n(x)| \to \infty$. We may even assume that $\rho_n = |y_n(0)| \to \infty$ and $|y_n(x)|/\rho_n \leq 2$. The function $z_n(\xi) = \rho_n^{-1} y_n(\rho_n^{-1/3}\xi)$ satisfies the equation

$$\frac{d^3z}{d\xi^3} + \rho_n^{-2/3}\frac{dz}{d\xi} = c_n^2/\rho_n^2 - \frac{1}{2}z^2 .$$

Since $|z_n| \leq 2$, by the above equation the derivatives $d^iz_n/d\xi^i$, $i \leq 4$ are uniformly bounded as $n \to \infty$. Hence there exists a subsequence z_{n_k} which converges uniformly on compact intervals to a bounded solution $z(\xi)$ of the equation $z''' = -\frac{1}{2}z^2$. The last equation has however only a trivial bounded solution which contradicts the fact that $|z(0)| = 1$.

For $c >> 1$ change in (1.5) the variables

$$z = y/(c\sqrt{2}), \quad \xi = x(c\sqrt{2})^{1/3}.$$

Eq. (1.5) then becomes

$$\frac{d^3z}{d\xi^3} + \varepsilon\frac{dz}{d\xi} = \frac{1}{2}(1-z^2), \quad \varepsilon = (c\sqrt{2})^{-2/3} << 1. \tag{2.1}$$

Again, as in eq. (1.5) the set of the bounded solutions of (2.1) for ε in a compact interval $\varepsilon \in [0,\varepsilon_*]$ is uniformly bounded.

Rewrite eq. (1.5) as a system

$$\frac{d\bar{y}}{dx} = (y_2, y_3, c^2 - y_2 - \frac{1}{2}y_1^2), \quad \bar{y} = (y_1,y_2,y_3). \tag{2.2}$$

Our next step is to show that the Conley index of the set of all bounded solutions of (2.2) is zero. Recall briefly (for details see [1]) some properties of this index.

a) Conley's index is a homotopy type of a pointed topological space.
b) For each isolated invariant set of a flow there is a corresponding index.
c) The index of a disjoint union of isolated invariant sets of a flow is a sum of their indices (i.e. the homotopy type of the wedge of the corresponding pointed topological spaces).
d) The index of an isolated invariant set does not change under a homotopy of the flow (provided the invariant set remains isolated under the homotopy).
e) The index of a hyperbolic critical point or of a hyperbolic periodic orbit is non-zero.

The flow in (2.2) could be extended by a two-parameter homotopy

$$\frac{d\bar{y}}{dx} = (y_2, y_3, t - sy_2 - \frac{1}{2}y_1^2), \; t \in [-c^2, c^2], \; s \in [0,1] \qquad (2.3)$$

to

$$\frac{d\bar{y}}{dx} = (y_2, y_3, -c^2 - \frac{1}{2}y_1^2).$$

The last system does not have bounded solutions (see [1], p. 12). Denote by $I(t,s) \in R^3$ the set of all points which belong to the bounded trajectories of (2.3). Again as with the system in (2.2) the set $I(t,s)$ is uniformly bounded for bounded t and s. Since the empty set $I(-c^2,0)$ has a zero index so is the index of $I(t,s)$. Note that system (2.2) has two hyperbolic critical points $\bar{y}_L = (\sqrt{2}c,0,0)$ and $\bar{y}_R = -\bar{y}_L$. Thus we have proved that the critical points or hyperbolic periodic orbits are not isolated components of the set of bounded solutions of (2.2). For a large c one can obtain a more detailed information regarding the bounded solutions of (2.2). Rewrite eq. (2.1) as a system

$$\frac{d\bar{z}}{d\xi} = (z_2, z_3, \frac{1}{2}(1-z_1^2) - \varepsilon z_2), \; \bar{z} = (z_1, z_2, z_3). \qquad (2.4)$$

For $\varepsilon = 0$ this system was firstly studied in [2] (see also [1] and [9]). It is known ([7]) that for $\varepsilon = 0$ eq. (2.2) has a unique bounded solution $\bar{z}(\xi)$. This solution connects the critical points $\bar{z}(-\infty) = \bar{z}_L = (1,0,0)$, $\bar{z}(+\infty) = \bar{z}_R = (-1,0,0)$, the function $z_1(\xi)$ is odd, vanishes only at zero and $dz_1/d\xi(0) < 0$. It is also known ([2]) that the two dimensional stable manifold $M_{st}(\bar{z}_R)$ of the critical point $\bar{z}_R$ and the two dimensional unstable manifold $M_{un}(\bar{z}_L)$ of the critical point $\bar{z}_L$ intersect transversally along the above bounded solution. Because of the transversality system (2.4) for small ε has a unique bounded solution $\bar{z}(\xi;\varepsilon)$ with the same properties as the $\bar{z}(\xi)$ above. The corresponding solution $v(x) = \int_0^x y(\tau)d\tau$ of eq. (1.4) for large c has conical form with a single maximum $v(0) = 0$ and the slopes $\pm c\sqrt{2}$ at $\mp\infty$. This solution has an interesting physical interpretation. The equation

$$u_t + u_{xxxx} + u_{xx} + \frac{1}{2}u_x^2 = c^2 \tag{2.5}$$

is a model equation (due to Sivashinsky) for a conical flame front on a Bunsen burner. Thus the above $v(x)$ represents a stationary Bunsen flame.

3. Periodic and quasi-periodic solutions

Consider eq. (1.5) for $0 < c << 1$. We are looking for periodic solutions with frequency $\omega = 1 + O(c)$. It is convenient to rescale the variables so that the period is independent of c. Change the variables $\xi = \omega x$, $z = y/\omega^3$. Then

$$\frac{d^3z}{d\xi^3} + \lambda\frac{dz}{d\xi} = \varepsilon^2 - z^2/2 \tag{3.1}$$

where

$$\varepsilon = c/\omega^3, \lambda = \omega^{-2} = 1 + O(\varepsilon) \tag{3.2}$$

Periodic solution of (3.1) with period 2π could be found by means of a power expansion in ε

$$z_{per} = \sum_{n=1}^{\infty} z_n(\xi)\varepsilon^n, \quad \lambda = 1 + \sum_{n=1}^{\infty} \lambda_n \varepsilon^n \tag{3.3}$$

Substitution of (3.3) into (3.1) gives

$$z_1''' + z_1' = 0, \quad z_2''' + z_2' = 1 - \lambda_1 z_1' - z_1^2/2$$

$$z_3''' + z_3' = - \lambda_1 z_2' - \lambda_2 z_1' - z_1 z_2 \quad \text{etc.}$$

Thus $z_1 = b_{10} + a_{11}\sin\xi + b_{11}\cos\xi$. Shifting ξ we may always assume that $b_{11} = 0$. In order to avoid resonance in the equation for z_2 one should assume that $\lambda_1 = 0$, $b_{10}\cdot a_{11} = 0$ and $b_{10}^2/2 + a_{11}^2/4 = 1$.

Here there are two possibilities. If $a_{11} = 0$ then $b_{10} = \pm\sqrt{2}$ and we recover the trivial solutions $z \equiv \pm\varepsilon\sqrt{2}$. If $b_{10} = 0$ then $a_{11} = \pm 2$ and as it follows from the subsequent equations

$$z_n(\xi) = \sum_{k=1}^{n} a_{nk} \sin k\xi \tag{3.4}$$

so that $z_{per}(\xi)$ is an odd function. The solutions with $a_{11} = 2$ and $a_{11} = -2$ are related by the shift $\xi \to \xi + \pi$. We select $a_{11} = -2$ so that for small $\varepsilon > 0$ $z'(0) < 0$. The first two terms of the expansion are

$$z_{per} = -2\varepsilon \sin\xi - \frac{\varepsilon^2}{6} \sin 2\xi + 0(\varepsilon^3),\ \lambda = 1 + \varepsilon^2/12 + 0(\varepsilon^4) \tag{3.5}$$

Actually all $\lambda_{2n+1} = 0$ and $a_{nk} = 0$ for odd n-k. The expansion in (3.4) could be justified rigorously using the Liapunov-Shmidt reduction. Using a computer we have calculated the first 100 terms of the expansion in (3.3). The radius of convergence is $|\varepsilon| \le R \approx 3.61$. By (3.2) one can also reconstruct the values of ω and c corresponding to ε. The ω,c curve is shown on Fig. 1. The maximal $c = c_{max} \approx 1.266$ corresponds to $\omega = 0.84$. The frequency $\omega_0 = \sqrt{2}/2$ as mentioned in Introduction is maximally amplified by the linear terms in (1.1). The corresponding value of $c^2 = 1.17$ on the graph is close to the mean propagation velocity $c_0^2 \approx 1.2$ of a turbulent flame as calculated in the numerical experiment [8]. The left end point P_4 of the graph corresponds to the boundary of the domain of convergence $\varepsilon = R$.

Next, we study the flow defined by (3.1) in a neighborhood of the above periodic solution. Change in (3.1) the variable

$$z \to z/\varepsilon \tag{3.6}$$

and rewrite the corresponding equation as a system

$$\frac{d\bar{z}}{d\xi} = f(\bar{z},\varepsilon) = (z', z'', \varepsilon - \lambda(\varepsilon)z' - \varepsilon z^2/2), \bar{z} = (z, z', z'') \tag{3.7}$$

The plane $z'' = 0$ intersects the periodic trajectory $\bar{z}_{per}$ at least at two points $\bar{z}_0 = (0, z'_{per}(0), 0)$ and $\bar{z}_1 = (0, z'_{per}(\pi), 0)$. (We use the old notation z_{per} for the rescaled z_{per}/ε). Denote by R^2 the plane $z'' = 0$, by $D^2 \subset R^2$ a small disk centered at $\bar{z}_0$ and by $P: D^2 \to R^2$ the corresponding Poincare map. Observe that the flow in (3.7) is volume preserving since div $f(\bar{z},\varepsilon) = 0$. Hence the map P preserves the measure $(\varepsilon - \varepsilon z^2/2 - \lambda(\varepsilon)z')dzdz'$. As far as $z'_{per}(0) < 0$ this measure is positive in a neighborhood of $\bar{z}_0$. For small ε,

$z'_{per}(0) = -2 + O(\varepsilon) < 0$. Our computations show that $z'_{per}(0)$ remains negative along the whole curve on Fig. 1. Besides the volume preservation, the flow in (3.7) is invariant under the change of variables $(z,z',z'') \to (-z,z',-z'')$, $\xi \to -\xi$.
As a result P satisfies the identity

$$JP = P^{-1}J \tag{3.8}$$

where $J = J^{-1}: R^2 \to R^2$ maps the pair (z,z') into $(-z,z')$. Consider the differential $dP(\bar{z}_0)$ of the map P at $\bar{z}_o$. Clearly

$$|\det dP(\bar{z}_0)| = 1. \tag{3.9}$$

In order to compute $dP(\bar{z}_0)$ one should solve the linearized equation

$$z''' + \lambda(\varepsilon)z' = -\varepsilon z_{per}\, z \tag{3.10}$$

An easy calculation shows that

$$dP(\bar{z}_0) = I + 2\pi\varepsilon \begin{pmatrix} 0 & 1 \\ -1 & 0 \end{pmatrix} + O(\varepsilon^2) \tag{3.11}$$

For small ε the eigenvalues of $dP(\bar{z}_0)$ are

$$\lambda_{1,2} = e^{\pm i\alpha_0(\varepsilon)}, \quad \alpha_0(\varepsilon) = 2\pi\varepsilon + O(\varepsilon^2) \tag{3.12}$$

i.e. they are complex conjugate and on the unit circle. Thus for small ε the periodic solution is elliptic. The matrix $dP(\bar{z}_0)$ could be expanded in power series with respect to ε. The radius of convergence of the series is the same as for z and λ in (3.3). We used this expansion in order to compute the parabolic points where $\lambda_1 = \lambda_2 = -1$ or $\lambda_1 = \lambda_2 = 1$. These points are indicated on Fig. 1 as P_1, P_2, P_3 and P_4. At P_1 $\lambda_1 = \lambda_2 = -1$ and $dP(\bar{z}_0)$ has a single eigenvector. The same result holds at P_2. At P_3 and P_4 $\lambda_1 = \lambda_2 = 1$ and $dP(\bar{z}_0)$ has a single eigenvector.

The periodic orbit is elliptic between P_0 and P_1, hyperbolic between P_1 and P_2, elliptic between P_2 and P_3 and again hyperbolic between P_3 and P_4. In the elliptic regions, based on Moser's twist map theorem ([10]), one would expect that there exists an infinite set of coelecial invariant tori surrounding the periodic orbit. Besides the measure preservation, Moser's theorem requires the map P to have a non-trivial Birkhoff normal form (see [13], pp. 158-159) with a rotation angle α_0 such that $n\alpha_0 \neq 0 (\mathrm{mod}\ 2\pi)$ for $1 \leq n \leq 4$. Using the explicit expansion in (3.5) we have indeed verified that the above conditions hold for a sufficiently small ε. The resulting tori are closures of quasiperiodic orbits of the flow. Using the symmetricity in (3.8) one can prove that the integrals of these orbits are quasi-periodic too. Thus we have found an infinite set of quasiperiodic solutions of (1.4).

4. <u>Numerical experiments</u>.

In order to gather more information about the set of bounded solutions of (1.5), especially for intermediate values of c, we have approximated (1.5) by a difference equation and solved it on a computer. The difference scheme employed was

$$(y_{j+3} - 3y_{j+2} + 3y_{j+1} - y_j)/\Delta x^3 + (y_{j+2} - y_{j+1})/\Delta x = c^2 - \frac{1}{4}(y_{j+2}^2 + y_{j+1}^2) \tag{4.1}$$

where y_j is the value of the grid function at $x_j = j\Delta x$, $j \in \mathbb{Z}$. In all computations we used $\Delta x = 0.05$. The scheme in (4.1) maintains the symmetry of the equation in (1.5) in the sense that it is invariant under the transformation $j \to -j$, $y \to -y$. We have investigated only the odd solutions of (4.1), i.e. those which satisfy the initial conditions

$$y_0 = 0, \; y_1 = -y_{-1} = \Delta x \cdot s, \tag{4.2}$$

where the slope s is a parameter. The values of y_j, $j \geq 2$ are then calculated by (4.1) until $|y_j|$ exceeds a certain large number y_{max} (we have used $y_{max} = 10$). Denote by $x_{max}(s)$ the point $x_j = j\Delta x$ where for the first time $|y_j| \geq y_{max}$. Recall that for $c \in [0,c_*]$ all bounded solutions of (1.5) lay in a strip $|y| \leq y_{max}$ where y_{max} depends only on c_*. The same is true for eg. (4.1). Thus for bounded solutions $x_{max}(s) = \infty$. It was a surprising empirical observation that with a few exceptions all local maxima of $x_{max}(s)$ have been infinite. Hence, for a sequence $s_1 < s_2 < s_3$, maximum of $x_{max}(s_i)$ at s_2 would imply that for some $s \in [s_1,s_3]$ the corresponding solution is bounded. Next, the interval $[s_1,s_3]$ was subdivided in n equial subintervals (usually n = 200), and if no additional maxima of $x_{max}(s)$ were found, it was assumed that there is only one bounded solution with the slope $s \in [s_1,s_3]$. The Golden Section method for a univalent function was then employed in order to converge to the above s. As suggested by the theory in Sections 2 and 3, only solutions connecting critical points were isolated. Note that these solutions are asymptotically unstable, and double precision on CDC (i.e. 30 decimal digits) was required in order to reach $x_{max} \sim 100$. On the other hand, if the "shooting" with n+1 slopes $s_{1i} = s_1 + i\Delta s, \Delta s = (s_3 - s_1)/n$, $i = 0,1,\ldots,n$ resulted in new local maxima at s_{1j_1}, s_{1j_2},.. etc, the above procedure was applied again to the relevant subintervals $[s_{1j} - \Delta s, s_{1j} + \Delta s]$ etc. Some of the new subintervals contained a single maximum of $x_{max}(s)$, others contained several maxima and (for some values of c) there were subintervals with $x_{max}(s) \equiv \infty$. After several iterations it was apparent that the subintervals of the second kind give rise to a Cantor type set of points C. Thus the set S of slopes s corresponding to bounded odd solutions of (4.1) is a union of three subsets: the set I of isolated points, the Cantor type set C and a set of closed intervals J . The set I corresponds to orbits connecting critical points, the set C - to chaotic orbits and the set J - to invariant tori. The numerical experiments lead as to the following conclusion: all non-isolated boundary points of S, in particular all points of C, are limit points of the subset I.

Let us describe the evolution of S as the parameter c decreases from ∞ to 0. For $c > c_1 \approx 1.283$ the set S consists of a single point $s_0 < 0$ corresponding to a solution $y^{(0)}$ which tends to $\pm c\sqrt{2}$ at $\mp \infty$ and has no zeros on the half line $x > 0$. As c decreases beyond c_1 towards $c_{max} \approx 1.2664$ there is sequence of bifurcation points $c_1 > c_2 > \ldots > c_n > \ldots$ tending to c_{max}. At c_n a new solution $y^{(n)}$ is "born" which splits into two solutions $y^{(n)}$ and $y^{(-n)}$ as c decreases beyond c_n. The functions $y^{(n)}$ and $y^{(-n)}$ have exactly n zeros in the domain $x > 0$ and the corresponding slopes s_n and s_{-n} have the sign $(-1)^{n-1}$. The sequences $\{s_{2n}\},\{s_{2n+1}\}$ are decreasing, $\{s_{-2n}\},\{s_{-2n-1}\}$ are increasing and at $c = c_{max}$ the limits $s_{ev} = \lim s_{2n} = \lim s_{-2n}$ and $s_{od} = \lim s_{2n+1} = \lim s_{-2n-1}$ are the slopes of the periodic solution y_{per} at $x = 0$ and $x = \pi$ respectively. At $c = c_{max}$ (which corresponds to the maximum poin P_3 on Fig. 1) the set S consists of the sequences $\{s_{\pm n}\}$ and the limit points s_{ev} and s_{od}. On Fig. 2 we have displayed the functions $y^{(1)}(x)$, $y^{(-1)}(x)$, $y^{(3)}(x)$ (with positive slope) and the functions $y^{(0)}(x)$, $y_{per}(x)$ (with negative slope) at $c = c_{max}$.

As c decreases beyond c_{max} there is a radical change in the set of bounded solutions. The periodic solution splits into two. The elliptic one is surrounded by a thin invariant torus while the hyperbolic one- by a Cantor set of bounded oscilating solutions. The set J consists of two intervals J_1 and J_2 centered at the slopes $y'_{per}(0)$ and $y'_{per}(\pi)$ of the elliptic periodic solutions. These intervals are very small, e.g. at $c = 1.2663$ (i.e. between the points P_3 and P_2)

$J_1 \approx [-3.0266, -3.0247]$. The Cantor-type set C is located in neighborhoods of the slopes $y'_{per}(0)$, $y'_{per}(\pi)$ of the hyperbolic periodic orbit. We conjecture that the Poincare map P has homoclinic points in a neighborhood of the hyperbolic periodic orbit. It is well known (e.g. see [11]) that in such a case there is a Cantor set in the plane $y'' = 0$ so that P acts on it as a Bernoulli shift. As mentioned earlier, all points of C and the boundary of J are the limit points of the subset I. In partical, in each "hole" of the set C there is a double sequence of points of I which tends to the endpoints of the "hole".

The above picture does not change qualitatively until the invariant torus disappears at $c \approx 1.266$. For $0.6 < c < 1.266$ both periodic solutions are hyperbolic and S is a union of C and I. On Fig. 3 one can see the graphs of two bounded solutions for $c = 0.8$ with the slopes which differ in the 15-th digit! One of them tends after $x \approx 80$ to the constant state while the second one still oscillates. The oscillations which have a typical wave length ≈ 12 apparently carry irregular modulations.

Approximately at $c = 0.6$ once again invariant tori appear. In our computations they manifest themselves as a set J of two or more intervals of slopes so that for $s \in J$ the solutions of (4.1), (4.2) remain bounded for a huge number of steps (we tried even 160000 steps). The set J is located in a neighborhood of the slopes $y'_{per}(0) \approx -1.5$ and $y'_{per}(\pi) \approx 0$ corresponding to the parabolic point P_4. The evolution of J is very interesting. At $c = 0.586$ J consists of an interval $J_1 \approx [-1.515, -1.495]$ and a similar interval J_2 near $s = 0$. As c decreases, first at $c = 0.5853$ J_1 splits into two intervals and later on in more smaller intervals. With each split in J, a Cantor type set is formed in a resulting "hole". At $c = 0.5$ we lost the trace of the set J. The same picture repeats itself as c increases until $c = 0.5865$.

In the domain $0.35 < c < 0.5$ the set J is empty. At the parabolic point P_1, $c \approx 0.3194$ and the slopes of periodic solution are $y'_{per}(0) \approx -0.6706$ and $y'_{per}(\pi) \approx 0.602$. The set J consists of an interval $J_1 \approx [-0.7723, -0.5219]$ and a similar interval J_2 near $s = 0.602$. As c increases, the above intervals first split at $c \approx 0.326$ and later disappear. On the other hand, as c decreases, the intervals J_1 and J_2 grow in size until they merge about $c = 0.1$ so that the set S consists of a single interval $J \approx [-0.3, 0.265]$.

As $c \to 0$ the set $S = J$ also tends to zero. Apparently, for small c the invariant set of the flow in (2.2) consists of a ball-like neighborhood of zero which includes, of course, the invariant tori described in Section 3.

We traced the solutions $y^{(0)}, y^{(-2)}$ and $y^{(-4)}$ (i.e. with the smallest negative slopes in S) as c decreased below c_{max}. On Fig. 4 their graphs are displayed for $c = 0.3$. The slopes s_0 and s_{-2} differ in the 4-th digit while s_{-2} an s_{-4} differ in the 9-th digit! The graphs of $y^{(-2)}, y^{(-4)}$ separate (visually) from $y^{(0)}$ at $x \approx 13$ while $y^{(-4)}$ separates from $y^{(-2)}$ at $x \approx 67$. The solution $y^{(-4)}$ disappeared somewhere between $c = 0.3$ and $c = 0.295$, $y^{(-2)}$ - between $c = 0.295$ and $c = 0.293$ and $y^{(0)}$ - around $c = 0.2$. On Fig. 5 the graphs of $y^{(0)}$ and y_{per} are shown for $c = 0.2$. We conjecture that each solution $y^{(n)}$ exists until some $c = c_n$ where it becomes a limit of a sequence of bounded oscilating solutions and disappears for $c < c_n$.

5. Conclusion. The above analytical and numerical study shows that the set of steady solutions of the Kuramoto-Sivashinsky equation is surprisingly complex. There are conic solutions which correspond to the Bunsen flames as well as periodic, quasi-periodic and chaotic solutions corresponding to a disturbed plane flames. For a high propagation velocity a single conical solution exists, while for a lower one all four types of solutions above do exist simultaneously. Our numerical study was limited to the odd solutions. The computations in [16] suggest that for moderate values of c eq. (1.5) has non-odd soliton-type solutions which tend to the same limit $c\sqrt{2}$ as $x \to \pm\infty$. Certainly, more computations are in place in order to understand the structure of the set of all steady solutions of (1.1) and its dependence on the propagation velocity c^2. Another important problem is the connection between the time dependent solutions of (1.1) and (2.5) and the above steady solutions. It was thought previously that the turbulence in the Kuramoto-Sivashinsky equation is primarily of a non-stationary orgin and is caused by a competition of a few spatial nodes. In view of the above results it is plausible that the set of steady solutions is an attractor for the time dependent problem. Note that for the (experimental) propagation velocity $c_0^2 \approx 1.2$ of a turbulent flame, both periodic orbits of (1.5) are hyperbolic and there is a plenty of spatial chaos in the set of bounded solutions of (1.5). Thus the turbulence in (1.1) may be attributed to the above "steady" chaos.

Bibliography

[1] C. Conley, Isolated Invariant Sets and the Morse Index, Conf. Bd. Math. Sci., No. 38, Amer. Math Soc. Providence, 1978.

[2] N. Kopell and L.N. Howard, Bifurcations and trajectories joining critical points, Advances in Math., v. 18 (1975), 306-358.

[3] Y. Kuramoto and T. Tsuzuki, Persistent propagation of concentration waves in dissipative media far from thermal equilibrium, Prog. Theor. Phys., v.55 (1976), 356-369.

[4] Y. Kuramoto and T. Yamada, Turbulent state in chemical reaction, Prog. Theor. Phys., v.56 (1976), 679.

[5] Y. Kuramoto, Diffusion induced chaos in reaction systems, Suppl. Prog. Theor. Phys., v. 64 (1978), 346-367.

[6] P. Manneville, Statistical properties of chaotic solutions of a one-dimensional model for phase turbulence, Phys. Lett. 84A (1981), 129-132.

[7] C.K. McCord, Uniqueness of connecting orbits in the equation $y^{(3)} = y^2-1$, preprint, Dept. of Math., Univ. of Wisconsin, Madison, 1983.

[8] D. Michelson and G. Sivashinsky, Nonlinear analysis of hydrodynamic instability in laminar flames, II. Numerical experiments, Acta Astronautica, 4 (1977), 1207-1221.

[9] M.S. Mock, On fourth-order dissipation and single conservation law, Comm. Pure Appl. Math., v. 29 (1976), 383-388.

[10] J. Moser, On invariant curves of area preserving mappings of an annulus, Nach. Akad. Wiss. Gottingen, Math. Phys. Kl (1962), 1-20.

[11] J. Moser, Stable and Random Motion in Dynamial Systems, with Special Emphasis on Celestial Mechanics, Princeton Univ. Press, 1973.

[12] B. Nicolaenko and B. Scheurer, Remarks on the Kuramoto-Sivashinsky equation, Proceedings of the Conference on Fronts, Interfaces and Patterns, CNLS, Los Alamos, May, 1983, special issue of Physica D.

[13] C. Siegel and J. Moser, Lectures on Celestial Mechanics, Springer, Grundlechren Bd. 187, 1971.

[14] G. Sivashinsky, Nonlinear analysis of hydrodynamic instability in laminar flames, I. Derivation of basic equations, Acta Astronautica, 4 (1977), 1117-1206.

[15] O. Thual and U. Frish, Natural boundary in Kuramoto model, presented at the workshop on Combustion, Flames a Fires, March 1-15, 1984, Les Houches.

[16] Yu. Tsvelodub, Izvest. Akad. Nauk, Mekh. Zhidk. Gaza No. 4 (1980), 142.

[17] T. Yamada and Y. Kuramoto, A reduced model showing chemical turbulence, Prog. Theor. Phys., v.56 (1976), 681.

Department of Mathematics, University of California, Los Angeles, CA 90024
and
Department of Mathematics, Hebrew University, Jerusalem, Israel

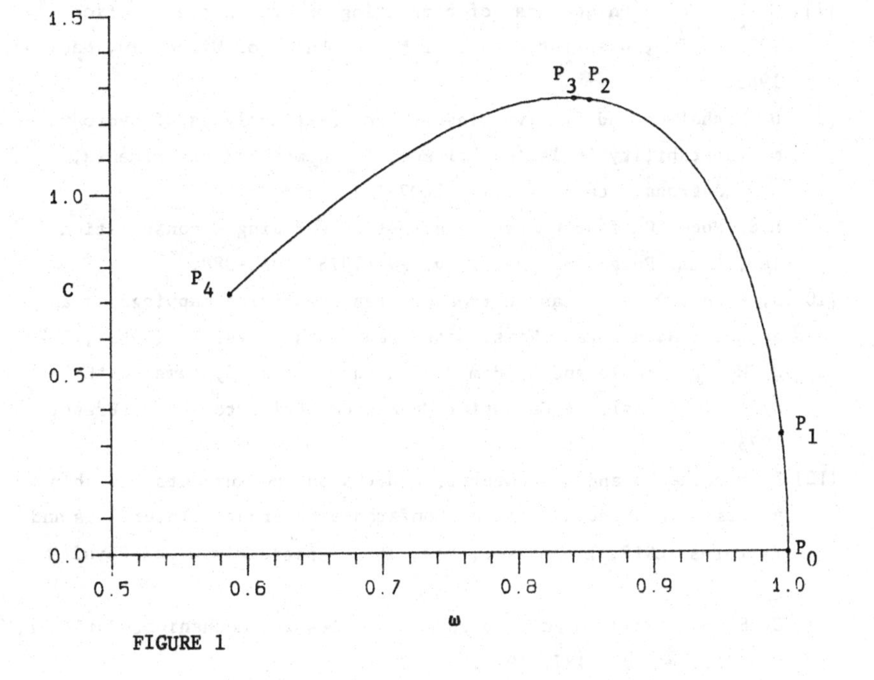

FIGURE 1

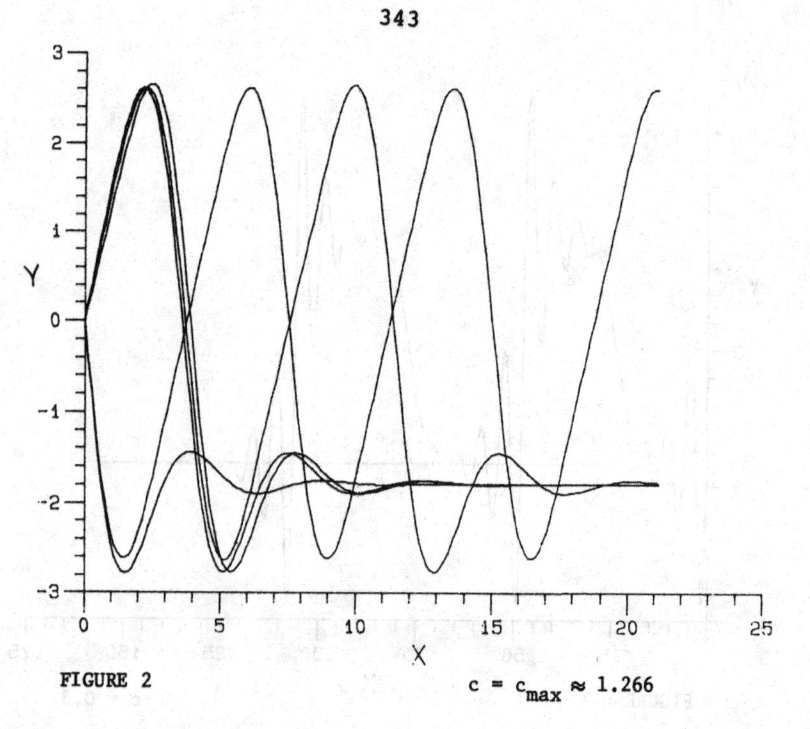

FIGURE 2 $c = c_{max} \approx 1.266$

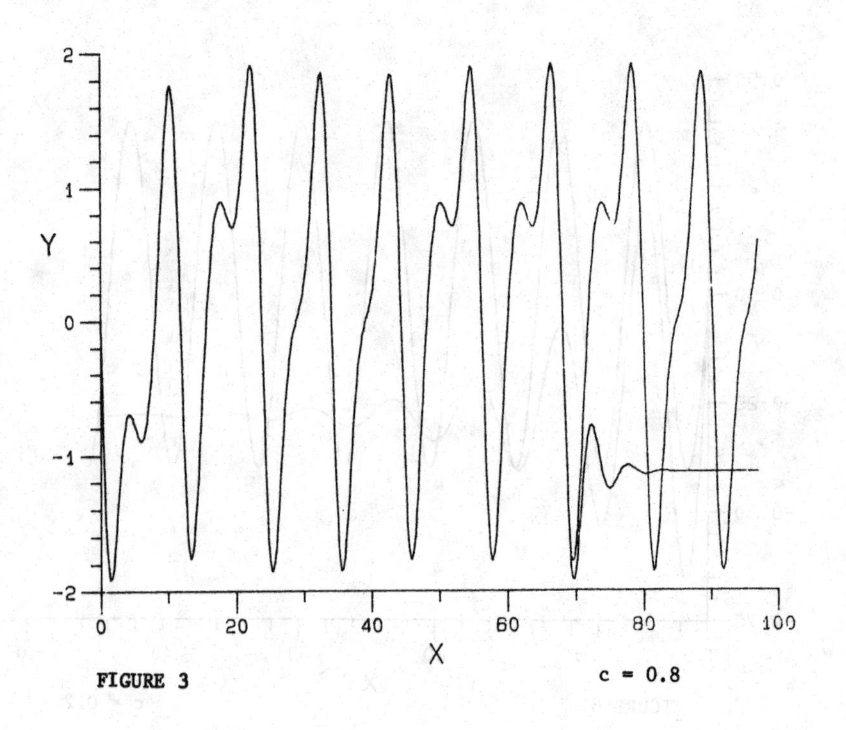

FIGURE 3 $c = 0.8$

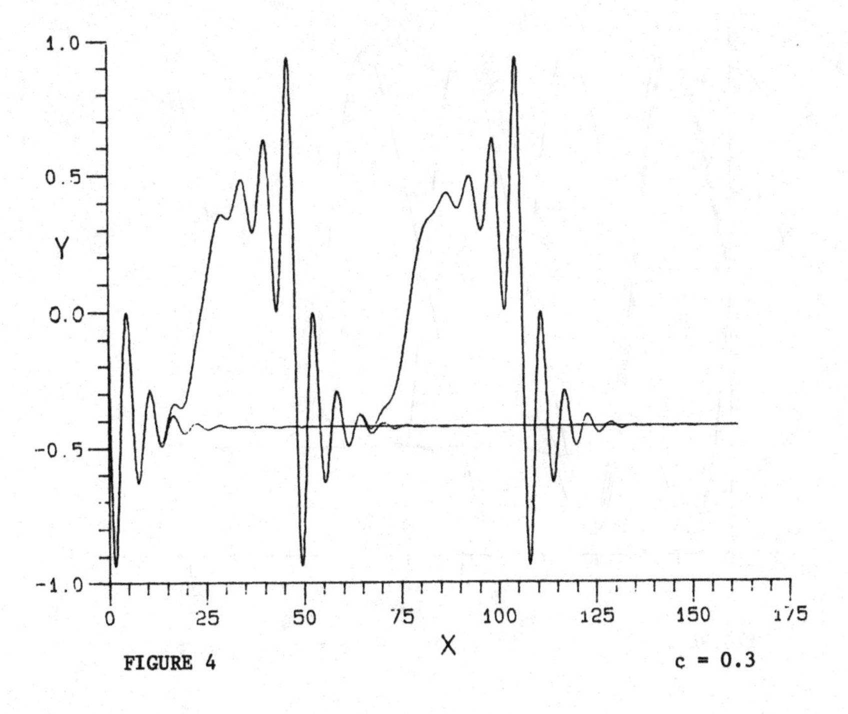

FIGURE 4 $c = 0.3$

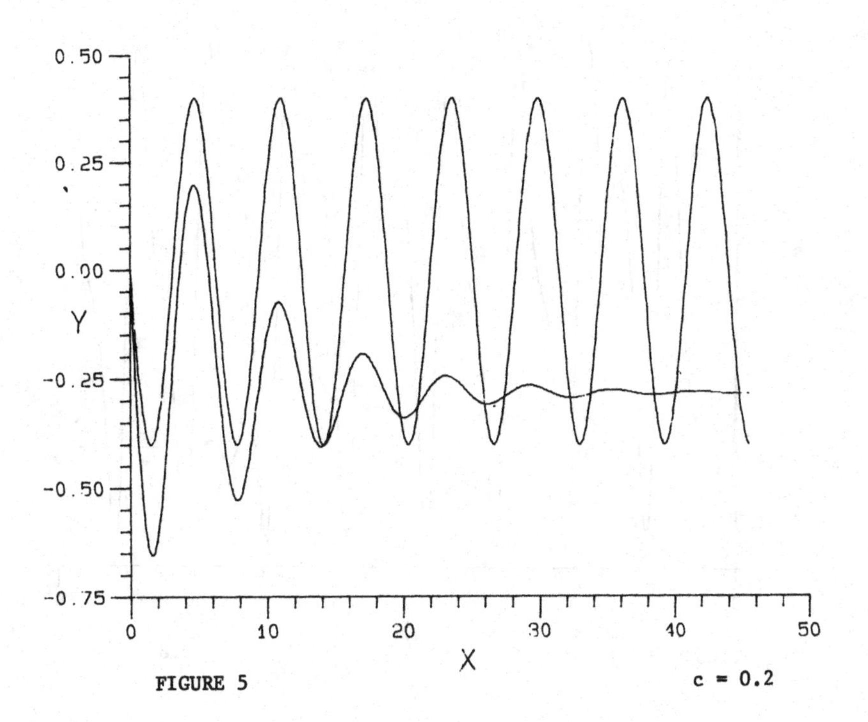

FIGURE 5 $c = 0.2$

Progress in Scientific Computing, Vol. 6
Proceedings of U.S.-Israel Workshop, 1984

INFORMATION CONTENT IN SPECTRAL CALCULATIONS

Saul Abarbanel and David Gottlieb

1. INTRODUCTION

It is well known that when one solves numerically, with the aid of higher order approximations, partial differential equations or systems one may get numerical oscillations. The oscillations may arise from different sources; e.g. incorrect treatment of the outflow boundaries in hyperbolic systems, mild non-linear instabilities, etc. One interesting class of numerical oscillations occurs when flows with extreme gradients, or local discontinuities, are simulated. The usual way of combating this manifestation of the Gibbs phenomenon is to introduce some kind of artificial viscosity (even TVD methods basically do this).

When the basic algorithm used in the computation has spectral accuracy the above strategy is not viable, since it will reduce the order of accuracy drastically. Even hybrid "switches" do not overcome this difficulty.

The approach taken in this paper is based on the premise that because of their very high accuracy spectral methods, regardless of the smoothness or lack of it on the computational grid, retain all the information* in the form of the solution interpolant in a hyperplane whose projection we see on the grid points.

In Section 1 we develop the theory for the case of a linear hyperbolic operator with constant coefficients and periodic boundary conditions. A numerical example is then presented which demonstrates the efficacy of the proposed procedure for extracting the piecewise smooth solution from the oscillatory raw data.

In Section 2 an analogous procedure is developed for the non-periodic case where the interpolating polynomials are not trigonometric but

---

* Here by information we do not mean information in the formal way as defined by Shannon [2], but rather by following Lax [3] - the ability to use the data to construct a piecewise smooth looking solution on the grid points.

Tchebyshev. The application is then tested on a physical problem - two dimensional supersonic wedge flow. The raw data in this example is the density as obtained by a pseudospectral steady state solution of the 2-D Euler equations. The postprocessing procedure is again found to be effective, particularly with regards to the (2-D) shock location and the magnitude of the density jump.

2. FOURIER METHODS

Consider the following initial-boundary value problem

$$u_t + u_x = 0 \qquad 0 < x < 2\pi , \qquad t > 0 \tag{2.1}$$

$$u(x,0) = g(x) \qquad 0 \leq x \leq 2\pi \tag{2.2}$$

$$u(x,t) = u(x+2\pi,t) \tag{2.3}$$

where $g(x)$ is piecewise smooth and has a discontinuity at a point $x = \overline{x}$, say. When problem (2.1) - (2.3) is solved by the pseudospectral Fourier method, [1], the initial condition $g(x)$ is replaced by its N-th order trigonometric interpolant, $g_N(x)$:

$$g_N(x) = \sum_{\ell=-N}^{N} \hat{g}_\ell e^{i\ell x} \tag{2.4}$$

where

$$g\,(x) = \frac{1}{2Nc_\ell} \sum_{j=0}^{2N-1} g(x_j) e^{-i\ell x_j} \tag{2.5}$$

$$x_j = \frac{\pi j}{N} \qquad j = 0,1,\ldots,2N-1 \tag{2.6}$$

and

$$c_\ell = \begin{cases} 1 & |\ell| < N \\ 2 & |\ell| = N . \end{cases} \tag{2.7}$$

Since (2.1) is linear with constant coefficients, the pseudospectral approximation, $u_N(x,t)$, satisfies the following problem

$$\frac{\partial u_N}{\partial t} + \frac{\partial u_N}{\partial x} = 0 \tag{2.8}$$

$$u_N(x,0) = g_N(x) \tag{2.9}$$

$$u_N(x,t) = u_N(x+2\pi,t) .$$

The solution to (2.8) is

$$u_N(x,t) = g_N(x-t) . \tag{2.10}$$

Since $g_N(x)$ collocates $g(x)$ at the points $0 \leq j \leq 2N-1$, the discreet values $g_N(x_j)$ convey visually the impression of a piecewise smooth function, which $g(x)$ indeed is. If however we examine $g_N(x)$ at points other than the collocation points x_j, the discreet values of $g_N(x)$ thus examined form an oscillatory pattern. Therefore if we evaluate $u_N(x_j,t) = g_N(x_j-t)$, at $t \neq \pi m/N$ (m an integer), then the solution (2.10) will indeed look oscillatory. If $g(x)$ were a smooth function, then these oscillations would be very small in magnitude and in fact would not be discerned by eye. The discontinuity at x_s, however, induces much larger oscillations, due to the Gibbs phenomenon. Note, however, that if $t = \pi m/N$ then the solution values at the collocation points are spectrally accurate. It is in this sense that we make the claim that the pseudospectral solution retains the same amount of information for problems with discontinuous initial data as for smooth initial conditions.

Our basic interest is in the "inverse problem": given the numerical results of a pseudospectral procedure applied to a hyperbolic problem (possibly even a non-linear one) which look oscillatory on the collocation points, $x_j = \pi j/N$ - how do we extract out of this data the correct piecewise smooth solution (i.e. the location of the discontinuity, the magnitude of the jump and the structure of the smooth part of the solution)?

The approach we shall use to recover the "lost" information is to consider first (2.1) - (2.3) with $g(x)$ being the saw-tooth function:

$$g(x) = AF(x,x_k) = \begin{cases} Ax & x < x_k \\ A(x-2\pi) & x_k \leq x , \quad (k \text{ integer}) \end{cases} \tag{2.11}$$

where $2\pi A$, then, is the magnitude of the jump. We use (2.11) because this saw-tooth function is the simplest periodic piecewise smooth function with a given discontinuity. We call (2.1) - (2.3) with $g(x)$ given by (2.11) the fundamental problem. The pseudospectral approximation to the solution of the fundamental problem satisfies, by analogy to (2.8) - (2.9) the following set:

$$\frac{\partial v_N}{\partial t} + \frac{\partial v_N}{\partial x} = 0 \tag{2.12}$$

$$u_N(x,0) = AF_N(x,x_k) \tag{2.13}$$

with $F_N(x,x_k)$ given by:

$$F_N(x,x_k) = \sum_{\ell=-N}^{N} a_\ell(k) e^{i\ell x} , \tag{2.14}$$

$$a_\ell(k) = \frac{1}{2Nc_\ell} \sum_{j=0}^{2N-1} e^{-i\ell x_j} \cdot F_N(x_j,x_k) \tag{2.15}$$

Performing (2.15) we get:

$$a_0(k) = \frac{\pi}{N} [k - N + .5] \tag{2.16}$$

$$a_\ell(k) = \frac{\pi}{2Nc_\ell} \left[2 \cdot \frac{1 - e^{-\frac{i\pi\ell(k+1)}{N}}}{1 - e^{-\frac{i\pi\ell}{N}}} + i \operatorname{ctn} \frac{\pi\ell}{2N} - 1 \right] \quad (\ell \neq 0) \tag{2.17}$$

The solution to (2.12) - (2.13) is

$$\begin{aligned} v_N(x,t) &= AF_N(x-t,x_k) \\ &= A \sum_{\ell=0}^{2N-1} a_\ell(k) e^{i\ell(x-t)} \\ &= A \left\{ \sum_{\ell=0}^{2N-1} a_\ell(k+\theta) e^{i\ell x} - t \right\} \\ &= A[F_N^{*}(x,x_k+t) - t] \end{aligned} \tag{2.18}$$

where

$$\theta = \frac{Nt}{\pi} \quad . \tag{2.19}$$

The last two equalities in (2.18) are obtained by manipulating (2.14) - (2.17).

We now use the pseudospectral approximation to the fundamental problem, $v_N(x,t)$ in order to remove the oscillations in (2.10). In fact, $\varphi_N(x,t) = u_N(x,t) - AF_N(x)$ is a continuous function satisfying the following p.d.e. and initial data:

$$\frac{\partial\varphi_N}{\partial t} + \frac{\partial\varphi_N}{\partial x} = 0 \tag{2.20}$$

$$\varphi_N(x,0) = g_N(x) - AF_N(x) \tag{2.21}$$

We note that $\varphi_N(x,0)$ is a continuous function because $g_N(x) - AF_N(x)$ does not have a jump anymore. It follows that $\varphi_N(x,t)$ converges to $\varphi(x,t)$ faster than $u_N(x,t)$ does to $u(x,t)$.

The above observations lead to the following procedure that extracts from $u_N(x_j,t)$ which is an oscillatory approximation to $u(x,t)$ a piecewise smooth representation of $u(x,t)$. Given $u_N(x,t) = \sum_{\ell=-N}^{N} \hat{u}_\ell e^{i\ell x}$, we try to find an unknown smooth and a saw-tooth function $AF_N(x-t,x_s)$ with an unknown jump $2\pi A$ at an unknown location x_s , such that

$$H = \sum_{j=0}^{2N-1} \left[u_N(x_j,t) - AF_N(x_j,x_s) - c - \sum_{\substack{\ell=-p \\ \ell\neq 0}}^{p} b_\ell e^{i\ell x_j} \right]^2 \tag{2.22}$$

is minimized. In (2.22) c is an unknown constant which when solved for will yield $t = c/A$ (see 2.18); $|p| < N$ and b_0 is taken to be zero. The inner sum in (2.22) represents the continuous function we seek. Note that we have $2p + 3$ unknowns in (2.2): A, x_s, c and $2p$ values of b_ℓ $(\ell\neq 0)$.

The conditions for local minima of H are found from the following $2p + 3$ equations:

$$\frac{\partial H}{\partial A} = 0 \Rightarrow \sum_{j=0}^{2N-1} \left[u_j F_j - AF_j^2 - cF_j - F_j \sum_{\substack{\ell=-p \\ \ell\neq 0}}^{p} b_\ell e^{i\ell x_j} \right] = 0 \tag{2.23}$$

where $F_j = F_N(x_j,x_s)$, $u_j = u_N(x_j,t)$. Also

$$\frac{\partial H}{\partial c} = 0 \Rightarrow \sum_{j=0}^{2N-1} \left[u_j - AF_j - c - \sum_{\substack{\ell=-p \\ \ell \neq 0}}^{p} b_\ell e^{i\ell x_j} \right] = 0 \qquad (2.24)$$

$$\frac{\partial H}{\partial s} = 0 \Rightarrow \sum_{j=0}^{2N-1} \left[F_j' u_j - AF_j' F_j - cF_j' - F_j' \sum_{\substack{\ell=-p \\ \ell \neq 0}}^{p} b_\ell e^{ix_j \ell} \right] = 0 \qquad (2.25)$$

where $F_j' = \partial F_N(x_j, x_s)/\partial s = \sum_{\ell=-N}^{N} \frac{\partial a_\ell(s)}{\partial s} \cdot e^{i\ell x_j}$;

and

$$\frac{\partial H}{\partial b_m} = 0 \Rightarrow b_m = \hat{u}_m - Aa_m , \quad |m| = 1,2,\ldots,p \qquad (2.26)$$

where $\hat{u}_m = \frac{1}{2Nc_m} \sum_{j=0}^{2N-1} u_N(x_j) e^{-i\ell x_j}$.

Substituting (2.26) into (2.23), (2.24) and (2.25) we get, respectively:

$$\hat{u}_0 - Aa_0 - c = 0 \qquad (2.27)$$

$$\sum_{|\ell|>p} c_\ell a_{-\ell} \hat{u}_\ell - A \sum_{|\ell|>p} c_\ell a_{-\ell} a_\ell = 0 \qquad (2.28)$$

$$\sum_{|\ell|>p} c_\ell a'_{-\ell} \hat{u}_\ell - A \sum_{|\ell|>p} c_\ell a'_{-\ell} a_\ell = 0 \qquad (2.29)$$

where $a'(s) = \partial a_\ell(s)/\partial s$. Next we combine (2.28) and (2.29) to get a single non-linear equation for s :

$$\sum c_\ell a'_{-\ell} \hat{u}_\ell \sum c_\ell a_{-\ell} a_\ell - \sum c_\ell a_{-\ell} \hat{u}_\ell \sum c_\ell a'_{-\ell} a_\ell = 0 \qquad (2.30)$$

where all sums run over $p < |\ell| \leq N$.

Equation (2.30) is solved iteratively for s . Having found s , one immediately obtains from (2.16) and (2.17) all the $a_\ell(s)$'s. Then from (2.26) we have the b_m's , and A from (2.28). Finally, having A we find c from (2.27).

The minimum thus obtained may be a local one while we are seeking a global minimum. This means that in practice one searches for the global minimum.

We now give an example that illustrates the efficacy of the procedure. We solve the following problem:

$$\frac{\partial u_N}{\partial t} + \frac{\partial u_N}{\partial x} = 0 \qquad 0 < x < 2\pi \ , \quad t > 0 \tag{2.31}$$

$$u_N(x,0) = \begin{cases} \sin \frac{x}{2} & 0 \leq x \leq \pi \\ -\sin \frac{x}{2} & \pi \leq x \leq 2\pi \end{cases} \tag{2.32}$$

$$u_N(0,t) = u_N(2\pi,t) \tag{2.33}$$

We ran the problem on several grids and exhibit here the numerical results for the case $N = 8$ (i.e. 16 subintervals in the domain $(0,2\pi)$). The unadulterated results at $t = \pi/2N$ on the grid points are shown in Figure 1, and the corresponding pointwise errors are listed in Table 1.

Table 1

j	exact solution	error 1 = |exact-unsmoothed|	error 2 = |exact-smoothed|	$\frac{\text{error 1}}{\text{error 2}}$
0	9.80×10^{-2}	5.86×10^{-5}	5.86×10^{-5}	1.00
1	9.80×10^{-2}	1.24×10^{-2}	5.86×10^{-5}	211
2	2.90×10^{-1}	2.57×10^{-2}	6.30×10^{-5}	408
3	4.71×10^{-1}	4.13×10^{-2}	7.33×10^{-5}	563
4	6.34×10^{-1}	6.15×10^{-2}	9.30×10^{-5}	661
5	7.73×10^{-1}	9.11×10^{-2}	1.31×10^{-4}	695
6	8.82×10^{-1}	1.43×10^{-1}	2.16×10^{-4}	662
7	9.57×10^{-1}	2.70×10^{-1}	4.42×10^{-4}	611
8	-9.95×10^{-1}	1.00×10^{0}	1.10×10^{-2}	91
9	-9.95×10^{-1}	2.68×10^{-1}	1.34×10^{-3}	200
10	-9.57×10^{-1}	1.42×10^{-1}	4.42×10^{-4}	321
11	-8.82×10^{-1}	9.07×10^{-2}	2.16×10^{-4}	420
12	-7.73×10^{-1}	6.12×10^{-2}	1.32×10^{-4}	464
13	-6.34×10^{-1}	4.11×10^{-2}	9.30×10^{-5}	442
14	-4.71×10^{-1}	2.55×10^{-2}	7.32×10^{-5}	348
15	-2.90×10^{-1}	1.22×10^{-2}	6.30×10^{-5}	194

We then postprocessed these $u_N(x_j,\pi/2N)$ values according to the procedure described above. The filtered values are shown on the same graph, and the errors are listed in the same table as the unprocessed ones. The dramatic improvement is evident.

In the next section we consider a two dimensional hyperobic system of non-linear equation, namely the Euler equations. Because the physical problem involves inflow, outflow and no-flow boundary conditions periodicity could not be imposed and we use the Tchebyshev, rather then Fourier, pseudospectral method.

3. TCHEBYSHEV METHODS - THE WEDGE FLOW PROBLEM

The physical problem is that of a wedge, inserted at a zero angle of attack, into a uniform supersonic flow of an ideal gas with $\gamma = 1.4$. An oblique shock develops in time and the flow reaches after a while a steady state.

The time dependent Euler equations in two space dimensions were discretized by the pseudospectral Tchebyshev method in space with an 8×8 grid and a modified Euler scheme was used for the time discretization (see [4]). Since we are interested in the steady state only, the accuracy fo the time integration is of little importance. In order to be sure that a steady state is reached, the code was run until all physical quantities did not change to 11 significant figures over a span of 100 time steps. The values of the density in the steady state at the grid points together with the grid points themselves are given in Table 2.

Table 2

	ρ									Y
	1.862	1.851	1.869	1.871	1.837	1.865	1.892	1.885	1.878	1.
	1.862	1.870	1.867	1.820	1.870	1.954	1.899	1.803	1.759	.961
	1.862	1.854	1.852	1.904	1.877	1.770	1.782	1.864	1.900	.853
	1.862	1.871	1.876	1.812	1.838	1.969	1.975	1.884	1.841	.691
	1.862	1.848	1.842	1.935	1.899	1.703	1.710	1.890	1.984	.5
	1.862	1.883	1.894	1.729	1.832	2.429	2.994	3.255	3.316	.308
	1.862	1.808	1.810	2.387	3.133	3.375	3.224	3.054	3.002	.146
	1.862	2.115	2.868	3.288	3.176	2.965	3.006	3.136	3.187	.038
	1.862	3.083	3.046	2.975	3.087	3.108	3.024	3.013	3.016	0
X	0	.038	.146	.308	.5	.691	.853	.961	1.	

Note that the raw data in Table 2 seems to indicate roughly the same y-shock location at $x_0 = 1$, $x_1 = .961$ and $x_2 = .853$, namely between the grid points $y_4 = .3086$ and $y_5 = .500$. This means that because of the coarse Tchebyshev grid the shock location cannot be resolved to better than 20% of the domain. In fact the correct shock locations at those x-stations are $y = .434$ for x_0, $y = .417$ for x_1 and $y = .370$ for x_2.

In the present case it is not necessary to employ a saw-tooth piecewise smooth function, as was done in the previous section, because there is no need to preserve periodicity. Instead we substract from the oscillatory data an expansion of the Heaviside function, $S(y,y_s)$:

$$S(y,y_s) = \begin{cases} d_1 + d_2 & -1 \leq y \leq y_s \\ d_1 & y_s \leq y \leq 1 \end{cases} \qquad (3.1)$$

where d_1, the state ahead of the shock, and d_2, the magnitude of the discontinuity, are constant. The description here of $S(y,y_s)$ as if independent of x has to do with the fact that the 2 dimensional results of the pseudospectral algorithm were post-processed at fixed x-stations. The expansion of $S(y,y_s)$ is given by

$$S_N(y,y_s) = \sum_{\ell=0}^{N} A_\ell(s) T_\ell(y) \qquad (3.2)$$

where $T_\ell(y)$ is the Tchebyshev polynomial of order ℓ, $T_\ell(y) = \cos[\ell \cos^{-1}(y)]$, and

$$A_0(s) = (s + \tfrac{1}{2})/N$$

$$A_\ell(s) = \sin\left[\frac{\pi\ell}{N}(s + \tfrac{1}{2})\right]/N \sin\frac{\pi\ell}{2N} \quad ; \quad 1 \leq \ell \leq N-1$$

$$A_N(s) = \sin\left[\,(s + \tfrac{1}{2})\right]/2N$$

If s is an integer than on the grid points, $y_j = \cos(\pi j/N)$,

$$S_N(y_j,y_s) = S(y_j,y_s) \qquad (3.3)$$

The L_2-norm which we wish to minimize is now, at any given x-station:

$$H = \sum_{j=0}^{N} \frac{1}{c_j} \left[\rho_N(y_j) - d_1 - d_2 S_N(y_j,y_s) - \sum_{\ell=1}^{p<N} b_\ell T_\ell(y_j)\right]^2 \qquad (3.4)$$

where

$$c_j = \begin{cases} 1 & 1 \leq j \leq N-1 \\ 2 & j = 0, N \end{cases} \tag{3.5}$$

Differentiating (3.5) with respect to the parameters d_1 , d_2 , s and $b_\ell (1 \leq \ell \leq p < N)$, using the orthogonality relations for the Tchebyshev polynomials and manipulations similar to those used in the previous section, we get $p + 3$ nonlinear algebraic equations which are completely analogous to (2.26) - (2.29). They are:

$$b_\ell = \hat{\rho}_\ell - d_2 A_\ell , \qquad \ell = 1, 2, \ldots, p. \tag{3.6}$$

$$\hat{\rho}_0 - d_2 A_0 - d_1 = 0 \tag{3.7}$$

$$\sum_{\ell=p+1}^{N} c_\ell A_\ell \hat{\rho}_\ell - d_2 \sum_{\ell=p+1}^{N} c_\ell A_\ell^2 = 0 \tag{3.8}$$

$$\sum_{\ell=p+1}^{N} c_\ell A_\ell' \hat{\rho}_\ell - d_2 \sum_{\ell=p+1}^{N} c_\ell A_\ell A_\ell' = 0 \tag{3.9}$$

where

$$\hat{\rho}_\ell = \frac{2}{N c_\ell} \sum_{j=0}^{N} \frac{1}{c_j} \rho(y_j) T_\ell(y_j) \tag{3.10}$$

$$A' = \frac{\partial}{\partial s} A_\ell(s) . \tag{3.11}$$

Again, we combine (3.8) and (3.9) into a single nonlinear equation for the shock location index, s :

$$\sum c_\ell A_\ell' \hat{\rho}_\ell \sum c_\ell A_\ell^2 - \sum c_\ell A_\ell \hat{\rho}_\ell \sum c_\ell A_\ell A_\ell' = 0 \tag{3.12}$$

where all the sums are from $\ell = p+1$ to $\ell = N$.

The procedure for extracting the shock location, jump magnitude and smooth part of the solution from the raw data $\rho(x,y_j)$ (given in Table 2) is exactly the same as described above for the Fourier problem.

For the wedge flow problem considered here, this procedure applied in the case of a coarse net $(N = 8)$, located the shock with an error only in the fourth significant figure. The smooth part was recovered to within 1% at the worst field point.

4. CONCLUSIONS

A procedure has been developed for extracting the piecewise smooth solution of hyperbolic system from the raw oscillatory data obtained by pseudospectral methods. The theory, so far, is precise only for the linear case with constant coefficients. Plausibily arguments point out that this procedure will also work for nonlinear operators when steady state has been achieved. The numerical results presented herein bear out this conjecture.

It remains to try and understand the phenomenon better in the nonlinear cases where waves are propogated with different speeds at different field points and at different times.

5. REFERENCES

[1] Gottlieb, D. and Orszag, S.A., "Numerical Analysis of Spectral Methods: Theory and Applications", CBMS Regional Conference Series in Applied Mathematics, 26, SIAM, 1977.

[2] Shannon, C.E., "A Mathematical Theory of Communication", Bell System Technical Journal, 27, pp. 379 - 423, 623 - 658, 1948.

[3] Lax, P.D., "Accuracy and Resolution in the Computations of Solutions of Linear and Nonlinear Equations" in Recent Advances in Dimensional Analysis, MRC University at Wisconsin, Academic Press, 1978, pp. 107 - 117.

[4] Gottlieb, D., "Spectral Methods for Compressible Flow Problems", ICASE Report No. 84-29, To appear in the Proceedings of the ICYNAFD.

School of Mathematical Sciences
Raymond and Beverly Sackler Faculty of Exact Sciences
Tel Aviv University
Tel Aviv, Israel

Research was supported in part by the Air Force Office of Scientific Research under Contract No. AFOSR 83-0089 and in part by the National Aeronautical and Space Administration under NASA Contract No. NAS1-17070 while the authors were in residence at ICASE, NASA Langley Research Center, Hampton, VA 23665.

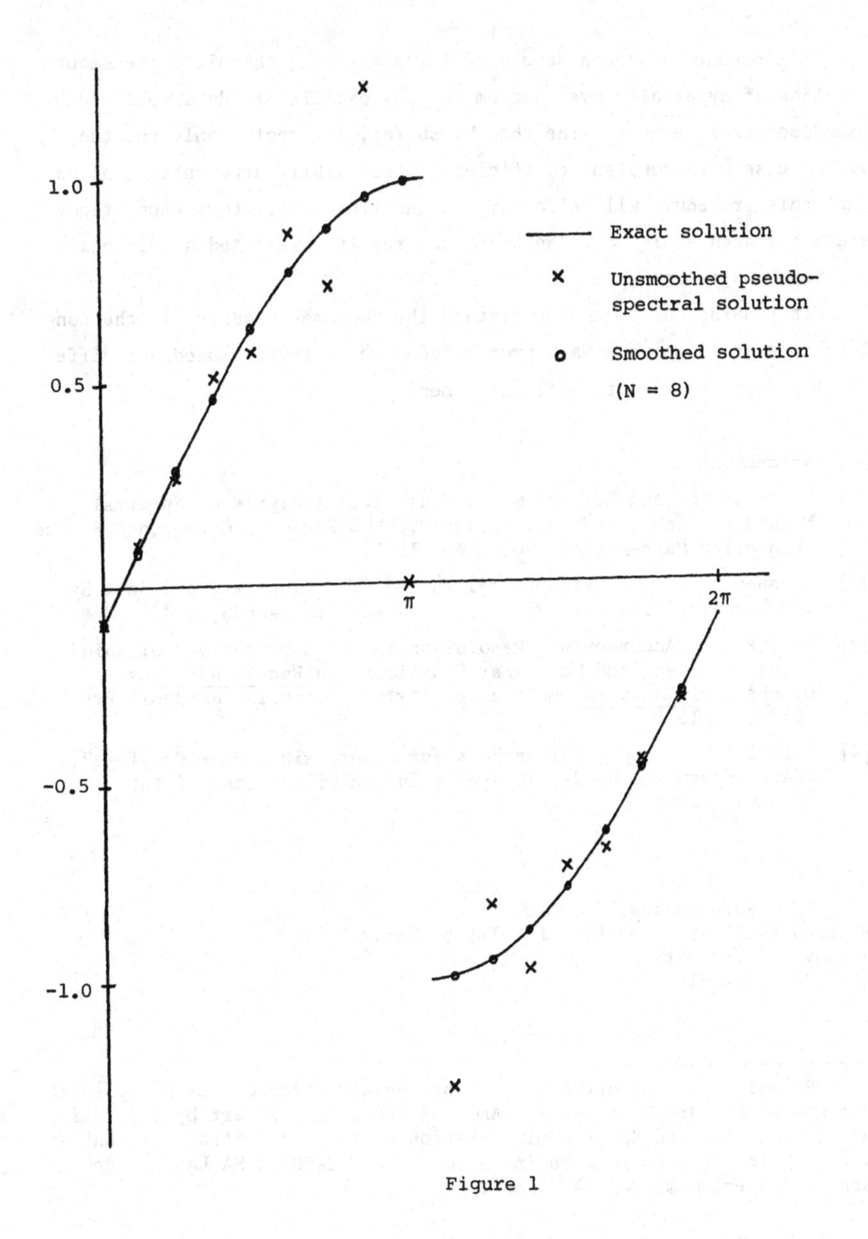
1.0
0.5
-0.5
-1.0
π
2π
Exact solution
Unsmoothed pseudo-
spectral solution
Smoothed solution
(N = 8)

Figure 1

Progress in Scientific Computing, Vol. 6
Proceedings of U.S.-Israel Workshop, 1984

RECOVERING POINTWISE VALUES OF DISCONTINUOUS DATA WITHIN SPECTRAL ACCURACY

David Gottlieb and Eitan Tadmor

1. INTRODUCTION

Let $f(x)$ be a bounded 2π-periodic function whose Fourier coefficients are given by

$$\hat{f}(k) = \frac{1}{2\pi}\int_{-\pi}^{\pi} f(y)e^{-ik\cdot y}dy \ , \quad -\infty < k < \infty \ . \tag{1.1}$$

It is well-known that whenever f is a smooth function, then its spectral approximation - consisting of the partial sums

$$S_N f(x) \equiv \hat{f}_N(x) = \sum_{|k|\le N} \hat{f}(k)e^{ik\cdot x} \ , \tag{1.2}$$

converges pointwise to $f(x)$. A typical error estimate in this case, asserts that for any x in the domain we have

$$|f(x) - \hat{f}_N(x)| \le C_s ||f||_{(s)}\cdot N^{-s+1} \ , \quad s > 1 \ . \tag{1.3}$$

Here and below, C_s stands for (possibly different) generic constant bounds, and $||f||_{(s)}$ denotes the largest maximum norm of f and its first s derivatives, the maximum taken over the whole domain.

We thus see that the decay rate of the truncation error on the left of (1.3), is restricted only by the degree of smoothness of the function f. In this sense, the spectral approximation is termed to be spectrally accurate. If, in particular, f is a C^∞-function, the truncation error is rapidly decaying, faster than any fixed ($\equiv$ independent of N) polynomial rate. Thus, the spectral approximation of C^∞-functions, enjoys the so called infinite order of accuracy; this is in constrast to the usually slower convergence rate due to a fixed degree polynomial accuracy.

Next, assume only the gridvalues $f_\nu = f(y_\nu)$ are known, at the $2N$ equidistant gridpoints $y_\nu = -\pi + \nu h$, $h = 2\pi/2N$, $\nu = 0,1,\ldots,2N-1$. Invoking the trapezoidal rule, the (exact) Fourier coefficients in (1.1) are approximated by discrete sums of these known gridvalues

$$\tilde{f}(k) = \frac{1}{2N} \sum_{\nu=0}^{2N-1} f_\nu e^{-ik\cdot y_\nu} , \quad -N \leq k \leq N . \tag{1.4}$$

The difference between the exact Fourier coefficients and their discrete approximation is also known to be spectrally small

$$|\hat{f}(k) - \tilde{f}(k)| \leq C_s ||f||_{(s)} N^{-s} , \quad s > 1 . \tag{1.5}$$

As a substitute to the (exact) Fourier coefficients appearing in the spectral approximation (1.2), $\hat{f}(k)$, let us use their discrete counterpart, $\tilde{f}(k)$. The resulting new approximation is found to be exact at the gridpoints $x = y_\nu$. In other words, we arrive at the trigonometric interpolant[(1)].

$$I_N f(x) \equiv \tilde{f}_N(x) = \sum_{|k|\leq N}{}'' \tilde{f}(k) e^{ik\cdot x} . \tag{1.6}$$

The two type of errors committed in this case - the original truncation error in (1.3) padded with the aliasing errors in (1.5) - both are spectrally small. Hence, if f is smooth over the whole domain, then its pseudo-spectral approximation (1.6) is spectrally accurate even in between the gridpoints

$$|f(x) - \tilde{f}_N(x)| \leq C_s ||f||_{(s)} N^{-s+1} , \quad s > 1 . \tag{1.7}$$

We also note that as in the spectral and pseudo-spectral cases (1.3) and (1.7), similar error decay is obtained with higher derivatives and in more space variables; the norm on the right hand-side of (1.3), (1.5) and (1.7) should be "raised" accordingly. Moreover, if the function is in particular analytic, then the spectral accuracy is further improved to be exponential: let $2\eta > 0$ be the width of analyticity strip with maximum modulus $||f||_\eta$ then an error bound of the form $C_\eta ||f||_\eta e^{-N\eta}$ follows, e.g. [7].

Unfortunately, the pointwise errors associated with the spectral or pseudospectral approximations, suffer from the limitation of being dependent on the smoothness of the function f over the whole domain (real or complex), and not just on its local behavior in the neighborhood of the point of interest. This dependence of the local convergence rate on the global smoothness, which is reflected by (though not a consequence of) the error estimates (1.3) and (1.7), is indeed inherent in both approximations. That is, the roughness of the function in one part of its domain, decelerates the convergence rate in the smoother part of it. Most notably is the case of piecewise smooth functions: not only that Gibbs phenomenon is recorded at points of discontinuity, but in addition, the spectral accuracy is lost at regions where the function is smooth.

In this paper we show how pointwise values of the function $f(x)$ can be recovered from the information contained in either its spectral or pseudo-spectral approximations, so that the accuracy solely depends on the local smoothness of f, that is, its smoothness in the neighborhood of the point of interest x. If, in particular, f is infinitely smooth in that neighborhood, then the value $f(x)$ is approximated within infinite order of accuracy. Most notably, we rocover pointwise values within spectral accuracy, despite the possible presence of discontinuities scattered in the domain.

For such pointwise recovery, we should dismantle the above local-global coupling limitation, associated with the (pseudo-) spectral approximations. To this end, we employ a regularization kernel which is convoluted against the (pseudo-) spectral approximation in the usual fashion. Our regularization kernel consists of the product of two terms: first we introduce a cut-off function to localize the kernel in the spirit advocated above; secondly, it is multiplied by the spectral approximation of the delta function (≡ Dirichlet kernel), so that spectral accuracy is guaranteed. Convolution with the resulting kernel has then the effect of (locally) smoothing the spectral and pseudo-spectral approximations.

The paper is organized as follows: in Section 2 we briefly discuss those fundamentals of Fourier summation which will be later needed. Smoothing of the spectral approximation is described in Section 3. In Section 4, we similarily treat the pseudo-spectral approximation. It should be emphasized that the latter case, directly involves only neighboring gridvalues, so that the construction of the pseudo-spectral approximation can be avoided altogether. In other words, (intermediate) pointvalues are recovered here, via a locally supported yet spectrally accurate interpolation recipe. We remark that more general orthogonal families - other than the treated above trigonometric one - can be used as well, to yield spectral smoothing: the notable examples of Legendre and Tchebyshev are brielfy sketched in Section 5. We conclude with numerical evidence which back up on theoretical considerations.

In [6], Mock and Lax have shown how to recover within polynomial accuracy, pointwise values of discontinuous solutions to linear hyperbolic equations. They have employed a locally suppoerted unit mass post-processing kernel with a finite number of vanishing higher moments. Our spectral smoothing is motivated by Mock and Lax discussion- indeed, our regularization kernel based on the Legendre spectral approximation is intimately related to their kernel. Majda and McDonough and Osher [5] on the other hand, extending their previous study [4] with regard to the same problem, have employed a spectrally accurate smoothing procedure by operating directly in the Fourier space. Our smoothing in the real space rather than in the transformed one seems to offer more robustness, resulting from the use of physical space localization; the latter is in fact the key element which enables us to apply our smoothing procedure to pseudo-spectral approximations. Moreover, it is also applicable in conjunction with orthogonal families other than the trigonometric one.

This work has been motivated by the numerical studies of (pseudo-) spectral simulation of shock waves. However, in this paper we restrict our attention to the level of approximation only; applications to P.D.E. will be discussed elsewhere.

2. PRELIMINARIES ON FOURIER SUMMATION

Given a 2π-periodic function Φ with Fourier coefficient $\hat{\Phi}(k) = (2\pi)^{-1} \int_{-\pi}^{\pi} \Phi(y)e^{-ik\cdot y}dy$, its spectral approximation $\hat{\Phi}_p(x)$ is

$$\hat{\Phi}_p(x) \equiv \sum_{|k|\leq p} \frac{1}{2\pi} \int_{-\pi}^{\pi} \Phi(y)e^{-ik\cdot y}dy e^{ik\cdot x} = \int_{-\pi}^{\pi} \Phi(x-y)D_p(y)dy \; ; \qquad (2.1a)$$

here $D_p(y)$ stands for the Dirichlet kernel

$$D_p(y) = \frac{1}{2\pi} \sum_{|k|\leq p} e^{ik\cdot y} = \frac{1}{2\pi} \frac{\sin (p+1/2)y}{\sin y/2} \qquad (2.1b)$$

normalized so that it has a unit mass

$$\int_{-\pi}^{\pi} D_p(y)dy = 1 . \qquad (2.1c)$$

An (a' postriori) bound on the truncation error is given by

$$\Big|\sum_{|k|>p} \hat{\Phi}(k)e^{ik\cdot x}\Big| \leqslant \sum_{|k|>p} |k|^s|\hat{\Phi}(k)||k|^{-s} \leqslant ||\Phi||_{(s)} \sum_{|k|>p} |k|^{-s} \leqslant$$

$$\leqslant ||\Phi||_{(s)}p^{-s+1} \qquad s > 1 , \qquad (2.2)$$

in agreement with (1.3), taking $(\Phi,p) = (f,N)$. Thus we have

$$\Big|\int_{-\pi}^{\pi} \Phi(y)D_p(x-y)dy-\Phi(x)\Big| \leqslant C_s||\Phi||_{(s)}p^{-s+1} , \qquad s > 1 ; \qquad (2.3a)$$

for later purpose, we quote here the special case $x = 0$,

$$\Big|\int_{-\pi}^{\pi} \Phi(y)D_p(y)dy - \Phi(0)\Big| \leqslant C_s||\Phi||_{(s)}p^{-s+1}, \qquad s > 1 . \qquad (2.3b)$$

The above error bound is not the sharpest bound possible: let $\omega(\ ; \ )$ denotes the function's modulus of continuity, then Kolmogorov's result yields an asymptotrially exact bound[(2)]

$$\Big|\int_{-\pi}^{\pi} \Phi(y)D_p(x-y)dy-\Phi(x)\Big| \leqslant \frac{2\ln p}{\pi^2 p^s}\int_0^{\pi/2} \omega\left(\frac{2\xi}{p};D^s\Phi\right)\sin\xi d\xi + O\left(p^{-s}\omega\left(\frac{1}{p}\right)\right). \qquad (2.4)$$

Turning to the pseudo-spectral approximation, we have encountered the additional source of aliasing errors, due to discretization of the (exact) Fourier coefficients' integral. Invoking the aliasing relations, e.g. [3],

$$\tilde{\Phi}(k) = \sum_{j=-\infty}^{\infty} \hat{\Phi}(k+2jp) , \qquad -p \leqslant k \leqslant p , \qquad (2.5)$$

($2p$ equidistant point's interpolant is assumed). The aliasing errors do not exceed

$$|\hat{\Phi}(k)-\tilde{\Phi}(k)| \leqslant \sum_{j\neq 0} |\hat{\Phi}(k+2jp)| \leqslant ||\Phi||_{(s)}\cdot\sum_{j\neq 0} |k+2jp|^{-s} \leqslant$$

$$\leqslant C_s||\Phi||_{(s)}p^{-s} , \qquad s > 1 , \qquad (2.6)$$

in agreement with (1.5), taking $(\Phi,p) = (f,N)$. Hence, the aliasing error $|\hat{\Phi}(k) - \tilde{\Phi}(k)|$, $k = -p,\dots,p$, adds up to a contribution similar to that of the truncation error, yielding in view of (2.5)

$$|\tilde{\Phi}_p(x)-\Phi(x)| \leq \Big|\sum_{|k|>p} \hat{\Phi}(k)e^{ik\cdot x}\Big| + \sum_{|k|\leq p} |\hat{\Phi}(k)| + \frac{1}{2}|\tilde{\Phi}(-p)+\tilde{\Phi}(p)| \leq$$

$$\leq C_s||\Phi||_{(s)}p^{-s+1}, \qquad s \geq 1, \tag{2.7}$$

in agreement with (1.7), taking $(\Phi,p) = (f,N)$. It should be noted, (e.g. [3]), that there is no qualitative difference between the spectral and pseudo-spectral approximations.

3. RECOVERING POINTWISE VALUES FROM THE SPECTRAL APPROXIMATION

In this section, we show how to extract highly accurate approximation to the point values of a discontinuous function from its first N Fourier coefficients in regions where the function is smooth. The basic idea is that these coefficients are moments of the functions and consequently, integral of any smooth function against the spectral approximation is highly accurate with that against the function itself. We therefore construct an auxilliary function such that when the spectral approximation is integrated against it, the desired original point value at a given point is recovered.

To do that, let $\rho(y)$ be a C^s-function vanishing outside the interval $(-\pi,\pi)$ and normalized to take the value one at the origin

$$\rho(y=0) = 1. \tag{3.1}$$

We recall that the Dirichlet kernel in (2.1b) is given by

$$D_p(y) \equiv \frac{1}{2\pi}\sum_{|k|\leq p} e^{ik\cdot y} = \frac{1}{2\pi}\frac{\sin\,(p+1/2)y}{\sin\,(y/2)}; \tag{3.2}$$

multiplying the two we obtain

$$\psi^{1,p}(y) = \rho(y)D_p(y). \tag{3.3}$$

We now set as our regularization kernel

$$\psi^{\theta,p}(y) \equiv \theta^{-1}\psi^{1,p}(\theta^{-1}y) = \theta^{-1}\rho(\theta^{-1}y)D_p(\theta^{-1}y) , \tag{3.4}$$

depending on a yet to be determined free parameter θ , $0 \leq \theta \leq 1$.

Given the spectral approximation, $\hat{f}_N$, we smooth its value via convolution with the above regularization kernel, computing

$$\hat{f}_N * \psi^{\theta,p}(x) = \int_{-\pi}^{\pi} \hat{f}_N(y')\psi^{\theta,p}(x-y')dy'. \tag{3.5}$$

In order to estimate the error, we decompose

$$\hat{f}_N * \psi^{\theta,p} - f = (\hat{f}_N-f) * \hat{\psi}_N^{\theta,p} + (\hat{f}_N-f) * (\psi^{\theta,p} - \hat{\psi}_N^{\theta,p}) + $$

$$+ (f * \psi^{\theta,p} - f). \tag{3.6}$$

The first term on the right vanishes in view of the orthogonality between the N-degree (trigonometric) polynomial $\hat{\psi}_N^{\theta,p}$ and the truncated sum $\hat{f}_N - f$,

$$(\hat{f}_N-f) * \hat{\psi}_N^{\theta,p} = 0 . \tag{3.7a}$$

Thus we are left with two sources of error in this case: the <u>truncation error</u> in the second term

$$T_N^{\theta,p} \equiv (\hat{f}_N-f) * (\psi^{\theta,p} - \hat{\psi}_N^{\theta,p}) \tag{3.7b}$$

and the <u>regularization error</u> in the third term,

$$R^{\theta,p} \equiv f * \psi^{\theta,p} - f . \tag{3.7c}$$

With regard to the truncation error $T_N^{\theta,p}$, Young inequality implies

$$||(\hat{f}_N-f) * (\psi^{\theta,p} - \hat{\psi}_N^{\theta,p})|| \leq ||\hat{f}_N-f||_{L^1}||\psi^{\theta,p} - \hat{\psi}_N^{\theta,p}|| , \tag{3.8a}$$

and in the view of (1.3) we conclude that this term is spectrally small

$$||T_N^{\theta,p}|| \equiv ||(\hat{f}_N-f) * (\psi^{\theta,p} - \hat{\psi}_N^{\theta,p})|| \leq$$

$$\leq C_s||\hat{f}_N-f||_{L^1}||\psi^{\theta,p}||_{(s)}N^{-s+1} . \tag{3.8b}$$

Turning to the regularization error, $R^{\theta,p}$, we compute at a given fixed point x

$$R^{\theta,p}_{(x)} \equiv f * \psi^{\theta,p}(x) - f(x) = \int_{x-\theta\pi}^{x+\theta\pi} f(y')\theta^{-1}\rho(\frac{x-y'}{\theta})D_p(\frac{x-y'}{\theta})dy' - f(x). \tag{3.9a}$$

Changing variables $y = \frac{x-y'}{\theta}$ and making use of (2.1c), the regularization error is simplified into

$$R^{\theta,p}_{(x)} \equiv f * \psi^{\theta,p}(x) - f(x) = \int_{-\pi}^{\pi} \Phi^{\theta,x}(y)D_p(y)dy \tag{3.9b}$$

where the auxiliary function $\Phi^{\theta,x}(y)$ is given by

$$\Phi^{\theta,x}(y) \equiv f(x-\theta y)\rho(y) - f(x) . \tag{3.9c}$$

In view of the normalization (3.1), $\Phi^{\theta,x}(y)$ vanishes at $y = 0$, and by appealing to the truncation error estimate quoted in (2.3b), we end up with

$$|R^{\theta,p}_{(x)}| \equiv |\int_{-\pi}^{\pi} \Phi^{\theta,x}(y)D_p(y)dy| \leq C_s||\Phi^{\theta,x}||_{(s)}p^{-s+1} . \tag{3.9d}$$

Added together, we have shown in (3.6) - (3.9) the following:

Proposition 3.1 (Main Error Estimate)

Let $\psi^{\theta,p}$ be the regularization kernel (3.4). Fix a point x in the domain, and set $\Phi^{\theta,x}$ to be the auxiliary function in (3.9c). Then, the following error estimate holds

$$|\hat{f}_N * \psi^{\theta,p}(x) - f(x)| \leq C_s||f||\cdot||\psi^{\theta,p}||_{(s)}N^{-s+1} + C_s||\Phi^{\theta,x}||_{(s)}p^{-s+1} . \tag{3.10}$$

The following two lemmas whose technical proofs are postponed to the end of this section, provide us with the necessary explicit bounds on the two terms appearing on the right of (3.10).

Lemma 3.2

Let $\psi^{\theta,p}$ denote the regularization kernel in (3.4). The following estimate holds

$$||\psi^{\theta,p}||_{(s)} \leq \theta^{-s-1}\cdot||\rho||_{(s)}\cdot(1+p)^{s+1} . \tag{3.11}$$

<u>Lemma 3.3</u>

<u>Let</u> $\Phi^{\theta,x}$ <u>denote the auxiliary function in</u> (3.9c). <u>The following estimate holds</u>

$$||\Phi^{\theta,x}||_{(s)} \leq (1+\theta)^s \cdot ||\rho||_{(s)} \cdot \underset{\substack{|y-x|\leq\theta\pi \\ 0\leq k\leq s}}{\text{Max}} |D^k f(y)| \,. \tag{3.12}$$

Choosing $p = N^\beta$, $0 < \beta \leq 1$, we conclude from (3.6), (3.8b), (3.9d) and the last two lemmas, the main result of this section, stating

<u>THEOREM 3.4</u>

<u>Let</u> f <u>be a bounded</u> 2π-<u>periodic function with a given</u> N-<u>degree spectral approximation</u> $\hat{f}_N$. <u>Setting the regularization kernel</u>

$$\psi^{\theta,N^\beta}(y) = \frac{1}{2\pi\theta}\,\rho(\theta^{-1}y)\,\frac{\sin\,(N^\beta + 1/2)y/\theta}{\sin\,(y/2\theta)}\,, \tag{3.13}$$

<u>then for any</u> x <u>in the domain, we have the pointwise error estimate</u>

$$|\hat{f}_N * \psi^{\theta,N^\beta}(x) - f(x)| \leq C_s||\rho||_{(s)} \cdot N^\beta [N\cdot\theta^{-s}||\hat{f}_N - f||_{L^1} \cdot N^{-(1-\beta)s} + \underset{\substack{|y-x|\leq\theta\pi \\ 0\leq k\leq s}}{\text{Max}} |D^k f(y)| \cdot N^{-\beta s}] \,. \tag{3.14}$$

Choosing $\theta = \beta = 1$ brings us back to the exactly same <u>global</u> error estimate we had in (1.3). Taking $\beta = 1/2$ on the other hand, the truncation and aliasing errors' contributions in (3.14) are balanced, and we are led to the following:

<u>Corollary 3.5</u> <u>(Spectral Smoothing)</u>

<u>Let</u> $\rho(y)$ <u>be a</u> C^{2s}-<u>function, supported in</u> $[-\pi,\pi]$ <u>and satisfying</u> (2.1). <u>Then, for any</u> x <u>in the domain, the value</u> $f(x)$ <u>can be recovered via the spectral smoothing of</u> $\hat{f}_N$, <u>which obeys the following error estimate</u>

$$|\hat{f}_N * \psi^{\theta,\sqrt{N}}(x) - f(x)| \leq C_s \cdot ||\rho||_{(2s)} \cdot [N\cdot\theta^{-2s} \cdot ||f|| + \underset{\substack{|y-x|\leq\theta\pi \\ 0\leq k\leq 2s}}{\text{Max}} |D^k f(y)|] \cdot N^{-s+1} \,, \quad s > 1. \tag{3.15}$$

In general, of course, the choices of the cut-off function, ρ , and the β-exponent, $0 < \beta < 1$, provide us with a whole variety of admissible kernels, for which we have:

Corollary 3.6 (Infinite Order of Accuracy)

Let $\rho(y)$ be a C^∞-function, supported in $[-\pi,\pi]$ and taking the value one at the origin. Assume the function f is C^∞ in the ε-neighborhood of a point x in the domain. Then the spectral smoothing

$$\hat{f}_N * \psi^{\theta=\varepsilon/\pi, N^\beta}(x) = \frac{1}{2\varepsilon}\int_{y=x-\varepsilon}^{x+\varepsilon} \hat{f}_N(y)\rho\left(\frac{\pi(x-y)}{\varepsilon}\right)\frac{\sin(N^\beta+\frac{1}{2})\frac{\pi(x-y)}{\varepsilon}}{\sin\frac{\pi(x-y)}{2\varepsilon}}\,dy,$$

$$0 < \beta < 1 , \qquad (3.16)$$

recovers the function value $f(x)$ within infinite order of accuracy.

Remarks

(i) Suppose f is known to be smooth in the asymmetric neighborhood of x , $(x - \varepsilon_L , x + \varepsilon_R)$, $0 < \varepsilon_L, \varepsilon_R \leq \pi$. Let ρ be a C^∞-function supported in the interval $[-\theta^{-1}\varepsilon_L, \theta^{-1}\varepsilon_R]$ inside of $[-\pi,\pi]$, such that $\rho(y = 0) = 1$. Then a nonsymmetric version of the above spectral smoothing reads

$$\hat{f}_N * \psi^{\theta,N^\beta}(x) = \frac{1}{2\pi\theta}\int_{x-\varepsilon_L}^{x+\varepsilon_R} \hat{f}_N(y)\rho\left(\frac{x-y}{\theta}\right)\frac{\sin(N^\beta+\frac{1}{2})(\frac{x-y}{\theta})}{\sin\frac{x-y}{2\theta}}\,dy, \qquad (3.17)$$

recovering $f(x)$ within spectral accuracy. The case $\varepsilon_L = \varepsilon_R = \varepsilon = \pi\theta$ coincides with Corollary 3.6.

(ii) The above error estimates concerning the spectral smoothing $\hat{f}_N * \psi^{\theta,p}$ still enjoy the further flexibility in choosing different s-orders in (3.10). This provides us with even further richness so as to tune the different free parameters to yield accurate results.

As promised, we conclude this section with the following:

Proof of Lemma 3.2

With the regularization kernel $\psi^{\theta,p}$ in (3.4), Liebnitz's rule gives us

$$||\psi^{\theta,p}||_{(s)} \leq \theta^{-s-1}||\rho(y)D_p(y)||_{(s)} \leq$$

$$\leq \theta^{-s-1}\sum_{j=0}^{s}\binom{s}{j}||\rho||_{(s-j)}||D_p||_{(j)} \; ; \qquad (3.18)$$

complemented by the maximum norm estimate

$$||D_p||_{(j)} \leq \frac{1}{2\pi} \sum_{|k|\leq p} |k|^j \leq \frac{1}{\pi(j+1)} p^{j+1} , \tag{3.19}$$

the desired result follows

$$||\psi^{\theta,p}||_{(s)} \leq \frac{1}{\pi} \theta^{-s-1} \sum_{j=0}^{s+1} \binom{s+1}{j} ||\rho||_{(s)} p^j \leq \theta^{-s-1} ||\rho||_{(s)} (1+p)^{s+1} . \tag{3.20}$$

Proof of Lemma 3.3

Let $\Phi^{\theta,x}(y) = f(x-\theta y)\rho(y) - f(x)$ be the auxiliary function in (3.9c) with $\rho(y)$ supported in $[-\pi,\pi]$. We observe that the only f-values participating in the definition of $\Phi^{\theta,x}$ are those from the $\theta\cdot\pi$ neighborhood of x , $|y - x| \leq \theta\cdot\pi$. Applying Liebnitz's rule restricted to that neighborhood, we find

$$||\Phi^{\theta,x}||_{(s)} \leq \sum_{j=0}^{s} \binom{s}{j} ||\rho||_{(j)} \theta^{s-j} \cdot \max_{\substack{|y-x|\leq\theta\pi \\ 0\leq k\leq s-j}} |D^k f| \leq$$

$$\leq (1+\theta)^s ||\rho||_{(s)} \cdot \max_{\substack{|y-x|\leq\theta\pi \\ 0\leq k\leq s}} |D^k f(y)| \tag{3.21}$$

as asserted.

4. RECOVERING POINTWISE VALUES FROM THE PSEUDOSPECTRAL APPROXIMATION

In this section we treat the case where the discrete gridvalues $f_\nu = f(y_\nu)$ are given, so that a peudo-spectral approximation $\tilde{f}_N$ collocating these gridvalues is uniquely determined, see (1.6). The key observation here is that the integrand $\hat{f}_N(y')\psi^{\theta,p}(x-y')$ in (3.5) is smooth over the whole domain, due to the kernel localization in the neighborhood of the point of interest, x . Hence, replacing the convolution integral with an appropriate trapezoidal sum, only an additional spectrally small aliasing error is committed. Thus, in analogy with (3.5), we smooth the pseudo-spectral approximation via the convolution sum

$$\frac{2\pi}{2N} \sum_{\nu=0}^{2N-1} \tilde{f}_N(y_\nu)\psi^{\theta,p}(x-y_\nu) \equiv \frac{2\pi}{2N} \sum_{\nu=0}^{2N-1} f_\nu \psi^{\theta,p}(x-y_\nu) . \tag{4.1}$$

Observe that since $\psi^{\theta,p}$ is supported in the neighborhood of x , only those neighboring gridvalues are taking part in the pseudo-spectral smoothing.

The computed error at a fixed point x , amounts to

$$\frac{2\pi}{2N}\sum_{\nu=0}^{2N-1} f_\nu \psi^{\theta,p}(x-y_\nu) - f(x) =$$

$$= \left(\frac{2\pi}{2N}\sum_{\nu=0}^{2N-1} f_\nu \psi^{\theta,p}(x-y_\nu) - f * \psi^{\theta,p}(x)\right) + \left(f * \psi^{\theta,p}(x) - f(x)\right). \tag{4.2a}$$

There are two sources of errors in this case: the <u>aliasing error</u> due to the use of the trapezoidal rule in the first difference

$$A_N^{\theta,p} \equiv \frac{2\pi}{2N}\sum_{\nu=0}^{2N-1} f_\nu \psi^{\theta,p}(x-y_\nu) - f * \psi^{\theta,p}(x) , \tag{4.2b}$$

and as before, the <u>regularization error</u> in the second difference

$$R^{\theta,p} = f * \psi^{\theta,p} - f . \tag{4.2c}$$

The aliasing error estimate in $(1.5)_0$ and the regularization error estimate in (3.9d) yield:

<u>Proposition 4.1</u> <u>(Main Error Estimate)</u>

<u>Let</u> $\psi^{\theta,p}$ <u>be the regularization kernel</u> (3.4). <u>Fix a point</u> x <u>in the domain and denote</u>

$$\chi^{\theta,p,x}(y) = f(y)\cdot\psi^{\theta,p}(x-y) . \tag{4.3}$$

<u>Also, let</u> $\Phi^{\theta,x}$ <u>be the auxiliary function in</u> (3.9c). <u>Then, the following error estimate holds</u>

$$\left|\frac{2\pi}{2N}\sum_{\nu=0}^{2N-1} f_\nu \psi^{\theta,p}(x-y_\nu) - f(x)\right| \le C_s||\chi^{\theta,p,x}||_{(s)}N^{-s} + C_s||\Phi^{\theta,x}||_{(s)}p^{-s+1} . \tag{4.4}$$

We observe that the newly introduced auxiliary function $\chi^{\theta,p,x}(y)$ is supported in the $\theta\cdot\pi$-neighborhood of x, where Liebnitz's rule yields

$$||\chi^{\theta,p,x}||_{(s)} \le \sum_{j=0}^{s} \binom{s}{j}||\psi^{\theta,p}||_{(j)}\cdot \max_{\substack{|y-x|\le\theta\pi \\ 0\le k\le s-j}} |D^k f(y)| ; \tag{4.5a}$$

invoking (3.11), the following bound is found

$$||\chi^{\theta,p,x}||_{(s)} \leq \sum_{j=0}^{s} \binom{s}{j}\theta^{-j+1}||\rho||_{(j)}(1+p)^{j+1}\cdot \underset{\substack{|y-x|\leq\theta\pi \\ 0\leq k\leq s}}{\mathrm{Max}} |D^k f(y)| \leq$$

$$\leq \theta(1+p)||\rho||_{(s)}\cdot \underset{\substack{|y-x|\leq\theta\pi \\ 0\leq k\leq s}}{\mathrm{Max}} |D^k f(y)|\left(\frac{1+\theta+p}{\theta}\right)^s . \quad (4.5b)$$

The last estimate on the aliasing part of the error, augmented with the previously derived estimate on the regularization error in Lemma 3.3, lead us to the main result of this section, stating:

THEOREM 4.2

Let f be a bounded 2π-periodic function with given gridvalues $f_\nu = f(y_\nu)$. Setting the regularization kernel

$$\psi^{\theta,N^\beta}(y) = \frac{1}{2\pi\theta}\rho(\theta^{-1}y)\frac{\sin (N^\beta+1/2)y/\theta}{\sin y/2\theta} , \quad (4.6)$$

then for any x in the domain, we have the pointwise error estimate

$$\left|\frac{2\pi}{2N}\sum_{\nu=0}^{2N-1} f_\nu\psi^{\theta,N^\beta}(x-y_\nu) - f(x)\right| \leq$$

$$\leq C_s||\rho||_s\cdot \underset{\substack{|y-x|\leq\theta\pi \\ 0\leq k\leq s}}{\mathrm{Max}} |D^k f(y)|\cdot[N^{-\beta s+1} + \theta^{-s}\cdot N^\beta\cdot N^{-(1-\beta)s}]. \quad (4.7)$$

Taking $\beta = 1/2$ to balance the two error's contributions, we find:

Corollary 4.3 (Pseudo-Spectral Smoothing)

Let $\rho(y)$ be a C^{2s}-function, supported in $[-\pi,\pi]$ and satisfying (2.1). Then for any x in the domain, the value $f(x)$ can be recovered via the pseudo-spectral smoothing of the neighboring gridvalues, f_ν , which obeys the following error estimate

$$\left|\frac{2\pi}{2N}\sum_{\nu=0}^{2N-1} f_\nu\psi^{\theta,\sqrt{N}}(x-y_\nu) - f(x)\right| \leq$$

$$\leq C_s||\rho||_{(2s)}\cdot \underset{\substack{|y-x|\leq\theta\pi \\ 0\leq k\leq 2s}}{\mathrm{Max}} |D^k f(y)|(1+\theta^{-2s})\cdot N^{-s+1}, \quad s > 1. \quad (4.8)$$

In analogy with Corollary 3.6, we also have:

Corollary 4.4 (Infinite Order of Accuracy)

Let $\rho(y)$ be a C^∞-function, supported in $[-\pi,\pi]$ and taking the value one at the origin. Assume the function f is C^∞ in the ε-neighborhood of a point x in the domain. Then the pseudo-spectral

<u>smoothing</u>

$$\frac{2\pi}{2N}\sum_{\nu=0}^{2N-1} f_\nu \psi^{\theta=\varepsilon/\pi, N^\beta}(x-y_\nu) = \frac{\pi}{2\varepsilon N}\sum_{y_\nu \geq x-\varepsilon}^{y_\nu \leq x+\varepsilon} f_\nu \rho\left(\frac{\pi(x-y_\nu)}{\varepsilon}\right)\frac{\sin(N^\beta+\frac{1}{2})\frac{\pi(x-y_\nu)}{\varepsilon}}{\sin\frac{\pi(x-y_\nu)}{2\varepsilon}},$$

$$0 < \beta < 1 \tag{4.9}$$

<u>recovers the function value</u> $f(x)$ <u>within infinite order of accuracy</u>.

In closing this section, we would like to emphasize another, slightly more global variant of the pseudo-spectral smoothing, based on integral convolution of the pseudo-spectral interpolant against the regularization kernel

$$\tilde{f}_N * \psi^{\theta,N^\beta} - f = \tilde{f}_N * \left(\psi^{\theta,N^\beta} - \tilde{\psi}_N^{\theta,N^\beta}\right) + \\ + \left(\tilde{f}_N * \tilde{\psi}_N^{\theta,N^\beta} - f * \psi^{\theta,N^\beta}\right) + \left(f * \psi^{\theta,N^\beta} - f\right). \tag{4.10}$$

The first term is spectrally small due to the interpolation error associated with the smooth regularization kernel as argued in Section 2; by the exactness of the trapezoidal rule applied to (trigonometric) polynomials of degree $\leq 2N$, we have

$$\tilde{f}_N * \tilde{\psi}_N^{\theta,N^\beta} = \frac{2\pi}{2N}\sum_{\nu=0}^{2N-1} f_\nu \psi^{\theta,N^\beta}(x-y_\nu) \tag{4.11}$$

and consequently, the second difference is spectrally small as argued above in relation to the aliasing errors. Finally the third difference is the spectrally small regularization error.

5. CONCLUDING REMARKS

The above arguments also apply to other orthogonal families. In conjunction with <u>Legendre polynomials</u>, we set as our regularization kernel

$$\psi^{\theta,p}(y) \equiv \theta^{-1}\rho(\theta^{-1}y)K_p(\theta^{-1}y) \; ; \tag{5.1a}$$

here $\rho(y)$ is C^s-function supported in the interval $[-1,1]$ such that $\rho(y=0) = 1$, and $K_p(y)$ stands for <u>Legendre spectral approximation</u> of the delta function

$$K_p(y) = \sum_{k=0}^{p} (k+\tfrac{1}{2}) P_k(y) P_k(0) \tag{5.1b}$$

normalized to have a unit mass

$$\int_{-1}^{1} K_p(y)\,dy = 1 \,. \tag{5.1c}$$

In view of the Christoffel-Durboux identity, we can rewrite

$$K_p(y) = \frac{p+1}{2} \frac{P_{p+1}(0)P_p(y) - P_p(0)P_{p+1}(y)}{y} \,. \tag{5.1d}$$

The resulting spectral smoothing via the above Legendre-type regularization kernel was introduced in [1], and is intimately related to Mock and Lax [6] post processing: indeed, $\psi^{\theta,p}$ serves as a locally supported kernel with vanishing higher moments and unit mass - modulo a negligble spectral error.

Similarily, we can use Tchebyshev orthogonal expansion where $K_p(y)$ in (5.1a) is replaced by

$$K_p(y) = \frac{2}{\pi} \sum_{k=0}^{p}{}' \, T_k(y) T_k(0) = \frac{(1-y^2)}{p\pi} T_p'(y) \tag{5.2}$$

We note that the (pseudo-) spectral smoothing done with the Tchebyshev kernel is not translated to the usual cut-off in the transformed space.

6. NUMERICAL EXAMPLES

In this section we demonstrate the efficacy of the smoothing procedure outlined above. As a test function we have chosen the piecewise C^∞-function

$$f(x) = \begin{cases} \sin \frac{x}{2} & 0 \leq x \leq \pi \\ -\sin \frac{x}{2} & \pi \leq x \leq 2\pi \,. \end{cases} \tag{6.1}$$

As before, denote its spectral approximation by $\hat{f}_N(x)$, and let $\tilde{f}_N(x)$ be the pseudo-spectral approximation to $f(x)$. It is evident from the first column of Tables I and III that $\hat{f}_N(y_\nu)$ - the spectral approximation sampled at $y_\nu = \nu\pi/N$ - do not approximate $f(y_\nu)$ within spectral accuracy. In fact, the error committed by $\hat{f}_{128}(y_\nu)$ is only half of that committed by $\hat{f}_{64}(y_\nu)$; this is in accordance with a suitably sharpened error estimate of type (1.3) - consult e.g. (3.4). Regarding the pseudo-spectral approximation, $\tilde{f}_N(x)$, it of course collocates the exact values at the sampling gridpoints, $\tilde{f}_N(y_\nu) = f(y_\nu)$; yet, in between these gridpoints, $\tilde{f}_N(y_{\nu+1/2} = (\nu + 1/2)\pi/N)$ approximate $f(y_{\nu+1/2})$ within first order accuracy only, as shown in the first column of Tables II and IV.

In order to construct our regularization kernel, we define the cut-off function $\rho(\xi) = \rho_\alpha(\xi)$ to be

$$\rho_\alpha(\xi) = \begin{cases} \exp \dfrac{\alpha\xi^2}{\xi^2-1} & |\xi| < 1 \\ 0 & \text{otherwise} \end{cases} \tag{6.2}$$

namely $\rho_\alpha(\xi)$ is a C^∞-function whose support is the interval $|\xi| < 1$. Our regularization kernel is now of the form (see (3.4))

$$\psi^{\theta,p}(y) = \frac{1}{2\pi\theta}\,\rho_\alpha(\theta^{-1}y)\,\frac{\sin\,(p+1/2)y/\theta}{\sin\,y/2\theta} \tag{6.3}$$

The post processing procedure of the spectral approximation $\hat{f}_N$ involves convoluting $\hat{f}_N$ against $\psi^{\theta,p}$, namely

$$f(x) \sim \frac{1}{2\pi\theta}\int_0^{2\pi} \hat{f}_N(y)\rho\left(\frac{x-y}{\theta}\right)\frac{\sin\,(p+1/2)(x-y)/\theta}{\sin\,(x-y)/2\theta}\,dy \tag{6.4}$$

where x is a fixed point of interest. (In practice we use the trapezoidal rule to evaluate the right-hand-side of (6.4) taking a large number of quaderture points.)

The parameter θ was chosen as

$$\theta = |x - \pi| ; \tag{6.5}$$

this guarantees that ψ is so localized that it does not interact with regions of discontinuity.

It should be noted, in this stage, that if θ was so chosen to be the same for each x , (and not as in (6.5)), the formula (6.4) admits a simpler form; that is, if

$$\psi^{\theta,p}(y) = \sum_{k=-\infty}^{\infty} \sigma_k e^{iky} \tag{6.6}$$

then

$$f(x) \sim \sum_{k=-N}^{N} \hat{f}(k)\sigma_k e^{ikx} \tag{6.7}$$

This procedure can be carried out efficiently in the Fourier space.

Next, we turn to the post-processing for the pseudo-spectral approximation $\hat{f}_N(x)$ which is simpler than (6.4). In fact, in this case

$$f(x) \sim \frac{2\pi}{2N} \sum_{\nu=0}^{2N-1} \tilde{f}(y_\nu)\psi^{\theta,p}(x-y_\nu) \tag{6.8}$$

Note that carrying out the smoothing procedure defined in (6.8) does not involve any extra evaluation of $\tilde{f}(y)$ in points other than y_ν, in contrast to spectral smoothing procedure in (6.4). As before, the parameter θ was chosen according to (6.5). We have yet to determine the parameters p and α. The parameter p must be equal to N^β for $0 < \beta < 1$ in view of (3.14), in order to assure infinite accuracy. (In our computations $\beta \approx .8$). Finally we feel that α is problem dependent and we chose $\alpha = 10$. We have not tuned the parameters to get optimal results; further tuning may improve the quality of our filtering procedure.

In Tables I, II, III, and IV we give the results of the smoothing procedure at several points in the domain. The pointwise values are now recovered with high accuracy. The first column in each table indicates the points in which the procedure was performed. We limited ourselves to four points in the interval $(0,\pi)$ because of the symmetry of of the function $f(x)$.

The second column gives either the spectral approximation $\hat{f}_N(x)$ or the pseudo-spectral approximation $\tilde{f}_N(x)$, $N = 128$ in Table I and II and $N = 64$ in Tables III and IV. The third column gives the smoothed results, when filtered by (6.4) on (6.8), at the same points as in column I.

The results indicate the dramatic improvement obtained by the smoothing procedure. Moreover, note that the error committed by $\tilde{f}_{128}$ (or $\hat{f}_{128}$) is better than the one committed by $\tilde{f}_{64}$ (or $\hat{f}_{64}$) only by a factor of 2 , whereas after the post-processing the error improves by a factor of 10^4.

Table I.

Results of smoothing of the spectral approximation of $f(x)$, $N = 128$.

$x_\nu = \frac{\pi\nu}{8}$ ν equals	$\lvert f(x_\nu) - \hat{f}_N(x_\nu)\rvert$	$\lvert f - \hat{f}_N * \psi\rvert$ at $x = x_\nu$
2	3.2 (-3)	5.8 (-10)
3	5.2 (-3)	7.9 (-10)
4	7.8 (-3)	6.3 (-10)
5	1.1 (-2)	1.1 (-10)

Table II.

Same as Table I for the pseudo-spectral approximation $\tilde{f}_N(x)$.

$x_{\nu+\frac{1}{2}} = \frac{\pi}{8}(\nu+\frac{1}{2})$ ν equals	$\lvert f(x_{\nu+\frac{1}{2}}) - \tilde{f}_N(x_{\nu+\frac{1}{2}})\rvert$	$\lvert f - \tilde{f}_N * \psi\rvert$ at $x = x_{\nu+\frac{1}{2}}$
2	5 (-3)	7 (-10)
3	8.1 (-3)	7.9 (-10)
4	1.2 (-2)	6.4 (-10)
5	1.8 (-2)	1.2 (-10)

Table III.

Results of smoothing of the spectral approximation of $f(x)$, $N = 64$.

$x_\nu = \frac{\pi\nu}{8}$ ν equals	$\lvert f(x_\nu) - \hat{f}_N(x_\nu)\rvert$	$\lvert f - \hat{f}_N * \psi\rvert$ at $x = x_\nu$
2	6.4 (-3)	4.8 (-6)
3	1 (-2)	5.9 (-6)
4	1.5 (-2)	7.7 (-6)
5	2.3 (-2)	8.9 (-6)

Table IV.

Same as Table III for the pseudospectral approximation, $\tilde{f}_N(x)$.

$x_{\nu+\frac{1}{2}} = \frac{\pi}{8}(\nu+\frac{1}{2})$ ν equals	$\lvert f(x_{\nu+\frac{1}{2}}) - \tilde{f}_N(x_{\nu+\frac{1}{2}})\rvert$	$\lvert f - \tilde{f}_N * \psi\rvert$ at $x = x_{\nu+\frac{1}{2}}$
2	1 (-2)	4.1 (-6)
3	1.6 (-2)	6 (-6)
4	2.4 (-2)	7.8 (-6)
5	3.6 (-2)	8.9 (-6)

7. ENDNOTES

[1] The single and double primed summations indicate halving the first and the last terms, respectively. It is used in this case to compensate for the use of <u>even</u> number of gridpoints.

[2] Referring to the convex case.

8. REFERENCES

[1] Gottlieb, D., "Spectral Methods for Compressible Flow Problems", ICASE Report No. 84-29.

[2] Gottlieb, D. and Orszag, S.A., "Numerical Analysis of Spectral Methods: Theory and Applications", CBMS, No. 26, SIAM, 1977.

[3] Kreiss, H.O. and Oliger, J., "Methods for the Approximate Solution of Time Dependent Problems", GARP Publications Series No. 10, 1973.

[4] Majda, A. and Osher, S., "Propagation of Error into Regions of Smoothness for Accurate Difference Approximations to Hyperbolic Equations", Comm. Pure Appl. Math., V. 30, 1977, p. 671-705.

[5] Majda, A. and Osher, S., "The Fourier Method for Nonsmooth Initial Data", Math. Comp., V. 32, 1978, p. 1041-1081.

[6] Mock, M.S. and Lax, P.D., "The Computation of Discontinuous Solutions of Linear Hyperbolic Equations", Comm. Pure Appl. Math., V. 31, 1978, p. 423-430.

[7] Tadmor, E., "The Exponential Accuracy of Fourier and Tchebyshev Differencing Methods", ICASE Report No. 84-40.

School of Mathematical Sciences
Raymond and Beverly Sackler Faculty of Exact Sciences
Tel Aviv University, Tel Aviv, Israel.

Both authors were supported in part by the National Aeronautical and Space Administration under NASA contract No. NAS1-17070 while the authors were in residence at ICASE, NASA LANGLEY Research Center, Hampton, Va. 23665. The first author was also supported in part by the Air-Force Office of Scientific Research under Contract No. AFOSR 83-0089.

Progress in Scientific Computing, Vol. 6
Proceedings of U.S.-Israel Workshop, 1984

Numerical problems connected with weather prediction

G. Browning* and Heinz-Otto Kreiss**

1. Introduction.

Numerical weather prediction has always been in the forefront of large scale computing and there is no doubt that this will continue. The reason is that the problems connected with weather prediction are so complex that they will tax the capacity of any size computer. The mathematical theory has advanced considerably and we want to summarize it in this paper.

2. Scaling of the basic equations.

We are mainly interested in short range weather prediction and assume that the heating and diffusion processes can to first approximation be neglected. In this case the atmospheric motions can be described by the gasdynamical equations. The adiabatic Eulerian equations in Cartesian coordinates x, y and z directed eastward, northward and upward, respectively, can be written as (see [1])

$$\begin{aligned} &ds/dt = 0 \, , \\ &dp/dt + \gamma p \nabla \cdot V = 0 \, , \\ &dV/dt + \rho^{-1} \nabla p + f(k \times V) + gk = 0 \, , \end{aligned} \tag{2.1}$$

*NCAR, Boulder, CO 80307, NCAR is sponsored by NSF.
**CALTECH, Pasadena, CA 91125, supported by NSF, Grant ATM8201207.

where t is time, $V = (u,v,w)^T$ is velocity, ρ is density, p is pressure and $s = \rho p^{-1/\gamma}$ is the reciprocal of the entropy. Also $f = f(y)$ is the Coriolis parameter, $g = 9.8ms^{-2}$ is the constant gravity acceleration, $\gamma = 1.4$ is the adiabatic exponent, and $k = (0,0,1)^T$ is the unit vector in the vertical direction. The total differential operator d/dt is given by

$$d/dt = \partial/\partial t + u\partial/\partial x + v\partial/\partial y + w\partial/\partial z \ .$$

We shall now introduce dimensionless variables to identify the relative magnitude of all terms in the equation. These are

$$\begin{aligned} x &= L_1x', \ y = L_2y', \ z = Dz', \ t = Tt', \\ u &= Uu', \ v = Vv', \ w = Ww' \ . \end{aligned} \tag{2.2}$$

Density and pressure can be written in the form

$$\begin{aligned} p &= P_0(p_0(z) + S_1p'), \ \rho = R_0(\rho_0(z) + S_1\rho'), \\ & \qquad 0 < S_1 \ll 1 \ , \end{aligned} \tag{2.3}$$

where

$$\begin{aligned} &P_0\partial p_0/\partial z + gR_0\rho_0 = 0 \ , \\ &P_0 = 10^5 kg \ m^{-1}s^{-2}, \quad R_0 = 1kg \ m^{-2} \ . \end{aligned}$$

(2.3) expresses the fact that a number of digits of pressure and density are independent of x,y and that p and ρ are to first approximation in hydrostatic balance. (2.3) implies also that

$$\begin{aligned} s &= R_0P_0^{-1/\gamma}\rho_0(z)(p_0(z))^{-1/\gamma}(1+S_1\rho'/\rho_0)(1+S_1p'/p_0)^{-1/\gamma} \\ &= R_0P_0^{-1/\gamma}s_0(z)(1+S_1s') \ , \\ s_0(z) &= \rho_0(z)(p_0(z))^{-1/\gamma}, \ s' = \rho'/\rho_0 - (1/\gamma)p'/p_0 \\ & \qquad + O(S_1) \ . \end{aligned} \tag{2.4}$$

We also scale the gravity and Coriolis force

$$g = Gg', \quad G = 10ms^{-2}, \quad f = 2\Omega f', \quad 2\Omega = 10^{-4}s^{-1} .$$

We assume that the scales in the x,y directions are the same and that $\partial u/\partial t$, $\partial v/\partial t$ balance the horizontal convection terms, i.e.

$$U = V, \quad L_1 = L_2 = L, \quad UT/L = 1 . \qquad (2.5)$$

Introducing the above relations into (2.1) gives us

$$\begin{aligned}
&ds'/dt' + (10S_1)^{-1}\tilde{S}_2\tilde{s}(z)(1 + S_1 s')w' = 0 , \\
&dp'/dt' + S_1^{-1}p_0(\gamma(1 + S_1 p'/p_0)d + \tilde{S}_2\tilde{p}(z)w') = 0 , \\
&d = u'_{x'} + v'_{y'} + S_2 w'_{z'} , \\
&du'/dt' + S_3\rho_0^{-1}(1 + S_1\rho'/\rho_0)^{-1}p'_{x'} - S_4 f'v' = 0, \qquad (2.6) \\
&dv'/dt' + S_3\rho_0^{-1}(1 + S_1\rho'/\rho_0)p'_{y'} + S_4 f'u' = 0 , \\
&dw'/dt' + S_1 S_5\rho_0^{-1}(1 + S_1\rho'/\rho_0)^{-1}L_1 = 0 , \\
&L_1 = p'_{z'} - D(\gamma D_0)^{-1}\tilde{p}(z)p' + D/D_0 g'\rho_0 s' + O(S_1) .
\end{aligned}$$

Here

$$\begin{aligned}
&d/dt' = \partial/\partial t' + u'\partial/\partial x' + v'\partial/\partial y' + S_2 w'\partial/\partial z' , \\
&S_2 = D^{-1}TW, \quad \tilde{S}_2 = D_0^{-1}TW, \quad D_0 = 10^4 m , \\
&S_3 = S_1 P_0(R_0 U^2)^{-1}, \quad S_4 = 2\Omega T , \\
&S_5 = TP_0(DR_0 W)^{-1} ,
\end{aligned}$$

and $\tilde{s}(z'),\tilde{p}(z')$ with $-3 < \tilde{s} < -1$, $\tilde{p} \sim -1.3$ are given smooth functions.

Now we choose the parameters according to the so called large scale dynamics

$$S_1 = 10^{-2}, \ L = 10^6 m, \ D = 10^4, \ U = V = 10ms^{-1}, \ W = 10^{-2}ms^{-1}.$$

Dropping the prime notation and neglecting terms of order

$O(S_1)$ we obtain

$$\begin{aligned}
&ds/dt + \tilde{s}(z)w = 0 , \\
&\varepsilon^2/(\gamma p_0)dp/dt + u_x + v_y + \varepsilon Lw = 0, \ \varepsilon = 10^{-1} , \\
&\varepsilon du/dt + \rho_0^{-1} p_x - fv = 0 , \\
&\varepsilon dv/dt + \rho_0^{-1} p_y + fu = 0 , \\
&\rho_0 \eta dw/dt - L^* p + g\rho_0 s = 0, \ \eta = 10^{-6} ,
\end{aligned} \tag{2.7}$$

where

$$\begin{aligned}
&d/dt = \partial/\partial t + u\partial/\partial x + v\partial/\partial y + \varepsilon w\partial/\partial z , \\
&Lw = w_z + \gamma^{-1}\tilde{p}(z)w, \quad L^* p = -p_z + \gamma^{-1}\tilde{p}(z)p .
\end{aligned}$$

There are three different time scales present in this system, the slow convection scale with time derivatives of order $O(1)$, the gravity wave scale with time derivatives of order $O(\varepsilon^{-1})$ and the soundwave scale with time derivatives of order $O(\eta^{-1/2}\varepsilon^{-1/2})$. We are only interested in the slow scale and shall discuss how to obtain the motions on that scale.

3. Unsymmetric hyperbolic systems.

Approximate d/dt by $d_H/dt = \partial/\partial t + u\partial/\partial x + v\partial/\partial y$ and assume that p,u,v are known smooth functions, i.e. p,u,v and a couple of its derivatives are of order $O(1)$. Then we can use the first and the last equation of (2.7) to determine s,w. For simplicity we set $\tilde{s}(z) = -1$ and $g = \rho_0 = 1$ and obtain

$$\begin{aligned}
&ds/dt - w = 0 \\
&\eta dw/dt + s = G, \quad G = L^* p .
\end{aligned} \tag{3.1}$$

(3.1) is a hyperbolic system but it is not symmetric. We assume that G and the initial data are periodic in x,y. For special initial data (3.1) has a smooth solution which can be obtained by an asymptotic expansion. Let

$$s = G + \eta\tilde{s}, \quad w = dG/dt + \eta\tilde{w} .$$

Then $\tilde{s},\tilde{w}$ satisfy

$$d\tilde{s}/dt - \tilde{w} = 0, \quad \eta d\tilde{w}/dt + \tilde{s} = -d^2G/dt^2 .$$

We can repeat the process and obtain

$$\begin{aligned} s_S &= G - \eta d^2G/dt^2 + O(\eta^2), \\ w_S &= dG/dt - \eta d^3G/dt^3 + O(\eta^2) . \end{aligned} \tag{3.2}$$

Assume now that the above solution is not compatible with the initial conditions. Then

$$s_H = s - s_S, \quad w_H = w - w_S ,$$

satisfy

$$\begin{aligned} ds_H/dt - w_H &= 0 , \\ \eta dw_H/dt + s_H &= 0 . \end{aligned} \tag{3.3}$$

Its solutions satisfy the following energy estimate

$$\|s_H(\cdot,t)\|^2 + \eta\|w_H(\cdot,t)\|^2 = \|s_H(\cdot,0)\|^2 + \eta\|w_H(\cdot,0)\|^2$$

i.e.

$$\|w_H(\cdot,t)\|^2 \sim \eta^{-1}(\|s_H(\cdot,0)\|^2 - \|s_H(\cdot,t)\|^2) .$$

Therefore we must expect that in general any disturbance from the smooth solution will introduce a component which will be amplified by $\eta^{-1/2} \sim 10^3$. This makes the numerical solution quite delicate and one has to calculate with very high accuracy.

There are two ways out of this difficulty.

1) Replace (3.1) by

$$s = G, \quad w = dG/dt .$$

If we are interested in the smooth solution only we make an error of order $O(\eta)$. For the full system (2.7) this means that we replace the last equation by the hydrostatic assumption

$$Lp^* = g\rho_0 s$$

and the first equation by

$$\frac{d}{dt}(\rho_0^{-1} Lp^*) + g\tilde{s}(z)w = 0 .$$

The resulting systems are the so called primitive equations.

2) Replace η by ε^2. The error we commit is $O(\varepsilon^2)$ which is acceptable. Also, disturbances are now amplified by $\varepsilon \sim 10$ which is manageable.

4. Instabilities due to shear layers.

In this section we want to examine the growth of perturbations. For that reason we linearize the system (2.7), i.e. we assume that

$$u = U + u', \; v = V + v', \; w = W + w', \; p = P + p',$$
$$s = S + s'$$

where U,V,W,P,S represent a smooth solution. Neglecting quadratic terms and dropping the prime sign we obtain

$$\begin{aligned}
&ds/dt + Q(S) + (\tilde{s}(z) + \varepsilon S_z)w = 0 \\
&(\varepsilon^2/\gamma p_0)dp/dt + (\varepsilon^2/\gamma p_0)Q(P) + u_x + v_y \\
&\qquad + \varepsilon(Lw + (\varepsilon^2/\gamma p_0)P_z w) = 0 \qquad (4.1) \\
&\varepsilon du/dt + \varepsilon Q(U) + \rho_0^{-1} p_x - fv + \varepsilon U_z w = 0 \\
&\varepsilon dv/dt + \varepsilon Q(V) + \rho_0^{-1} p_y + fu + \varepsilon V_z w = 0 \\
&\rho_0 \eta dw/dt + \rho_0 \eta Q(W) - L^* p + g\rho_0 s + \rho_0 \eta \varepsilon W_z w = 0 .
\end{aligned}$$

Here

$$\frac{d}{dt} = \frac{\partial}{\partial t} + U\frac{\partial}{\partial x} + V\frac{\partial}{\partial y} + \varepsilon W\frac{\partial}{\partial z}, \quad Q(H) = H_x u + H_y v \, .$$

We tolerate slow exponential growth $e^{\alpha t}$ where α does not depend on $\varepsilon^{-1}, \eta^{-1}$ or the frequencies. One can show that the Q-terms and $\rho_0 \eta W_z w$ only produce slow growth and therefore we neglect these terms. Also, for simplicity only we replace γp_0 and ρ_Q by 1, $\tilde{s}(z) + \varepsilon S_z$ by -1, $Lw + (\varepsilon^2/\gamma p_0)P_z w$ by w_z, $-L^* p$ by p_z. Then (4.1) becomes

$$\begin{aligned}
&ds/dt - w = 0 \, , \\
&\varepsilon^2 dp/dt + u_x + v_y + \varepsilon w_z = 0 \, , \\
&\varepsilon du/dt + p_x - fv + \varepsilon^2 U_z w = 0 \, , \\
&\varepsilon dv/dt + p_y + fu + \varepsilon^2 V_z w = 0 \, , \\
&\eta dw/dt + p_z + s = 0 \, .
\end{aligned} \tag{4.2}$$

Now freeze coefficients and Fourier transform the above system. Denoting by $\hat{h} = (\hat{s},\hat{p},\hat{u},\hat{v},\hat{w})^T$ the transformed and by $(\omega_1,\omega_2,\omega_3)$ the dual variables we obtain

$$\frac{d\hat{h}}{dt} + ((iU\omega_1 + iV\omega_2 + i\varepsilon W\omega_3)I + \hat{H}_1)\hat{h} =: \hat{H}\hat{h}$$

where

$$\hat{H}_1 = \begin{pmatrix} 0 & 0 & 0 & 0 & -1 \\ 0 & 0 & i\varepsilon^{-2}\omega_1 & i\varepsilon^{-2}\omega_2 & i\varepsilon^{-1}\omega_3 \\ 0 & i\varepsilon^{-1}\omega_1 & 0 & -\varepsilon^{-1}f & \varepsilon U_z \\ 0 & i\varepsilon^{-1}\omega_2 & \varepsilon^{-1}f & 0 & \varepsilon V_z \\ \eta^{-1} & i\eta^{-1}\omega_3 & 0 & 0 & 0 \end{pmatrix} .$$

Let $\tau = \kappa - i(U\omega_1 + V\omega_2 + \varepsilon W\omega_3)$ then the characteristic equation is given by

$$0 = \mathrm{Det}|\hat{H} - I\kappa| = \mathrm{Det}|\hat{H}_1 - I\tau| =$$

$$\begin{vmatrix} -\tau & 0 & 0 & 0 & -1 \\ 0 & -\tau & i\varepsilon^{-2}\omega_1 & i\varepsilon^{-2}\omega_2 & i\varepsilon^{-1}\omega_3 \\ 0 & i\varepsilon^{-1}\omega_1 & -\tau & -\varepsilon^{-1}f & \varepsilon U_z \\ 0 & i\varepsilon^{-1}\omega_2 & \varepsilon^{-1}f & -\tau & \varepsilon V_z \\ \eta^{-1} & i\eta^{-1}\omega_3 & 0 & 0 & -\tau \end{vmatrix} =$$

$$-\tau\begin{vmatrix} -\tau & i\varepsilon^{-2}\omega_1 & i\varepsilon^{-2}\omega_2 & i\varepsilon^{-1}\omega_3 \\ i\varepsilon^{-1}\omega_1 & -\tau & -\varepsilon^{-1}f & \varepsilon U_z \\ i\varepsilon^{-1}\omega_2 & \varepsilon^{-1}f & -\tau & \varepsilon V_z \\ \eta^{-1}i\omega_3 & 0 & 0 & -\tau \end{vmatrix}$$

$$+ \eta^{-1}\begin{vmatrix} -\tau & i\varepsilon^{-2}\omega_1 & i\varepsilon^{-2}\omega_2 \\ i\varepsilon^{-1}\omega_1 & -\tau & -\varepsilon^{-1}f \\ i\varepsilon^{-1}\omega_2 & \varepsilon^{-1}f & -\tau \end{vmatrix} =$$

$$(\tau^2 + \eta^{-1})\begin{vmatrix} -\tau & i\varepsilon^{-2}\omega_1 & i\varepsilon^{-2}\omega_2 \\ i\varepsilon^{-1}\omega_1 & -\tau & -\varepsilon^{-1}f \\ i\varepsilon^{-1}\omega_2 & \varepsilon^{-1}f & -\tau \end{vmatrix}$$

$$+ i\omega_3\eta^{-1}\tau\begin{vmatrix} i\varepsilon^{2}\omega_1 & i\varepsilon^{-2}\omega_2 & i\varepsilon^{-1}\omega_3 \\ -\tau & -\varepsilon^{-1}f & \varepsilon U_z \\ \varepsilon^{-1}f & -\tau & \varepsilon V_z \end{vmatrix} =$$

$$= -\tau\{\tau^4+\varepsilon^{-1}\eta^{-1}(\varepsilon+\omega_3^2+\varepsilon^{-2}\eta(\omega_1^2+\omega_2^2) + \varepsilon^{-1}\eta f^2)\tau^2 +$$
$$+ \varepsilon^{-1}\eta^{-1}(U_z\omega_3\omega_1+V_z\omega_2\omega_3)\tau$$
$$+ \varepsilon^{-2}\eta^{-1}(\varepsilon^{-1}(\omega_1^2+\omega_2^2+f^2\omega_3^2)+f^2-\omega_1\omega_3 fV_z+\omega_2\omega_3 fU_z)\} .$$

To first approximate the solutions of the characteristic equations are

$$\tau_{1,2} = \pm i\sqrt{\varepsilon^{-1}\eta^{-1}(\varepsilon+\omega_3^2+\varepsilon^{-2}\eta(\omega_1^2+\omega_2^2)+\varepsilon^{-1}\eta f^2)}$$
$$=: \pm i(\varepsilon\eta)^{-1/2}a \sim \pm 3000i\sqrt{\varepsilon + \omega_3^2} ,$$
$$\tau_{3,4} = -\frac{1}{2}\frac{U_z\omega_3\omega_1+V_z\omega_2\omega_3}{a^2} \pm i\varepsilon^{-1}\sqrt{\frac{\omega_1^2+\omega_2^2+f^2\omega_3^2}{a^2}} \tag{4.3}$$
$$\tau_5 = 0 .$$

$\tau_{1,2}$ represent the sound waves. By using the primitive equations $(\eta = 0)$ we make their speed infinite. If we instead modify the system, i.e. set $\eta = \varepsilon^2$, we decrease the soundspeed and obtain $\tau_{1,2} \sim \pm i\varepsilon^{-3/2} a \sim \pm 30i\sqrt{\varepsilon + \omega_3^2}$.

$\tau_{3,4}$ represent the gravity waves. Note that in general Real $\tau_{3,4} \neq 0$. In fact for $\omega_3 = O(1)$

$$\max_{\omega_1, \omega_2} |\text{Real } \tau_{3,4}| = O(\varepsilon\eta^{-1/2}) \sim 10^2 .$$

Thus the amplitude of the gravity waves can grow rapidly. An explanation that this instability has not much effect on the global flow might be the following. Due to $|\text{Real } \tau_{3,4}| = O(\varepsilon)|\text{Im } \tau_{3,4}|$ the speed of the gravity waves is large compared with the exponential growth. Therefore, if U_z, V_z change sign sufficiently rapidly the total exponential growth might be small.

If we use the primitive equations we make the exponential growth worse. In fact the initial value problem for the primitive equations is mathematically ill posed because

$$\text{Real } \tau_{3,4} = -\frac{1}{2} \frac{U_z \omega_3 \omega_1 + V_z \omega_3 \omega_2}{\varepsilon + \omega_3^2}$$

can become arbitrarily large.

If one uses the primitive equations for numerical calculations one has to add diffusion terms and time filters to avoid "exponential explosions".

Again these problems go away by setting $\eta = \varepsilon^2$. In this case $|\text{Real } \tau_{3,4}|$ is bounded independently of ε^{-2} and ω_j.

$\tau_5 = 0$ or $\kappa = i(U\omega_1 + V\omega_2 + W\omega_3)$ corresponds to the slow scale (Rossby waves) and is not effected by a change of η.

5. The modified system.

In the last section we have shown that the modified system $(\eta = \varepsilon^2)$ has better mathematical properties than both the primitive equations and the full system. In this section we want to investigate the error which we commit by replacing η by ε^2. Let U,V,W,P,S denote a smooth solution of the system (2.7) and $\tilde{u},\tilde{v},\tilde{w},\tilde{p},\tilde{s}$ the solution of the modified system. To first approximation the differences

$$u = \tilde{u} - U,\ v = \tilde{v} - V,\ w = \tilde{w} - W,\ p = \tilde{p} - P,\ s = \tilde{s} - S$$

satisfy the first four equations of (4.1) while the last equation is replaced by

$$\rho_0\varepsilon^2 dw/dt + \rho_0\varepsilon^2 Q(w) - L^* p + \rho_0 s + \rho_0\varepsilon^2 W_z w = -\varepsilon^2 G,$$
$$G = g\rho_0(1 - \eta/\varepsilon^2)dW/dt .$$

We can neglect the same terms as in section 4 and make the same simplifications. Then we obtain instead of (4.2)

$$\begin{aligned} &ds/dt - w = 0 ,\\ &\varepsilon^2 dp/dt + u_x + v_y + \varepsilon w_z = 0 ,\\ &\varepsilon du/dt + p_x - fv + \varepsilon^2 U_z w = 0 ,\\ &\varepsilon dv/dt + p_y + fu + \varepsilon^2 V_z w = 0 ,\\ &\varepsilon^2 dw/dt + p_z + s = -\varepsilon^2 G . \end{aligned} \qquad (5.1)$$

We transform (5.1) into a symmetric hyperbolic system by introducing new variables

$$s' = s + \varepsilon^2 G,\ p' = \varepsilon^{1/2}p,\ u' = u,\ v' = v,\ w' = \varepsilon w$$

and obtain

$$
\begin{aligned}
&ds'/dt - \varepsilon^{-1}w' = -\varepsilon^2 dG/dt \,, \\
&dp'dt + \varepsilon^{-3/2}(u'_x + v'_y + w'_z) = 0 \,, \\
&du'/dt + \varepsilon^{-3/2}p'_x - \varepsilon^{-1}fv' + U_z w' = 0 \,, \\
&dv'/dt + \varepsilon^{-3/2}p'_y + \varepsilon^{-1}fu' + V_z w' = 0 \,, \\
&dw'/dt + \varepsilon^{-3/2}p'_z + \varepsilon^{-1}s' = 0 \,.
\end{aligned} \tag{5.2}
$$

We now choose the initial conditions for the modified system such that

$$s' = p' = u' = v' = w' = 0 \quad \text{for} \quad t = 0 \,. \tag{5.3}$$

Specifying periodic boundary conditions we can estimate the solution of (5.2), (5.3) in the usual way by the energy method.

$$
\begin{aligned}
&\|s\|^2 + \|\varepsilon^{1/2}p\|^2 + \|u\|^2 + \|v\|^2 + \|\varepsilon w\|^2 = \\
&\|s'\|^2 + \|p'\|^2 + \|u'\|^2 + \|v'\|^2 + \|w'\|^2 = O(\varepsilon^4) \,.
\end{aligned} \tag{5.4}
$$

To obtain an improved estimate for p and w we observe that the total time derivatives satisfy - except for lower order terms - the same system with dG/dt replaced by d^2G/dt^2. Also (5.2), (5.3) show that at $t = 0$

$$
\begin{aligned}
ds'/dt = -\varepsilon^2 dG/dt,\ dp'/dt &= du'/dt = dv'/dt \\
&= dw'/dt = 0 \,.
\end{aligned}
$$

Thus for every fixed t

$$
\begin{aligned}
&\|ds'/dt\|^2 + \|dp'/dt\|^2 + \|du'/dt\|^2 + \|dv'/dt\|^2 \\
&\quad + \|dw'/dt\|^2 = O(\varepsilon^4) \,.
\end{aligned}
$$

Using the differential equations (5.2) we ob'ain now that

$$\|w'\| = O(\varepsilon^3),\ \|p'\| = O(\varepsilon^{5/2}) \,.$$

Thus we have proved

<u>Theorem 5.1</u>. The error we commit by replacing η by ε^2 is of the order $O(\varepsilon^2)$.

6. The initial boundary value problem.

Already J. Oliger and A. Sundström [3] have shown that the initial boundary value problem for the primitive equations is not well posed even if no shear is present. In this section we shall prove that the initial boundary value problem for the modified system is well posed and that its solutions are continuous uniformly in ε up to the boundary provided certain compatibility conditions are satisfied. All results are "in the small" and therefore we need only to consider the halfspace problem for the perturbation equation which we simplify again slightly. This time we neglect also the shear terms because for the modified system these are genuinely of lower order. Thus we consider the system

$$\begin{aligned} &ds/dt - w = 0, \; d/dt = \partial/\partial t + U\partial/\partial x + V\partial/\partial y + \varepsilon W\partial/\partial z \,, \\ &\varepsilon^2 dp/dt + u_x + v_y + \varepsilon w_z = 0 \,, \\ &\varepsilon du/dt + p_x - fv = 0 \,, \\ &\varepsilon dv/dt + p_y + fu = 0 \,, \\ &\varepsilon^2 dw/dt + p_z + s = 0 \end{aligned} \tag{6.1}$$

for

$$t > 0, \; x > 0, \; -\infty < y < \infty, \; -\infty < z < \infty \,.$$

For simplicity only, we assume that U,V,W are constant. We give initial data for $t = 0$ and boundary conditions at $x = 0$. We consider only the case of inflow, i.e. $U > 0$. Then four characteristics point into the region of integration and we have to specify four boundary conditions

$$u + \alpha\varepsilon^{1/2} p = h_0, \; v = v_0, \; w = w_0, \; s = s_0 \quad \text{for } x = 0 \tag{6.2}$$

where we choose α such that

$$\frac{U}{2}(\alpha^2\varepsilon^2 + 1) - \alpha\varepsilon^{-1} < -\frac{1}{2}\varepsilon^{-1} .$$

Remark. The first boundary condition $u + \alpha\varepsilon^{1/2}p = h_0$ seems to be strange. We choose it instead of the more natural condition $u = u_0$ for technical reasons. It makes the boundary conditions dissipative (see (6.4)).

For smooth solutions we cannot choose the boundary data arbitrarily because

$$p_z = -s + O(\varepsilon^2),\ p_x - fv = O(\varepsilon),\ p_y + fu = O(\varepsilon),$$

imply

$$-s_y + fu_z = O(\varepsilon) .$$

Therefore we obtain for $x = 0$

$$\begin{aligned} h_{0z} &= u_z + \alpha\varepsilon^{1/2}p_z = \frac{1}{f}s_y - \alpha\varepsilon^{1/2}s + O(\varepsilon) \\ &= \frac{1}{f}s_{0y} - \alpha\varepsilon^{1/2}s_0 + O(\varepsilon) , \\ 0 = ds/dt - w &= s_t - Up_{zx} + Vs_y - w + O(\varepsilon) \\ &= s_{0t} - Ufv_{0z} + Vs_{0y} - w_0 + O(\varepsilon) . \end{aligned}$$

If these relations are satisfied then we can find smooth functions $\bar{s},\bar{p},\bar{u},\bar{v},\bar{w}$ which satisfy the boundary conditions and

$$\begin{aligned} &\bar{u}_x + \bar{v}_y + \varepsilon\bar{w}_z = O(\varepsilon^{3/2}) , \\ &\bar{p}_x - f\bar{v} = O(\varepsilon),\quad \bar{p}_y + f\bar{u} = O(\varepsilon) , \\ &\bar{p}_z + \bar{s} = O(\varepsilon^2) . \end{aligned}$$

Introducing new variables

$$s' = s - \bar{s},\ p' = \varepsilon^{1/2}(p - \bar{p}),\ u' = u - \bar{u},\ v' = v - \bar{v},$$
$$w' = \varepsilon(w - \bar{w}) ,$$

we obtain an inhomogenous symmetric hyperbolic system

$$\begin{aligned} ds'/dt - \varepsilon^{-1}w' &= \varepsilon G_1 , \\ dp'/dt + \varepsilon^{-3/2}(u'_x + v'_y + w'_z) &= G_2 , \\ du'/dt + \varepsilon^{-3/2}p'_x - \varepsilon^{-1}fv' &= G_3 , \\ dv'/dt + \varepsilon^{-3/2}p'_y + \varepsilon^{-1}fu' &= G_4 , \\ dw'/dt + \varepsilon^{-3/2}p'_z + \varepsilon^{-1}s' &= \varepsilon G_5 , \end{aligned} \tag{6.3}$$

with homogenous boundary conditions. For its solutions we obtain an energy estimate in the usual way. Let $\{(f,g),\|f\|^2\}$ and $\{(f,g)_B,\|f\|_B^2\}$ denote the L_2-scalar product and norm over the halfspace $x > 0$ and the boundary $x = 0$ respectively. Integration by parts gives us

$$\begin{aligned} \frac{\partial}{\partial t} (\|s'\|^2 + \|p'\|^2 + \|u'\|^2 + \|v'\|^2 + \|w'\|^2) < \\ -\varepsilon^{-1}\|p'\|_B^2 + 2((s',\varepsilon G_1) + (p',G_2), + (u',G_3) \\ + (v',G_4) + (w',\varepsilon G_5) . \end{aligned} \tag{6.4}$$

Thus the problem is well posed. We want to show that the solution is smooth up to the boundary. We proceed in the same way as in [2]. Assume that the initial conditions are properly initialized, i.e. that a number of time derivatives at $t = 0$ are bounded independently of ε and that the initial data are compatible with the homogenous boundary conditions. Then we can estimate the time derivatives and the derivatives in the tangential variables y,z because they satisfy equations of the same type. We want to estimate the x-derivatives. The second and third equation of (6.3) show that

$$\begin{aligned} \|u'_x\| = O(1), \ \|p'_x\| &= O(\varepsilon^{1/2}), \\ \|p'_x - \varepsilon^{1/2}fv'\| &= O(\varepsilon^{3/2}), \\ \|u_x + v_y + w_z\| &= O(\varepsilon^{3/2}) . \end{aligned} \tag{6.5}$$

We can also estimate the derivatives of u_x,p_x with respect to t,y,z. Thus we can estimate the 1-dimensional

L_2-norms $\|u_x\|_L, \|p_x\|_L$ on every line $y = z = \text{const}$. The same is true for any variable. If we have an estimate over the whole halfspace then we obtain also an estimate over every line $y = z = \text{const}$. (6.4) implies for every interval $0 < x < A$, $y = \text{const.}$, $z = \text{const.}$

$$(p_y' + \varepsilon^{1/2}u')_x + \varepsilon^{1/2}w_z' = O(\varepsilon^2) ,$$

i.e.

$$p_y' + \varepsilon^{1/2}fu' = C(y) - \varepsilon^{1/2} \int_0^x w_z' d\tilde{x} + O(\varepsilon^2) .$$

The first equation of (6.3) tells us

$$\int_0^x w_z' dx = \varepsilon \int_0^x s_t' + Vs_y' + Ws_z' dx + \varepsilon Us'(x) = O(\varepsilon) .$$

Thus

$$p_y + \varepsilon^{1/2}fu = C(y) + O(\varepsilon^{3/2}) .$$

We want to show that $C(y) = O(\varepsilon^{3/2})$. Using the fourth equation of (6.3) we obtain

$$\begin{aligned} C(y) &= - \int_0^\infty e^{-x}C(y)dx = - \int_0^\infty e^{-x}(p_y + \varepsilon^{1/2}fu)dx \\ &\quad + O(\varepsilon^{3/2}) = \varepsilon^{3/2} \int_0^\infty e^{-x}dv'/dtdx + O(\varepsilon^{3/2}) \\ &= \varepsilon^{3/2} \int_0^\infty e^{-x}Uv_x'dx + O(\varepsilon^{3/2}) \\ &= \varepsilon^{3/2} \int_0^x e^{-x}Uv'dx + O(\varepsilon^{3/2}) = O(\varepsilon^{3/2}) . \end{aligned}$$

Thus $\|p_y' + \varepsilon^{1/2}fu'\|_A = O(\varepsilon^{3/2})$ where $\|f\|_A^2 = \int_0^A \int_{-\infty}^{+\infty} \int_{-\infty}^{+\infty} |f|^2 dxdydz$ and therefore

$$\begin{aligned} U\|v_x'\|_A &< \varepsilon^{-3/2}\|p_y' + \varepsilon^{1/2}fu'\|_A \\ &\quad + \|dv'/dt - Uv_x'\|_A = O(1) . \end{aligned} \tag{6.6}$$

Now differentiate the last four equations of (6.3) and combine them into

$$\begin{aligned}(1 + \varepsilon^{3/2}U^2)p'_{xx} + p'_{yy} + p'_{zz} - \varepsilon^{1/2}f(v'_x - u'_y) \\ = \varepsilon^{3/2}(d^2p'/dt^2 - U^2p'_{xx}) + \varepsilon^{3/2}\tilde{G} .\end{aligned} \quad (6.7)$$

By (6.4) and the corresponding relations for the y,z derivatives it follows that $\varepsilon^{-1/2}p$ is a smooth function of y,z for $x = 0$. Therefore (6.7) shows that

$$\|\varepsilon^{1/2}p'\|_A + \|\varepsilon^{-1/2}p'_x\|_A + \|\varepsilon^{-1/2}p'_{xx}\|_A = O(1) . \quad (6.8)$$

Now we can estimate w'_x, s'_x. Introducing into the first and the last equation of (6.3) new variables by

$$\begin{aligned}s' &= q + \tilde{s},\ q = \varepsilon^2 G_5 - \varepsilon^{-1/2}p'_z, \\ w' &= \varepsilon(\varepsilon G_1 + dq/dt) + \tilde{w} ,\end{aligned}$$

gives us

$$\begin{aligned}d\tilde{s}/dt - \varepsilon^{-1}\tilde{w} &= 0 , \\ d\tilde{w}/dt + \varepsilon^{-1}\tilde{s} &= -\varepsilon^2 dG_1/dt + \varepsilon d^2q/dt^2 .\end{aligned} \quad (6.9)$$

By assumption the initial data are such that

$$ds'/dt = O(1), \quad dw'/dt = O(1) ,$$

i.e.

$$\tilde{s} = O(\varepsilon), \quad \tilde{w} = O(\varepsilon), \quad \text{for} \quad t = 0 . \quad (6.10)$$

Also, at the boundary,

$$\tilde{s} = O(\varepsilon^2), \quad \tilde{w} = O(\varepsilon^2) \quad \text{for} \quad x = 0 . \quad (6.11)$$

Then (6.8) implies $\|d^2q/dt^2\|_A^2 = O(1)$ and therefore there exist a time interval $0 \le t \le T$, T proportional to A independent of ε, such that

$$\|\tilde{s}\|_{A/2} + \|\tilde{w}\|_{A/2} = O(\varepsilon),\ \|\tilde{s}_x\|_{A/2} + \|\tilde{w}_x\|_{A/2} = O(1),\quad 0 \le t \le T,$$

i.e.

$$\|w'\|_{A/2} = O(\varepsilon),\ \|w_x'\|_{A/2} = O(1),\ \|s_x'\|_{A/2} = O(1),\quad 0 \le t \le T. \tag{6.12}$$

For $0 \le t \le T$ we have now obtained bounds for the first x-derivatives near the boundary for all variables. In the interior we obtain the bounds by a standard "partition of unity" argument. Let $\phi(x) \in C_0^\infty$ be a monotone function with $\phi(0) = 0$, $\phi(x) \equiv 1$ for $x \ge 1$. Differentiate the equations (6.3) with respect to x and multiply them by $\phi(x)$. Then we obtain an inequality for

$$\frac{\partial}{\partial t}(s_x',\phi s_x') + (p_x',\phi p_x') + (u_x',\phi u_x') + (v_x',\phi v_x') + (w_x',\phi w_x')$$

which provide us with the desired estimates for $0 \le t \le T$. At $t = T$ we can restart the whole process and therefore we obtain in any finite time interval

$$\|u_x'\| + \|v_x'\| + \|\varepsilon^{-1/2}p_x'\| + \|s_x'\| + \|w_x'\| = O(1)\ . \tag{6.13}$$

For the original variables (6.13) translates into

$$\|u_x\| + \|v_x\| + \|p_x\| + \|s_x\| + \varepsilon\|w_x\| = O(1)\ . \tag{6.14}$$

To show that also $\|w_x\| = O(1)$ we have to use higher derivatives. Differentiate the equations (6.3) with respect to x. By the same argument as before we obtain now also bounds for $\|\varepsilon^{-1/2}p_{xxx}\|_A$. Therefore also $\|d^3q/dt^3\|_A = O(1)$ and we obtain the next term in the asymptotic expansion of the solution of (6.9) by

introducing $\tilde{s} + q_1 = \approx{s}$, $\approx{w} = \tilde{w} - \varepsilon dq_1/dt$,
$q_1 = \varepsilon^2(dG_1/dt + d^2q/dt^2)$ as new variables.

Then it follows that $\approx{s} = O(\varepsilon^2)$, $\approx{w} = O(\varepsilon^2)$, $\approx{w}_x = O(\varepsilon)$ and therefore

$$\|s'_{xx}\| = O(1), \ \|w'_x\| \doteq O(\varepsilon), \quad \text{i.e.} \qquad (6.15)$$
$$\|w_x\| = O(1), \ \|s_{xx}\| \doteq O(1) .$$

We have now estimated all the first and second x-derivatives except w_{xx}. We can also estimate their y and z derivatives and therefore continuity of the variables up to the boundary follows. Thus we have proved

Theorem 6.1. If the initial data are suitably initialized and the boundary data satisfy certain compatibility conditions then the solutions of (6.1) are continuous up to the boundary.

7. References

[1] Kasahara, A., Various vertical coordinate systems used for numerical weather prediction, Mon. Wea. Rev., 1974, Vol. 102, p. 509-522.

[2] Kreiss, H.-O., Problems with different time scales for partial differential equations, Comm. Pure Appl. Math., 1980, Vol. 33, p. 399-439.

[3] Oliger, J. and Sundström, A., Theoretical and practical aspects of some initial-boundary value problems in fluid dynamics, SIAM J. Appl. Math., 1978, Vol. 35, p. 839-866.

Progress in Scientific Computing, Vol. 6
Proceedings of U.S.-Israel Workshop, 1984

ORDER OF DISSIPATION NEAR RAREFRACTION CENTERS

Michael Sever
Department of Mathematics
The Hebrew University, Jerusalem
Israel

1. General discussion.

We consider the approximation of weak solutions of hyperbolic systems of conservation laws,

$$(1.1)\qquad u_t + f(u)_x = 0, \quad -\infty < x < \infty, \; t > 0 \; ; \; u(\cdot,0) \text{ given},$$

by projection methods of finite-difference, finite-element, spectral, etc. type (as opposed to methods such as those of Godunov [2] or Glimm [1]). These methods all contain a dissipation term or mechanism, as needed for the generation of entropy in the presence of shocks. (Throughout this discussion, we specialize to systems for which there exists a convex entropy function U, with corresponding entropy flux F [3].) In the presence of shocks, the magnitude of the required dissipation is determined by the requirement that entropy be generated at the correct rate, given a discrete shock profile uniformly bounded (i.e. without excessive overshooting) and confined to a width of $O(h)$, where h is the mesh size. For example, a regularized form of (1.1) such as

$$(1.2)\qquad u_t + f(u)_x = h(A(u,hu_x)u_x)_x$$

if often used on the basis of a discretization procedure, for a suitably chosen viscosity matrix $A(\cdot,\cdot)$.

The purpose of this note is to suggest that in the neighborhood of rarefraction centers, the dissipation term is preferably one factor of h smaller than that shown in (1.2), for example proportional to h^2u_{xx} or htu_{xx}, $0 < t \le 1$, where the rarefraction center occurs at $t = 0$. Considering the lack of generation of entropy, it is conceivable that dissipation should be avoided entirely near rarefraction centers. However, we shall describe some computational results to the effect that this is not the case. Next we observe that dissipation appropriate for shocks is too strong near rarefraction centers, in the sense that it leads to accuracy of only first order in the mesh size, even in distribution sense. We shall make this precise. Finally we shall show thatin a special case, much better accuracy is achieved using a method with the weaker dissipation.

Computations were made for the scalar case of (1.1), using initial data corresponding to a single rarefaction center. The discretization of (1.1) was of the form

$$(1.3) \qquad \left(\frac{u^{m+1}-u^m}{\Delta t} + f\left(\frac{u^{m+1}+u^m}{2} \right)_x, \phi \right) = 0 \quad \text{for all} \quad \phi \in X, \quad m = 0,1,\ldots,$$

in which $u^m \in X$ is our approximation to $u(\cdot, m\Delta t)$. In (1.3), X is the space of piecewise linear functions on a uniform mesh of size h, and (,) is the real L_2 inner product on R. Choosing $\phi = u^{m+1} + u^m$ in (1.3), it is clear that this method has zero dissipation in L_2.

Several variations of the method were used, corresponding to different initial data values, different flux functions f, preprocessing of the initial data as in [5], lumping the mass matrices, etc. The results were most disappointing. A "sawtooth" instability generally appeared, giving a solution uniformly bounded intime but with variation growing apparently unboundedly with time. In those cases where this effect was not present, at least noticeably after a reasonable number of time steps, the accuracy obtained was still low, no better than first order in distribution sense.

As against this, if dissipation strong enough for shocks is included, the obtained accuracy will be reduced to first order. Initial data corresponding to a rarefaction center is similar enough to that for a shock, or to that for a general Riemann problem, that excess entropy will be dissipated at a finite rate (bounded in magnitude from below, independently of h) initially. This will continue for a time of at least $O(h)$, for the time derivatives are no greater than $O(h^{-1})$ and hence the initial discontinuity cannot be removed in any less time. Thus for any $t \geq O(h)$, we expect

$$(1.4) \qquad \int_{-\infty}^{\infty} \eta(x) \, (U(u(x,t)) - U(u_h(x,t))) \, dx \geq ch$$

where u is the correct weak solution, u_h the computed approximation, and $\eta \in C_0$ is nonnegative and equal to unity in an interval containing the rarefaction wave. In (1.4) and below c is a generic constant.

Although (1.4) implies a limit of first order accuracy in L_1, it does not necessarily limit the accuracy in distribution sense. The same phenomenon arises, for example, in linear hyperbolic systems with smooth coefficients but discontinuous initial data, in which case higher order accuracy can still be readily obtained [4,5]. However, in the case of rarefraction waves the solution u is continuous, and we have the following result.

Theorem 1: Suppose that (1.4) holds, that

$$(1.5) \qquad \| u(\cdot,t) - u_h(\cdot,t)\|_{L_1} \leq ch,$$

and that within the support of η, $u(\cdot,t)$ is continuous and $u_x(\cdot,t)$ of bounded variation; then there exists $\theta \in C_0^\infty$ (θ is vector-valued in the case of systems) such that

$$(1.6) \qquad |(u(\cdot,t) - u_h(\cdot,t), \theta)| \geq ch.$$

The proof is given in section 2. Thus either the obtained accuracy is worse than first order in L_1, which seems unlikely for methods typically used, or else no better than first order in distribution sense.

Finally we present an example, for a very simple problem, of how better results can be obtained with an apparently weaker dissipation term. We specialize to the case of a single conservation law, with initial data corresponding to a single rarefraction center. For simplicity of notation, we set $f = \frac{1}{2}u^2$; these results are readily extended to general flux function f.

We consider the problem with the single independent variable $z = x/t$, thus obtaining

$$(1.7) \qquad u'(u-z) = 0, \quad a < z < b; \quad u(a) = u_-, \; u(b) = u_+,$$

in which prime denotes differentiation with respect to z and $a < u_- < u_+ < b$. The solution, of course, is

$$(1.8) \qquad u(z) = \begin{cases} u_-, & a \leq z \leq u_- \\ z, & u_- < z < u_+ \\ u_+, & u_+ \leq z \leq b. \end{cases}$$

We discretize the problem (1.7) by a simple finite element method. Let X be the space of piecewise linear functions on a uniform mesh of size h, in the interval $a < z < b$. Let X_0 be the restriction of X to functions vanishing at the endpoints. We seek $u_h \in X$ satisfying the boundary conditions $u_h(a) = u_-$, $u_h(b) = u_+$, and

$$(1.9) \qquad (u_h'(u_h - z), \phi) = 0 \quad \text{for all} \quad \phi \in X_0.$$

Choosing ϕ to be a tent function, centered at z_j, we obtain the discrete equation

$$(1.10) \qquad \left(\frac{u_{j+1}-u_{j-1}}{2h}\right)\left(\frac{u_{j+1}+u_j+u_{j-1}}{3} - z_j\right) = \frac{1}{6}\,(u_{j+1} - 2u_j + u_{j-1})$$

for each interior mesh point z_j, where $u_i = u_h(z_i)$ are the point values for our approximation u_h. A dissipative term proportional to $h^2 u_{zz}$ is observed in the right side of (1.10). The results quoted in Theorem 2 below remain valid if the coefficient of this term is increased.

If the values u_-, u_+ correspond exactly to mesh points, then the method (1.9) is exact, as in this case the solution $u \in X$. More generally, however, we have the following, the proof of which is given in section 3:

Theorem 2: There exists a solution to the system generated by (1.9); furthermore for any such solution u_h,

$$(1.11) \qquad ||u - u_h||_{L_2} \leq ch^{3/2},$$

and for all $\theta \in C_0^\infty$

$$(1.12) \qquad |(u - u_h, \theta)| \leq ch^{5/2}.$$

These are good results. The L_2 estimate in quasi-optimum; the power of h cannot be improved for this space X. The distribution sense estimate (1.12) is better than that obtained for the interpolate of u in X, for which we obtain only $O(h^2)$ error in distribution sense. The obtained power of h is that obtained for the error in L_2 plus the order of the space X minus the order of the differential equation, which is often the best possible for finite-element methods [6,7].

The exact solution is unique in the case that u_-, u_+ fall on mesh points, and we conjecture that the approximate solution u_h is unique in general.

2. Proof of theorem 1.

Throughout this section $u = u(\cdot,t)$ $u_h = u_h(\cdot,t)$ and $(,)$ denotes also the complex innder product on R in Fourier transform variables. From the convexity of U we have

$$(2.1) \quad (\eta, U'(u) \cdot (u-u_h) - U(u) + U(u_h)) \geq 0,$$

so that from (1.4) and (2.1)

$$(2.2) \quad (\eta, U'(u) \cdot (u-u_h)) \geq ch.$$

Since u is continuous and u_x of bounded variation, we have

$$(2.3) \quad |\widehat{\eta^{1/2}U'(u)} \ (\xi)| \leq c/\xi^2$$

where $\widehat{\ }$ denotes Fourier transform and ξ is the Fourier transform variable.

Let $g(x) = \dfrac{\sin kx}{x}$, k to be determined below, and set

$$(2.4) \quad \theta(x) = \eta^{1/2}(x) \ ((U'(u)\eta^{1/2}) o g) \ (x),$$

then

$$(\theta, u-u_h) = (\widehat{\eta^{1/2}(u-u_h)}, \ \hat{g} \ \overline{\widehat{U'(u)\eta^{1/2}}})$$
$$= \int_{-k}^{k} \widehat{\eta^{1/2}(u-u_h)} \ \overline{\widehat{U'(u)\eta^{1/2}}} \ d\xi$$
$$= (\eta^{1/2}(u-u_h), \ \overline{U'(u)\eta^{1/2}}) - \int_{-\infty}^{-k} \widehat{\eta^{1/2}(u-u_h)} \ \overline{\widehat{U'(u)\eta^{1/2}}} \ d\xi$$
$$(2.5) \quad - \int_{k}^{\infty} \widehat{\eta^{1/2}(u-u_h)} \ \overline{\widehat{U'(u)\eta^{1/2}}} \ d\xi,$$

where $\overline{}$ denotes complex conjugate. The first right hand side term in (2.5) is bounded from below by ch, from (2.2); for the other two terms, we use (1.5) and (2.3),

$$\left| \int_{k}^{\infty} \widehat{\eta^{1/2}(u-u_h)} \ \overline{\widehat{U'(u)\eta^{1/2}}} \ d\xi \right| \leq c \ ||\widehat{u-u_h}||_{L_\infty} \int_{k}^{\infty} |\widehat{U'(u)\eta^{1/2}}| \ d\xi$$

$$\leq c\, ||u-u_h||_{L_1}/k$$

$$\leq c\, h/k; \tag{2.7}$$

from (2.5) and (2.7), choosing k sufficiently large, (1.6) follows.

3. Proof of Theorem 2.

We first prove convergence. Let n be the number of interior mesh points, i.e. $z_0 = a$ and $z_{n+1} = b$. Let u_I denote the interpolate of u in X, and for any $v \in X$ set

$$T_j(v) = \left(\frac{v_{j+1}-v_{j-1}}{2h}\right)\left(\frac{v_{j+1}+v_j+v_{j-1}}{3} - z_j\right) - \frac{1}{6}(v_{j+1}-2v_j+v_{j-1}), \quad j = 1,2,\ldots,n. \tag{3.1}$$

Let p,q be such that $z_p \leq u_- < z_{p+1}$, $z_q \leq u_+ < z_{q+1}$; then using (1.8), we have

$$T_j(u_I) = \begin{cases} O(h), & j = p,p+1,q,q+1 \\ 0, & \text{otherwise.} \end{cases} \tag{3.2}$$

Let $e \in X_0 = u_I - u_h$; we observe

$$\begin{aligned}(u_I'(u_I-z) - (u_h'(u_h-z),e) &= (u_I'e + u_he' - ze',e) \\ &= (u_I' + \tfrac{1}{2},e^2) - \tfrac{1}{2}(u_h',e^2) \\ &= (u_I' + \tfrac{1}{2},e^2) - \tfrac{1}{2}(u_I' - e',e^2) \\ &= \tfrac{1}{2}(u_I' + 1,e^2) \\ &\geq c\,||e||_{L_2}^2 \\ &\geq ch \sum_{j=1}^{n} e_j^2 \end{aligned} \tag{3.3}$$

Since u_I is non-decreasing. Since (1.10) is exactly what one gets from (1.9) by choosing a tent function for ϕ, we may infer from (3.3) that

$$(3.4)\qquad \sum_{j=1}^{n} (T_j(u_I)-T_j(u_h))\, e_j \geq c \sum_{j=1}^{n} e_j^2$$

without the necessity of a tedious computation of summations by parts and collecting the various terms.

From (3.2) and (3.4)

$$\sum_{j=1}^{n} e_j^2 = c \sum_{j=1}^{n} e_j\, T_j(u_I)$$

$$\leq ch\, \Big(\sum_{j=1}^{n} e_j^2\Big)^{1/2}$$

$$(3.5)\qquad \leq ch^2,$$

so that

$$||u_I - u_h||_{L_2}^2 \leq ch \sum_{j=1}^{n} e_j^2$$

$$(3.6)\qquad \leq ch^3$$

and since $||u-u_I||_{L_2} \leq ch^{3/2}$, we have proved the L_2 estimate (1.11). For the distribution sense estimate, set $r = u-u_h$, and consider the expression

$$E = (u'(u-z)-u_h'\,(u_h-z),\theta)$$

$$= (u'r + u_h r' - zr',\theta)$$

$$= ((ur)' - rr' - zr',\theta)$$

$$= -\,(r, u\theta' - (z\theta)') + \frac{1}{2}\,(r^2,\theta')$$

$$(3.7)\qquad = ((z-u)\theta' + \theta, r) + O(h^3 ||\theta||_{C^1})$$

using (1.11) in the last step.

Using (1.7) and (1.9) we choose $\phi \in X$ to be the interpolate of θ, obtaining

$$E = ((ur)' - rr' - zr', \theta-\phi)$$

$$= (r, \theta-\phi) + \frac{1}{2}(r^2, (\theta-\phi)') + (r, (z-u)(\theta-\phi)')$$

$$\le ch^{3/2} ||\theta-\phi||_{L_2} + ch^3 ||\theta-\phi||_{C^1} + ch^{3/2} \quad (3.8)$$

$$||(z-u)(\theta'-\phi')||_{L_2}$$

again using (1.11).

Given a fixed $\chi \in C_0^\infty$, we choose θ so that

$$(z-u)\theta' + \theta = \chi \quad \text{, i.e.} \quad (3.9)$$

$$\theta(z) = \begin{cases} \chi(u_-) + \frac{1}{u_- - z} \int_z^{u_-} (\chi(y) - \chi(u_-))dy, & z < u_- \\ \chi(z) = & u_- \le z \le u_+ \\ \chi(u_+) + \frac{1}{z-u_+} \int_{u_+}^{z} (\chi(y) - \chi(u_+))dy, & z > u_+ . \end{cases} \quad (3.10)$$

From (3.10) we see that θ is continuous, but the first derivative θ' has jump discontinuities at $z = u_\pm$. Thus

$$||\theta-\phi||_{L_2} \le ch^{3/2}$$

$$||\theta-\phi||_{C^1} \le c$$

$$||\theta||_{C^1} \le c$$

$$||(z-u)(\theta'-\phi')||_{L_2} \le ch, \quad (3.11)$$

the final estimate in (3.11) depending on the fact that $z-u$ is $O(h)$ in the two zones in which $\theta'-\phi'$ is $O(1)$. Combining (3.7), (3.8), (3.9), and (3.11), we get $(r,\chi) = O(h^{5/2})$, which is equivalent to (1.12).

Finally, we prove existence of an approximate solution u_h, using (3.6) as a priori estimate. From (3.1), we have a mapping $T : R^n \to R^n$, determined by

$$T(v) = w, \; w_j = T_j(v), \quad j = 1,\dots,n, \quad v_0 \equiv u_-, \; v_{n+1} \equiv u_+ ;$$

a discrete solution u_h corresponds to v such that $T(v) = 0$. Using (3.6) as an a priori estimate for such solutions, it is clear that for a sufficiently large open ball $B \subset R^n$, $\deg(T,B,0)$ is defined and homotopy invariant, where $\deg(\ ,\ ,\ )$ is the ordinary finite dimensional topological degree. We deform the mapping T so that $u_- = u_+$ (alternatively, so that u_-, u_+ fall exactly on mesh points). Then $T_j(u_I) = 0$ for all j, and from (3.5) we see that the exact (trivial) solution is unique in this case. Finally from (3.3), replacing u_h by an arbitrary function in X satisfying the endpoint conditions, it follows that the Jacobian matrix $\partial T(v)/\partial v$ cannot be singular at $v = u_I$; thus $\deg(T,B,0)$ is either $+1$ or -1, which completes the existence proof.

4. References

1. J. Glimm, "Solutions in the large for nonlinear hyperbolic systems of equations", Comm. Pure Appl. Math., 18 (1965), pp. 697-715.
2. S. K. Godnnov, "Bounds on the discrepancy of approximate solutions constructed for the equations of gas dynamics", J. Computational Math. and Math. Physics. 1 (1961), pp. 623-637 (in Russian).
3. P. D. Lax, "Shock waves and entropy", Proc. Symp. at Univ. of Wisc. (1971) E. H. Zarontonello, ed., pp. 603-634.
4. A. Majda and S. Osher, "Propagation of error into regions of smoothness for accurate difference approximations to hyperbolic equations," Comm. Pure Appl. Math. 30 (1977), pp. 671-706.
5. M. S. Mock and P. D. Lax, "The computation of discontinuous solutions of linear hyperbolic equations", Comm. Pure Appl. Math. 31 (1978), pp. 423-430.
6. J. Nitsche, "Ein Kriterium fur die Quasi-Optimalitat des Ritzschen Verfahrens", Numer. Math. 11 (1968), pp. 346-348.
7. G. Strang and G. Fix, "An analysis of the finite element methods". Prentice-Hall, Englewood Cliffs , N.J., (1973).

PROGRESS IN SCIENTIFIC COMPUTING ALREADY PUBLISHED

PSC 1 *The Finite Element Method in Thin Shell Theory*
Application to Arch Dam Simulations
M. Bernadou, M. M. Boisserie
ISBN 3-7643-3070-8

PSC 2 *Numerical Treatment of Inverse Problems in Differential and Integral Equations*
P. Deuflhard, E. Hairer, editors
ISBN 3-7643-3125-9

PSC 3 *Lanczos Algorithms for Large Symmetric Eigenvalue Computations Vol. 1, Theory*
Jane K. Cullum, Ralph A. Willoughby
ISBN 0-8176-3058-9
ISBN 3-7643-3058-9

PSC 4 *Lanczos Algorithms for Large Symmetric Eigenvalue Computations Vol. 2, Programs*
Jane K. Cullum, Ralph A. Willoughby
ISBN 0-8176-3294-8
ISBN 3-7643-3294-8
ISBN for two-volume set 0-8176-3295-6, 3-7643-3295-6

PSC 5 *Numerical Boundary Value ODEs*
Uri M. Ascher, Robert D. Russell, editors
ISBN 0-8176-3302-2
ISBN 3-7643-3302-2